Mathematics

Macmillan Master Series

Accounting
Advanced English Language
Advanced Pure Mathematics
Arabic
Basic Management
Biology
British Politics
Business Administration
Business Communication
C Programming
C++ Programming
Chemistry
COBOL Programming
Communication
Counselling Skills
Database Design
Desktop Publishing
Economic and Social History
Economics
Electrical Engineering
Electronic and Electrical Calculations
Electronics
English Grammar
English Language
English Literature
Fashion Styling
French
French 2
Geography
German
Global Information Systems
Internet
Italian
Italian 2
Java
Marketing
Mathematics
Microsoft Office
Microsoft Windows, Novell NetWare and UNIX
Modern British History
Modern European History
Modern World History
Networks
Pascal and Delphi Programming
Philosophy
Photography
Physics
Psychology
Shakespeare
Social Welfare
Sociology
Spanish
Spanish 2
Statistics
Systems Analysis and Design
Visual Basic
World Religions

Macmillan Master Series
Series Standing Order ISBN 0–333–69343–4
(outside North America only)

You can receive future titles in this series as they are published by placing a standing order. Please contact your bookseller or, in case of difficulty, write to us at the address below with your name and address, the title of the series and the ISBN quoted above.

Customer Services Department, Macmillan Distribution Ltd
Houndmills, Basingstoke, Hampshire RG21 6XS, England

Mastering

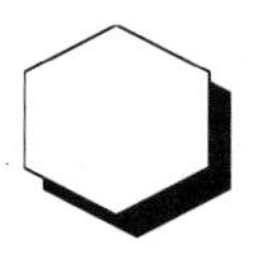

Mathematics

Second Edition

Geoff Buckwell

First edition 1991
Reprinted four times
Second edition 1997

Published by
MACMILLAN PRESS LTD
Houndmills, Basingstoke, Hampshire RG21 6XS
and London
Companies and representatives
throughout the world

ISBN 0–333–66508–2

A catalogue record for this book is available
from the British Library.

This book is printed on paper suitable for recycling and
made from fully managed and sustained forest sources.

10 9 8 7 6 5 4 3 2
06 05 04 03 02 01 00 99 98

Copy-edited and typeset by Povey–Edmondson
Tavistock and Rochdale, England

Printed and bound in Great Britain by
Biddles Ltd
Guildford and King's Lynn

Contents

Introduction

This book covers basic mathematical principles in a clear, straightforward style. The purpose of this book is to help those of you who are finding some (or all!) aspect of mathematics difficult. Very often, it is a small point that needs explanation, and many books do not cover everything in sufficient detail. *Mastering Mathematics* is designed to overcome these problems. It includes pencil and paper techniques as well as the correct use of a calculator. It also covers a wide range of topics so that if you are following a GCSE, BTEC, college course, or just 'doing mathematics for fun', you will almost certainly find the topics you need.

Scattered throughout the book are a number of Investigations which are designed to help you explore topics in a more practical way. Also as you progress through the book, there are Examples and Exercises to see if you have indeed *mastered* those subjects covered in that chapter. None of these investigations requires any specialist equipment. It is important to remember when studying mathematics at any time that methods sometimes need to be read several times. Perseverance will bring its reward, particularly if the exercises are tried and checked against the answers at the back. If you get anything wrong, go back to the text and read the relevant section.

Geoff Buckwell

Acknowledgements

I am indebted to Colin Prior for his illustrations which I hope give some light relief, to Josie Buckwell for her work in preparing and checking the manuscript, and Matthew Buckwell for his work in producing the diagrams.

The author and publishers wish to thank the following for permission to use copyright material:

Express Newspapers plc for articles 'Quiet for God's Sake' and 'No Camping', *Daily Star*, 4 March 1991. The London East Anglian Group: Midland Examining Group; Northern Examining Association comprising Associated Lancashire Schools Examining Board, Joint Matriculation Board, North Regional Examinations Board, North West Regional Examinations Board and Yorkshire & Humberside Regional Examinations Board; Southern Examining Group, University of Cambridge Local Examinations Syndicate, and the Welsh Joint Education Committee for questions from past examination papers. Times Newspapers Ltd for 'Britain will honour troops with Gulf victory parade' by Nicholas Wood and Ruth Gledill, *The Times*, 4 March 1991.

Every effort has been made to trace all the copyright-holders, but if any have been inadvertently overlooked the publishers will be pleased to make the necessary arrangement at the first opportunity.

Geoff Buckwell

Working with whole numbers

1.1 Number systems

The number system that we use today has taken thousands of years to develop. The symbols used are 0, 1, 2, 3, 4, 5, 6, 7, 8, 9. Each symbol is called a **digit**. Our system involves counting in tens. This type of system is called a **denary** system, and 10 is called the **base** of the system. It is possible to use a number other than 10 as the base, see Section 1.6. For example, computer systems use base 2 (the **binary** system) or base 16 (the **hexadecimal** system).

Numbers are combined together, using the four **arithmetic operations**:

addition (+), subtraction (−), multiplication (×) and division (÷)

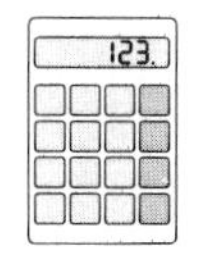

Calculators are frequently used nowadays by most people. However, this book will try to ensure that you can use 'pencil and paper' techniques as well. Sections where the use of a calculator is needed will be clearly marked. There are many types of calculators available, and you must *always refer to the manual for detailed instructions*. Throughout this book a keyboard such as that shown in Figure 1.1 will be used. When using a calculator, make sure that the display shows zero, and the memory is empty, before starting a new calculation. For the keyboard shown, [AC] will clear any calculation, and [AC] [M in] will clear the memory. An M on the screen indicates that something is stored in the memory.

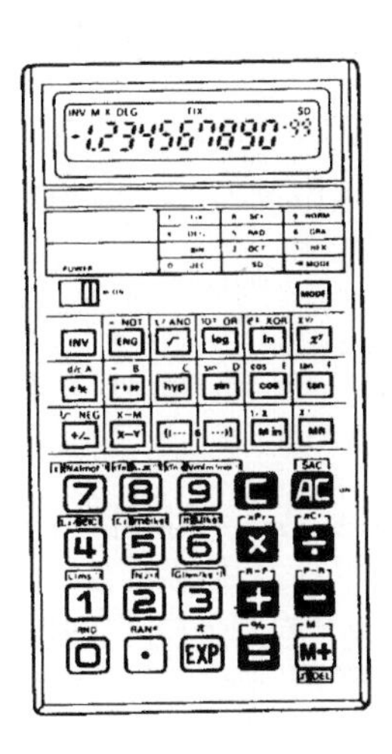

Figure 1.1

(i) Whole numbers

The numbers 1, 2, 3, 4, ... which we use for *counting* are called **whole numbers** or **natural numbers**. In higher level mathematics, they are referred to as **integers**.

(ii) Powers

Repeated multiplication by the same number is known as raising **to a power**.

$8 \times 8 \times 8 \times 8$ is written 8^4 (8 to the power 4).

Writing numbers like this is called **index notation**.

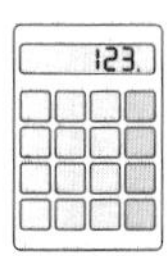

If you have a power button $\boxed{x^y}$ on your calculator, the answer to a power becomes easy to find.

On the calculator, 8^4 is entered as follows:

				Display
8	x^y	4	=	4096.

(iii) Place value

Once a number contains more than one digit, the idea of **place value** is used to tell us its worth. In the numbers 285 and 2850, the 8 stands for something different. In 285, the 8 stands for 8 'tens'. In 2850, the 8 stands for 8 'hundreds'.

The following table shows the names given to the first seven places.

millions	*hundred thousands*	*ten thousands*	*thousands*	*hundreds*	*tens*	*units*
4	0	8	7	0	2	6

The number shown is 4087026, which is four million eighty-seven thousand and twenty six.

We could write this in index notation.

Since $100 = 10 \times 10 = 10^2$, $1000 = 10 \times 10 \times 10 = 10^3$ and so on, our table could be written:

10^6	10^5	10^4	10^3	10^2	10^1	1
4	0	8	7	0	2	6

Hence $4087026 = 4 \times 10^6 + 8 \times 10^4 + 7 \times 10^3 + 2 \times 10 + 6 \times 1$

This type of notation will be met again when we look at standard form (see p. 42).

1.2 The four rules of working with numbers

(i) Addition

A number obtained by adding two numbers is called the **sum** of the numbers. It is important when carrying out addition to keep every digit in its correct column.

Example 1.1

Find the sum of 216, 4083 and 95.

Solution

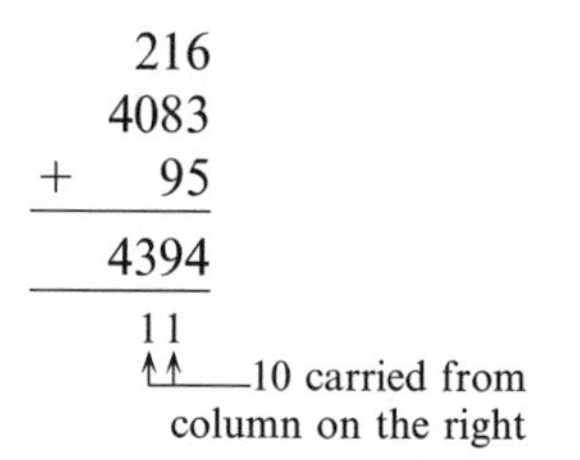

Columns are added from the right (units first). If the sum of a column exceeds 9, the excess number of tens is carried to the next column. Here the units column adds up to 14, and so 1 (one ten) is carried to the tens column and so on.

The sum is 4394.

Using a calculator:

Display

216 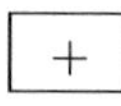 4083 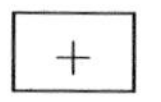 95 = 4394.

Notice that a decimal point is shown at the end of the number.

(ii) Subtraction

If you subtract one number from another, you are taking away one number from another.

Example 1.2

Find (a) 846 – 723 (b) 54 – 27 (c) 1011 – 826.

Solution

(a)

$$\begin{array}{r} 846 \\ -723 \\ \hline 123 \\ \hline \end{array}$$

Each digit in 723 is less than the digit above it, so the subtraction is straightforward.

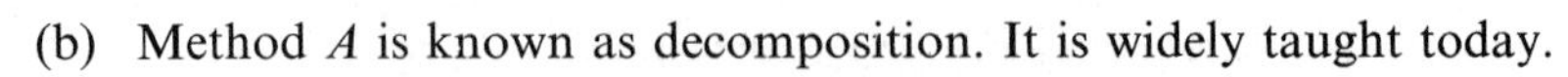

(b) Method A is known as decomposition. It is widely taught today.

A:

$$\begin{array}{r} {\scriptstyle 4\ 14} \\ \not{5}\ \not{4} \\ -2\ 7 \\ \hline 2\ 7 \\ \hline \end{array}$$

Since 7 is less than 4, we borrow 10 from the next column to the left.Then we subtract 7 from 14. Then we reduce the 5 to 4.

The alternative, method B, used to be the main method learned, and is called equal addition.

B:

$$\begin{array}{r} {\scriptstyle 14} \\ 5\ \not{4} \\ {\scriptstyle 3\ \ } \\ -\not{2}\ 7 \\ \hline 2\ 7 \\ \hline \end{array}$$

Since 7 is less than 4, we add ten to the 4 making 14, and add one to the 2 (the lower number) in the next column (this is in fact adding 10).

(c) A:

$$\begin{array}{r} {\scriptstyle 9\ 10\ 11} \\ \not{1}\ \not{0}\ \not{1}\ \not{1} \\ -\ \ 8\ 2\ 6 \\ \hline 1\ 8\ 5 \\ \hline \end{array}$$

In this example, we have had to borrow 10 from the tens column to make 11, and 100 from the hundreds column leaving 9(00).

B:

$$\begin{array}{r} {\scriptstyle 10\ 11\ 11} \\ 1\ \not{0}\ \not{1}\ \not{1} \\ {\scriptstyle 1\ \ 9\ \ 3\ \ \ } \\ -\ \ \not{8}\ \not{2}\ 6 \\ \hline 1\ 8\ 5 \\ \hline \end{array}$$

It is probably sensible to stick to the method you know already.

Using a calculator, each of these problems is just as easy. Hence part (b) would become:

				Display
54	–	27	=	27.

(iii) Multiplication

The result of multiplying two numbers together is called the **product** of the two numbers.

Example 1.3

Work out 242 × 36.

Solution

$$\begin{array}{r} 242 \\ \times \quad 36 \\ \hline 7260 \\ 1452 \\ \hline 8712 \\ \hline \end{array}$$

7260 ← this row is from multiplying 242 by 30

1452 ← this row is from multiplying 242 by 6

8712 ← the answer is obtained by adding the two rows above.

It is not surprising that people prefer to multiply using a calculator. Quite simply, we have:

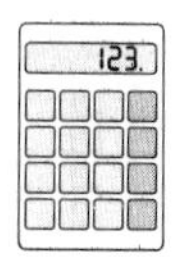

				Display
242	×	36	=	8712.

(iv) Division

The number obtained by dividing one number by another is called the **quotient**. If the division is not exact, you will be left with a **remainder**. Division can be carried out by *long* or *short* division.

Example 1.4

(a) Find $8478 \div 9$.
(b) Find the remainder when 827 is divided by 7.
(c) Find $8675 \div 25$ using the process of long division.

Solution

(a)

$$\begin{array}{r} 9\;4\;2 \\ 9\overline{)8 4^3 7^1 8} \end{array}$$

After dividing 8400 by 9, it divides 900 times leaving a remainder 300. The 300 is then added to 70 to give 370. 9 is then divided into 370. It divides 40 times with a remainder of 10. The 10 is then added to 8 to give 18. 9 divides into 18 twice.

The answer or *quotient* is 942. This process is called **short division**.

(b)

$$\begin{array}{r} 1\;1\;8 \text{ r } 1 \\ 7\overline{)8^1 2^5 7} \end{array}$$

On dividing 7 into 57, it divides 8 times ($7 \times 8 = 56$) and there is a remainder of 1.

Hence the remainder when 827 is divided by 7 is 1.

(c)

$$\begin{array}{r} 347 \\ 25\overline{)8675} \\ \underline{75}\downarrow \\ 117 \\ \underline{100}\downarrow \\ 175 \end{array}$$

25 divides into 86, 3 times, $3 \times 25 = 75$ which is put under 86. $86 - 75 = 11$. (This gives the remainder.) The 7 is brought down to give 117. 25 divides into 117, 4 times. $4 \times 25 = 100$ which is put under 117. $117 - 100 = 17$ (the remainder). The 5 is brought down to give 175. 25 divides into 175 exactly 7 times.

Hence the quotient is 347.

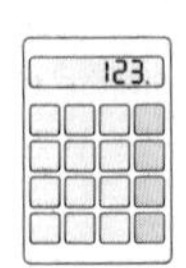

One area where the calculator is difficult to use is finding a remainder when you are dividing. The first example is straightforward.

(a) **Display**

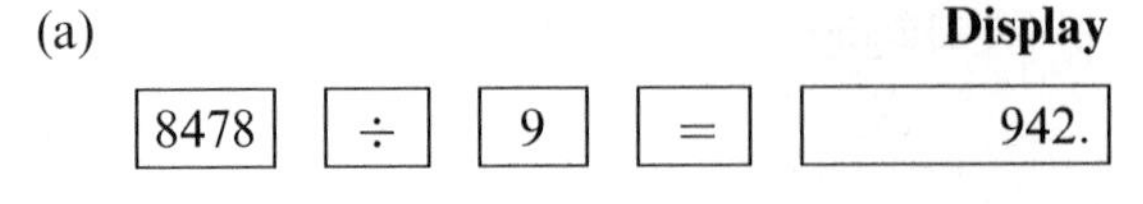

(b) **Display**

827 ÷ 7 = 118.1428571

It divides 118 times, but it is certainly not clear that the remainder is one.

To find the remainder, you would have to multiply 118 by 7:

Display

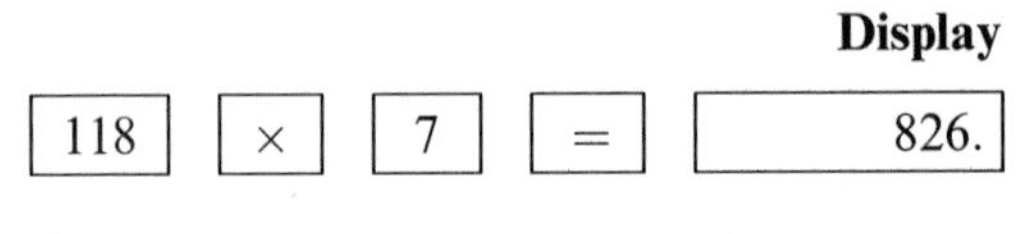

The remainder = 827 − 826 = 1.

The following Investigation is based simply around whole numbers, although it could be extended. You may find part (v) difficult at this stage.

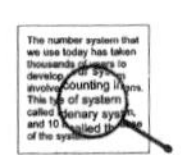

Investigation I: Arithmogons

In Figure 1.2 (i), can you find the missing numbers in the circles, so that the numbers in the squares are each equal to the sum of the two numbers in the circles on either side?

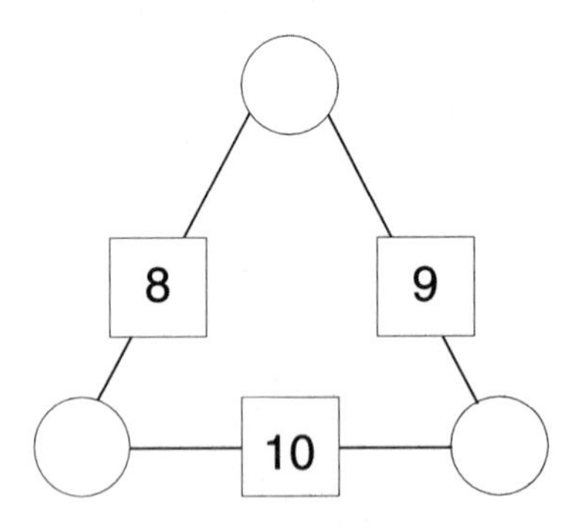

Figure 1.2(i)

When you have solved this problem, try the same thing with Figure 1.2 (ii) and 1.2(iii).

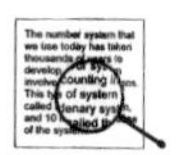

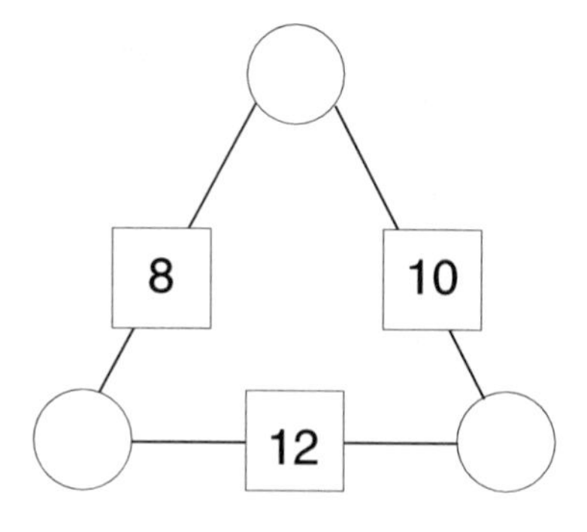

Figure 1.2(ii)

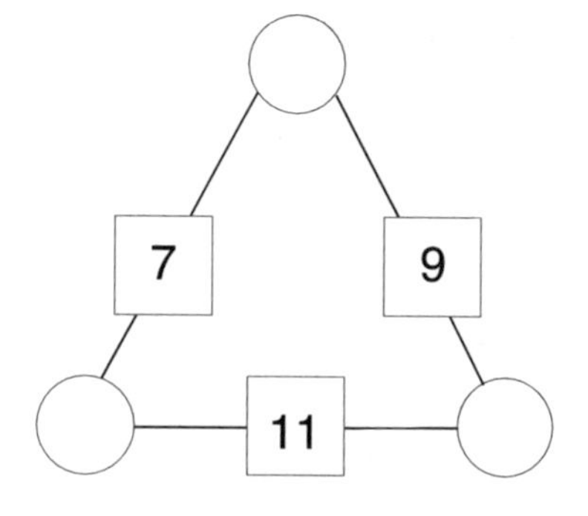

Figure 1.2(iii)

Can you detect any pattern?

You should now be trying to find a system for predicting what the numbers are. Hence, starting with Figure 1.2(iv), could you find the missing numbers easily?

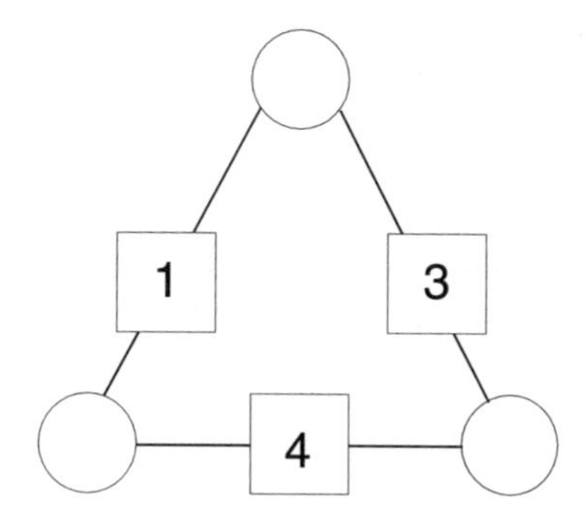

Figure 1.2(iv)

In the case of Figure 1.2(v) can you find a general way of predicting the missing numbers?

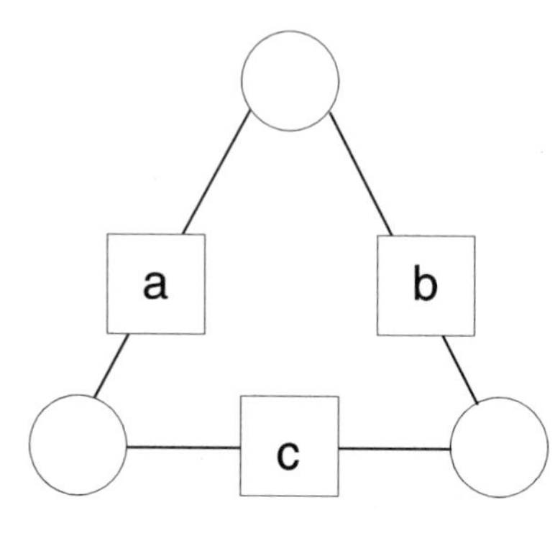

Figure 1.2(v)

Extension. Suppose you use a different operation than addition. Will it work?

1.3 The use of brackets

When a calculation involves more than one operation, it is important to make clear the *order* of operations. Brackets ( ) are used to indicate the operations that are carried out first.

Example 1.5

Find (a) $(8+6)\times(15\div 3)$
(b) $18-(6\times 2)$.

Solution

(a) $(8+6)\times(15\div 3)=14\times 5=70$
[*Note:* Each bracket is worked out separately.]

(b) $18-(6\times 2)=18-12=6$.

If brackets have not been included, the order in which operations are carried out can be summarised by the word BODMAS. The letters stand for 'brackets, of, divide, multiply, add, subtract' in that order. The word 'of' is usually only used in context such as if we ask 'What is the cost of five of these apples?'. The answer would be found by multiplying the cost of one apple by five. Hence 'of' is represented by $\times$ on the calculator. The following example shows how the rule would work.

Example 1.6

Work out (a) $8\times 4+12\div 2$
(b) $3+5$ of $8-10\div 5$.

Solution

You will notice that brackets have not been included. The operations are now carried out following the BODMAS rule.

(a) $8\times 4+12\div 2=8\times 4+6$ (i) divide
$=32+6$ (ii) multiply
$=38$ (iii) add

(b) $3+5$ of $8-10\div 5=3+40-10\div 5$ (i) (of means $\times$)
$=3+40-2$ (ii) divide
$=43-2$ (iii) add
$=41$. (iv) subtract

Most calculators follow the BODMAS rule, but if you are unsure put brackets into an expression to make it clear. If your calculator has buttons for brackets, (and), then you can proceed immediately. If not, you may have to use the memory. These ideas are illustrated in the following example:

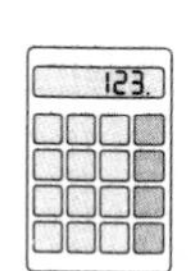

Example 1.7

Use a calculator to evaluate: (a) $6 \times (8 + 23)$ (b) $148 - 12 \times 7$.

Solution

(a) *Method 1.* (If your calculator does not have brackets)

Display

8 + 23 = × 6 = 186.

You have to work out the contents in the bracket first, and then multiply by 6.

Method 2. (If you have brackets)

Display

6 × (8 + 23) = 186.

(b) *Method 1.* (Without brackets)

Display

12 × 7 = M in 148 − MR = 64.

Method 2. (With brackets)

Display

148 − (12 × 7) = 64.

Method 3.
If you are sure that your calculator will multiply before subtracting, then part (b) becomes

Display

148 − 12 × 7 = 64.

1.4 Multiples and factors

We know that $7 \times 5 = 35$. We say that 35 is a **multiple** of 5 or a multiple of 7. Since 7 and 5 divide exactly into 35, then 7 and 5 are said to be **factors** of 35.

The number 13 has no factors except 1 and 13. We say that 13 is a **prime** number. The smallest prime number is 2, it only has factors 1 and 2. Note that 1 is *not* regarded to be a prime number.

Example 1.8

Express the following numbers as a product of their prime factors (a) 210 (b) 180.

Solution

(a) $210 = 2 \times 105$
$= 2 \times 3 \times 35$
$= 2 \times 3 \times 5 \times 7$

(b) $180 = 2 \times 90$
$= 2 \times 2 \times 45$
$= 2 \times 2 \times 3 \times 15$
$= 2 \times 2 \times 3 \times 3 \times 5.$

It would be more convenient to use the index notation for this example. Hence $180 = 2^2 \times 3^2 \times 5$.

(i) Highest common factor (HCF)

The largest number that divides exactly into two or more given numbers is the **highest common factor (HCF)** of those numbers.

Example 1.9

Find the HCF of (a) 28 and 40 (b) 120, 80 and 90.

(a) $28 = 2 \times 14 = 2 \times 2 \times 7 = 2^2 \times 7$
$40 = 2 \times 20 = 2 \times 2 \times 10 = 2 \times 2 \times 2 \times 5 = 2^3 \times 5.$

First express each number as a product of its prime factors. The prime factor 2 appears in both numbers. You choose the lowest power of this common prime factor which is 2^2.

Hence the HCF of 28 and 40 is $2^2 = 4$.

(b) $120 = 2 \times 60 = 2 \times 2 \times 30 = 2 \times 2 \times 2 \times 3 \times 5$

$= 2^3 \times 3 \times 5$

$80 = 2^4 \times 5$

$90 = 2 \times 3^2 \times 5.$

The only factors common to *all three* numbers are 2 and 5.

The lowest power of 2 in all the numbers is just 2^1.

The lowest power of 5 in all the numbers is just 5^1.

Hence the HCF $= 2^1 \times 5^1 = 10$.

(ii) Lowest common multiple (LCM)

The smallest number that two or more numbers divide into exactly is called the **lowest common multiple (LCM)** of these numbers.

Example 1.10

Find the LCM of 8, 10 and 12.

Solution

List the multiples of each number.

8: 8, 16, 24, 32, 40, 48, 56, 64, 72, 80, 88, 96, 104, 112, **120**, . . .

10: 10, 20, 30, 40, 50, 60, 70, 80, 90, 100, 110, **120**, . . .

12: 12, 24, 36, 48, 60, 72, 84, 96, 108, **120**, . . .

It can be seen that 120 is the lowest common multiple of 8, 10, 12. Although this method is adequate in this case, it is rather lengthy for larger numbers. A better method is to use the prime factors.

$8 = 2^3, \qquad 10 = 2^1 \times 5^1, \qquad 12 = 2^3 \times 3^1.$

Take the *highest* power of *every* prime number that occurs, i.e. 2^3, 3^1 and 5^1.

The LCM is $2^3 \times 3^1 \times 5 = 120$.

Example 1.11

An electric sign contains three flashing lights, red, blue and green. The red light flashes on every 10 seconds, the blue light flashes on every 15 seconds, and the green light flashes on every 18 seconds. The lights are on together at 0800. How many times are they on together between 0800 and 0900 inclusive?

Figure 1.3

Solution

The question looks as if it might involve a 'hit and miss' method. In fact, it is a problem about lowest common multiples.

The multiples of the three numbers are as follows:

10, 20, 30, 40, 50, 60, 70, 80, **90**
15, 30, 45, 60, 75, **90**
18, 36, 54, 72, **90**

The LCM is 90.

Hence the lights flash on together every 90 seconds, i.e. every $1\frac{1}{2}$ minutes.

We now need to find how many $1\frac{1}{2}$-minute intervals there are in 60 minutes.
Now 0800 to 0900 is 1 hour = 60 minutes.

$60 \div 1\frac{1}{2} = 40.$

The answer is *not* 40, because it says 0800 to 0900 *inclusive*, and they are on together at 0800.

Hence the correct answer is 41.

Exercise 1(a)

1 Write as a power:
(i) $3 \times 3 \times 3$ (ii) $2 \times 2 \times 2 \times 2 \times 2$ (iii) 100×100.

2 Express in words:
(i) 483 (ii) 4096 (iii) 801 469.

3 Write as a denary number:
(i) eight thousand nine hundred and eight
(ii) forty million six hundred thousand and seventy-nine.

4 Work out:
(i) 26×10 (ii) 306×100 (iii) $4090 \div 10$ (iv) $18000 \div 100$.

5 Find the sum of 968, 29 and 603.

6 Subtract 89 from 211.

7 From the product of 12 and 18, subtract the sum of 18 and 25.

8 What number must be added to 895 to give:
(i) 1946 (ii) 8084?

9 Find the value of:

(i) $8 \times (6 + 3)$ (ii) $25 \times (12 \div 3)$ (iii) $8 \times 15 - 3 + 6$

(iv) 7 of $8 \div 4$ (v) $3 \times 15 \div 5 - 2$.

10 The sum $6 + 3 \times 15 \div 3 + 2 = 47$ is correct, but the brackets have been omitted. Insert brackets in the right places to make it correct.

11 Write the following numbers as a product of prime factors:

(i) 36 (ii) 75 (iii) 140 (iv) 290 (v) 175.

12 Find the highest common factor (HCF) of the following numbers:

(i) 40 and 60 (ii) 75 and 90 (iii) 27, 45 and 54 (iv) 84, 126 and 196.

13 Find the lowest common multiple (LCM) of the following numbers:

(i) 12 and 18 (ii) 4, 8 and 12 (iii) 25 and 30 (iv) 6, 22 and 121.

14 The ten tiles shown in Figure 1.4 are part of a set of tiles used in a game of Scrabble.

Figure 1.4

These ten tiles can be used to make words and the numbers on the tiles are added together.

For example JUG scores $8 + 1 + 2 = 11$.

(i) Calculate the score for JAPE.

(ii) Brian uses four of the ten tiles to make a word which scores 17. Three of the letters in his word are A, C and J. What is the fourth letter?

(iii) Claire uses five of the ten tiles to make a word which scores 20. Three of the letters in her word are C, A and U. What are the other two letters?

(iv) Davinder picks up four of the ten tiles. These four tiles have a total of 27. What are the four letters? [Note that they will not make a word].

(v) Erica picks up five of the ten tiles. What is her lowest possible total?

A word can be played on the game board in such a way that the number on one tile is trebled before the other numbers are added on. This total is then doubled to give the score.

(vi) The word 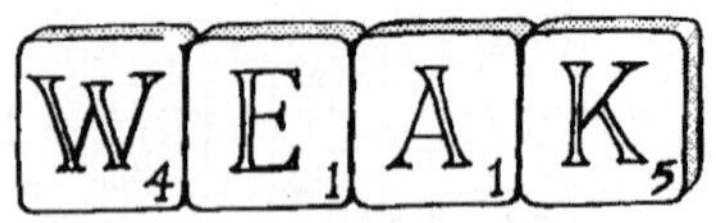

is played in this way.

What is:

(a) The highest score possible,
(b) The lowest score possible?

(vii) A five letter word is played in the same way

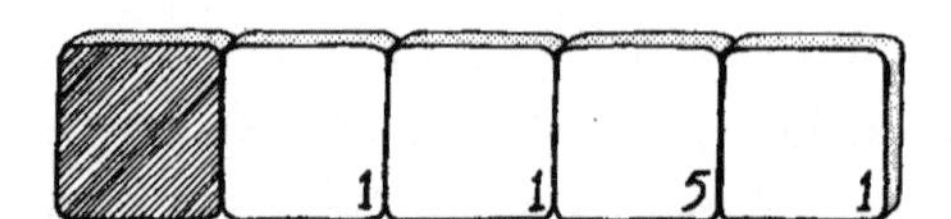

= score of 76

What is the number on the shaded tile if that is the one which is trebled?

[MEG]

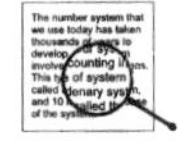

Investigation II: Fun with numbers

In the following questions, each sum is correct, but many of the digits have been replaced by a letter. Can you find in each case what digits the letters stand for (each question is separate). There may be more than one solution.

(i)
$$\begin{array}{r} 1ab \\ +\ \ ab \\ \hline 1b8 \\ \hline \end{array}$$

(ii)
$$\begin{array}{r} cdc \\ \times\ \ \ cc \\ \hline 6e6ed \\ 6e6e\ \\ \hline f11de \\ \hline \end{array}$$

(iii)
$$\begin{array}{r} gk \\ 97\overline{)95hi} \\ \underline{kj3}\ \\ jji \\ \underline{jji} \\ \underline{0} \end{array}$$

(iv)
$$\begin{array}{r} pqr \\ \times\ \ \ st \\ \hline pqru \\ rvt\ \\ \hline qswt \\ \hline \end{array}$$

1.5 Negative numbers

Using ordinary positive numbers, it is not possible to work out 8 – 11. The number system has to be extended to include **negative** numbers (numbers below zero).

Hence 8 – 11 = ⁻3 (negative 3).

To distinguish between a positive and a negative number, we sometimes write $^{+}3$ to mean positive 3. Zero, which is represented by the digit 0, is *neither* positive nor negative, but is included in the set of all integers.

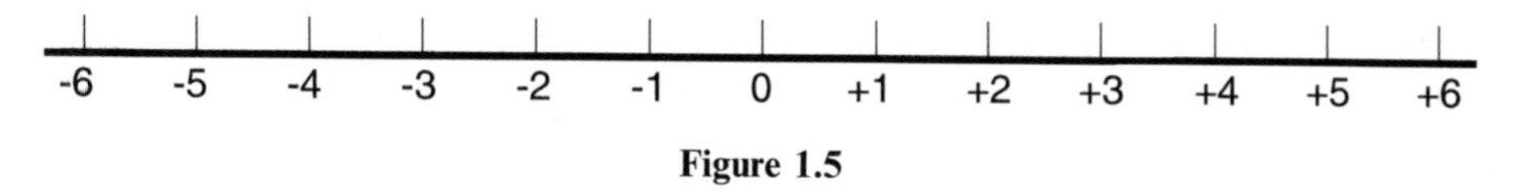

Figure 1.5

Negative numbers are commonly used in temperature measurements (see Example 1.13, p. 18).

Every whole number can be represented by a point on a straight line called the **number line**.

Any number on the line is *less* than any number to the right of it, and more than any number to the left of it.

The symbol $<$ is used to mean less than

Hence $^{+}2 < ^{+}8$ (positive 2 is less than positive 8).

$^{-}4 < ^{-}1$ (negative 4 is less than negative1).

The symbol $>$ is used to mean greater than.

Hence $^{+}8 > ^{-}4$ (positive 8 is greater than negative 4).

(i) Addition

The addition of a positive number to any given number of the number line, produces an answer that is further to the right. Addition of a negative number produces a result that is further to the left.

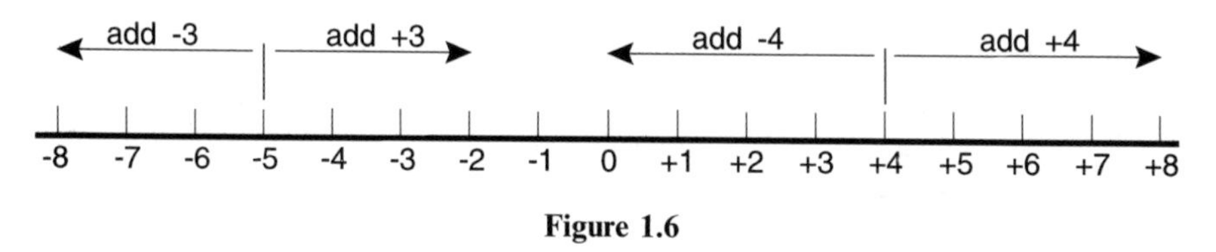

Figure 1.6

Figure 1.6 illustrates the following results:

$$^{-}5 + ^{+}3 = ^{-}2, \qquad ^{-}5 + ^{-}3 = ^{-}8$$

$$^{+}4 + ^{+}4 = ^{+}8, \qquad ^{+}4 + ^{-}4 = 0.$$

(ii) Subtraction

The subtraction of a positive number from any given number on the number line produces an answer that is further to the left. Subtraction of a negative number produces a result that is further to the right.

This last point is frequently forgotten. It is the same as saying that

subtraction of a negative number becomes addition of that number

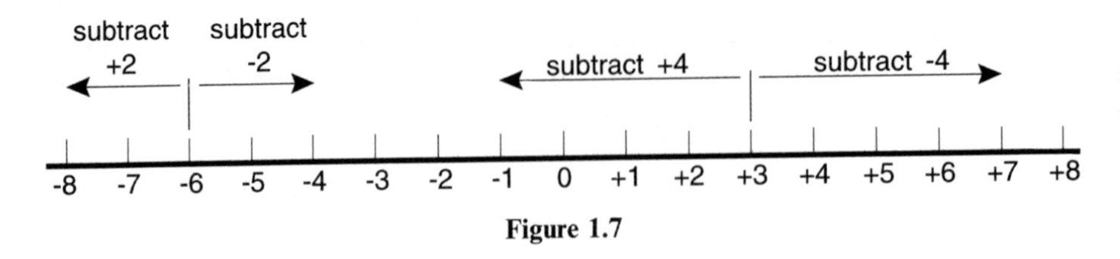

Figure 1.7

Figure 1.7 illustrates the following results:

$$^{-}6 - {}^{+}2 = {}^{-}8, \qquad {}^{-}6 - {}^{-}2 = {}^{-}4$$
$$^{+}3 - {}^{+}4 = {}^{-}1, \qquad {}^{+}3 - {}^{-}4 = {}^{+}7.$$

(iii) Multiplication

$^{+}2 \times {}^{-}4 = {}^{-}4 + {}^{-}4 = {}^{-}8$ (since multiplying by 2 is the same as adding the same quantity twice).

Hence:

positive × negative = negative

Now when we multiply two numbers together, it does not matter in what order we do the multiplication, so for example $8 \times 7 = 7 \times 8$. We would not want this to alter now.

Hence

$$^{-}4 \times {}^{+}2 = {}^{-}8$$

So that

negative × positive = negative

Also we have:

positive × positive = positive

To complete this, we need to decide what negative × negative becomes. We complete the pattern if the answer is positive.

So

negative × negative = positive

For example,

$$^{-}4 \times {}^{-}6 = {}^{+}24.$$

(iv) Division

Since $^{-}4 \times {}^{+}2 = {}^{-}8$, dividing by $^{-}4$, we have $^{+}2 = {}^{-}8 \div {}^{-}4$

i.e.

negative ÷ negative = positive

In a similar way to multiplication,

negative ÷ positive = negative
positive ÷ negative = negative

and clearly

positive ÷ positive = positive

Numbers that have a + or − sign in front of them are known generally as **directed numbers**. From now on, the + sign will be omitted from positive numbers, and the negative sign will be at the same level as the subtraction sign.

So $^{+}8 - {}^{-}2$ will be written $8 - -2$.

These results are summarised in the following table:

Rule

÷ or ×	+	−
+	+	−
−	−	+

Negative numbers are entered on the calculator by means of the [+/−] button.

Example 1.12

Use a calculator to find:

(a) $8 - -14$ (b) $(-3)^3$ (c) $28 - (-16 \div 4)$.

Solution

(a) **Display**

(b) If you do not have a power button:

[3] [+/–] [×] [3] [+/–] [×] [3] [+/–]

Display

[=] [– 27.]

If you do have a power button:

Display

[3] [+/–] [x^y] [3] [=] [– 27.]

(c) Without brackets:

[16] [+/–] [÷] [4] [=] [M in] [28] [–] [MR]

Display

[=] [32.]

With brackets:

Display

[28] [–] [(] [16] [+/–] [÷] [4] [)] [=] [32.]

Example 1.13

The midday temperatures on a given day at nine different cities throughout the world are as follows:

London 8°C, Moscow –4°C, Bombay 21°C, Quebec –9°C, Cape Town 18°C, Edinburgh –3°C, Paris 12°C, Mexico 25°C, New York 0°C.

(a) Find the average of these temperatures.
(b) What is the difference in temperature between the hottest and coldest places?
(c) How many places are within 10°C of the average?

Solution

This is a question about directed numbers.

(a) The average $= (8 + -4 + 21 + -9 + 18 + -3 + 12 + 25 + 0) \div 9$ (see Chapter 9).

[*Note*: Although one of the numbers is zero, you still divide by 9.]

When entering negative numbers on the calculator, remember to use the button [+/–]. So the first part of the calculation would be:

[8] [+] [4] [+/–] [+] [21] [+] [9] [+/–] . . .

$= 68 \div 9 = 7.6$

Hence the average temperature is 7.6°C correct to 1 decimal place (1 d.p.) (see Section 2.3).

(b) The hottest temperature is 25°C

The coldest temperature is −9°C

The difference $= 25 - -9$

$= 25 + 9 = 34$°C.

(c) 10°C from the average means

$7.6 + 10 = 17.6$°C

$7.6 - 10 = -2.4$°C

The cities in this range are New York, London, Paris.

The answer is three.

Exercise 1(b)

In this exercise, work out each question without a calculator first, and then check your answers using the calculator.

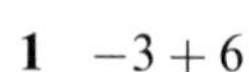

1 $-3 + 6$

2 $8 + -8$

3 $-8 - 4$

4 $4 + -9$

5 $-2 + -6$

6 $13 - -7$

7 $3 - -2 - 1$

8 $-1 - -4 + 2$

9 $27 + -2 + 3$

10 $2 - -10 + -4 - 7$

11 $-15 + 33 - 12 + -11$

12 $47 - -9 + -7 - 8$

13 $-10 - 15 + 10 - -5 + 5$

14 2×-3

15 -4×6

16 -2×-1

17 -8×0

18 -16×-5

19 $10 \div -2$

20 $-18 \div 3$

21 $-12 \div -4$

22 $(-24 \div -3) \times -2$

23 $(18 \div -9)$

24 $(-2 \times -12) \div (8 \div -8)$

25 $(6 + -3) \times (-2 - -3)$

26 $(5 - 30) \div (8 - 3)$

27 $12 \div (4 - 8)$

28 $28 \times (6 - -4)$

29 $(120 - 30) \div (-42 - 3)$

30 $1000 \div (125 \times -2)$

31 $2^3 + 3^3 + 4^3$

32 $12^4 \div 6^3$

33 $(-4)^3 \times 2^2$

34 $88 \div 22 \times -66$

35 This table gives the temperature in Sheffield during one week in January 1987

Day	*Sun*	*Mon*	*Tues*	*Wed*	*Thurs*	*Fri*	*Sat*
Noon	−3°C	−2°C	1°C	−3°C	2°C	3°C	−2°C
Midnight	−8°C	−8°C	−6°C	−10°C	−6°C	−3°C	−5°C

(i) What is the lowest temperature in the table?
(ii) On which day was there the biggest drop in temperature between noon and midnight?
(iii) How much was this drop?
(iv) What was the least rise in temperature between midnight one day and noon the following day?
(v) On the next Sunday the temperature was 8° higher at noon than at midnight the previous night. What was the temperature at noon?

[MEG]

1.6 Other number bases

The denary system which was introduced at the beginning of this chapter has ten digits and the columns are powers of ten. We say that ten is the base of the system. Any other positive integer could be used as the **base** of a number system. If eight is used, we refer to it as the **octal** system, if two is used as the base we have the **binary** system, commonly used in electronic circuits and computers. Also used in computers is the **hexadecimal** system or **hex** for short, which uses 16 as the base, and requires new symbols to represent 10, 11, 12, etc. (see Example 1.16). To indicate that a base other than 10 is being used, a small subscript is added after the number. Hence 23_8 indicates that the number has been written in base 8.

Example 1.14

Change the following numbers into denary numbers:

(a) 23_8 (b) 11011_2 (c) 4031_5.

Solution

(a)

8	units
2	3

$23_8 = 2 \times 8 + 3 \times 1 = 16 + 3 = 19.$

(b)

2^4 16	2^3 8	2^2 4	2	units
1	1	0	1	1

$11011_2 = 1 \times 16 + 1 \times 8 + 1 \times 2 + 1 = 27.$

(c)

5^3 125	5^2 25	5	units
4	0	3	1

$4031_5 = 4 \times 125 + 3 \times 5 + 1$
$= 500 + 15 + 1 = 516.$

You will notice that the headings to the columns reading from the right are just the powers of the base being used.

There are certain calculators that have the facility to change from one system to another in certain number bases, usually base 2, base 8 and base 16, where

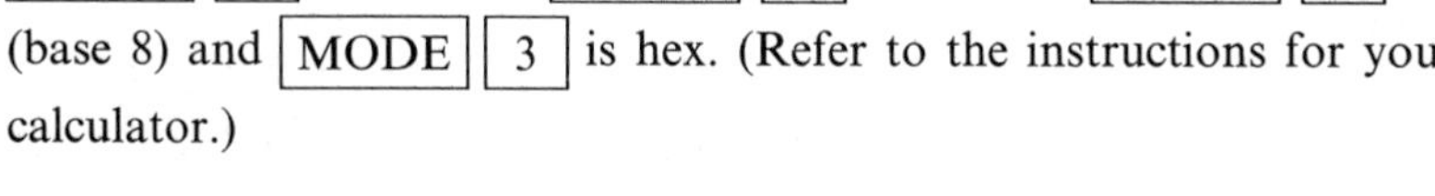

[MODE] [0] is denary, [MODE] [1] is binary, [MODE] [2] is octal (base 8) and [MODE] [3] is hex. (Refer to the instructions for your own calculator.)

In part (a) of Example 1.14, to change octal to denary:

					Display
MODE	2	23	MODE	0	19.

In part (b), to change binary to denary:

					Display
MODE	1	11011	MODE	0	27.

In order to change from denary into a different base, a technique involving *successive division* is used. The given denary number is divided repeatedly by the appropriate base number, and the remainder is noted at each stage.

Example 1.15

Write the denary number 176, in (a) base eight (b) base five (c) base two.

Solution

(a)

		remainder
8)	176	
8)	22	0 ↑
8)	2	6
	0	2

(b)

		remainder
5)	176	
5)	35	1 ↑
5)	7	0
5)	1	2
5	0	1

The answer is obtained by reading upwards the remainders, so that $176 = 260_8$ and $176 = 1201_5$.

(c)

		remainder
2)	176	
2)	88	0 ↑
2)	44	0
2)	22	0
2)	11	0
2)	5	1
2)	2	1
2)	1	0
	0	1

You will notice that the smaller the base number, the longer the answer becomes.

Here $176 = 10110000_2$.

(i) The hexadecimal system

A fairly common notation on a calculator that can work in hex, is as follows:

10 is A	13 is d
11 is b	14 is E
12 is c	15 is F

Example 1.16

(a) Convert 2 *bc* hex into a denary number
(b) Convert the denary number 879 into hex.

Solution

(a)

16^2 256	16	units
2	*b*	*c*

$$2\,bc = 2 \times 256 + 11 \times 16 + 12$$
$$= 512 + 176 + 12 = 700.$$

(b)

	remainder
16)879	
16) 54	15 (*F*)
16) 3	6
0	3

Hence 879 hex = 36*F*.

Investigation III: Ancient number systems

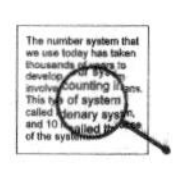

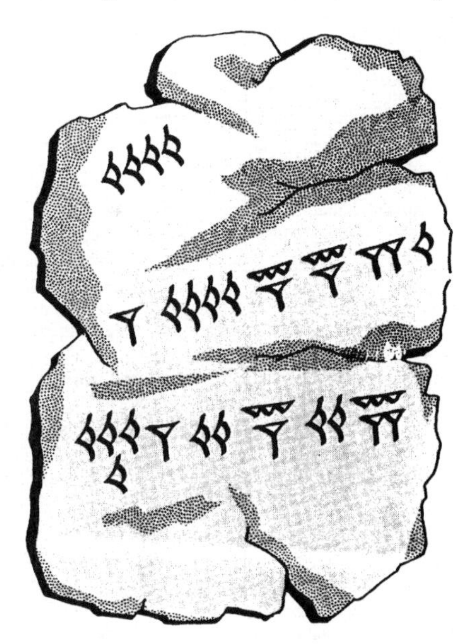

The illustration shows part of an old Babylonian clay tablet. It has three sets of numbers on it.

stands for 40

1 44 4 2 10

What do you think the third set of numbers stands for?

Question: Do you think this is a good system? Try and devise a system of your own. Can you see why ancient people had difficulty in counting efficiently?

(ii) Calculation in other number bases

The rules for adding, subtracting, multiplying and dividing in other number bases are exactly the same as for denary numbers, all you have to remember is that the base is not 10. Most people would use a scientific calculator which has a base 2, 8 or 16 facility on it. But just in case you mislay yours or it stops working, the following example should indicate how you proceed.

Example 1.17

Carry out the following calculations in the stated base:

(a) $43_8 + 25_8$ (b) $1011_2 \times 11_2$ (c) $111_4 - 23_4$ (d) $1604_8 \div 44_8$

Solution

(a)
```
   43
  +25
  ---
   70   The answer is 70₈.
  1↑
   5 + 3 = 8 so carry 1
   as the base is 8
```

(b)
```
     1011
  ×    11
  -------
    10110
     1011
  -------
   100001   The answer is 100001₂.
    111
```

(c) *Method A* (decomposition)
```
    4
  0 3 5
  1 1 1
 -  2 3
  -----
    2 2   The answer is 22₄.
```

Method B (equal addition)
```
  0 5 5
  1 1 1
    1 3
 -  2 3
  -----
    2 2   The answer is 22₄.
```

Now 1 becomes 5 if you borrow 4. The process is identical to working in base 10, as long as you remember you are borrowing 4 and not 10.

(d) This question is not for the faint-hearted.

$$\begin{array}{r} 31 \\ 44\overline{)1604} \\ 154 \\ \hline 44 \\ \hline \end{array}$$

Remember here that 160 is *not* a denary number

Hence the answer is 31_8.

On many scientific calculators, as mentioned earlier, it is possible to work directly in other number bases. On the keyboard used in this book, [MODE] [1] would produce base 2, [MODE] [2] would produce base 8, and [MODE] [3] would produce base 16.

Example 1.18

Use the calculator to carry out the following calculations in the given base:

(a) $1010100_2 \div 111_2$ (b) $53_8 \times 67_8$ (c) $49_{hex} \times 5A_{hex}$.

Solution

(a) [MODE] [1] [1010100] [÷] [111] [=] Display: 1100.

(b) [MODE] [2] [53] [×] [67] [=] Display: 4475.

(c) [MODE] [3] [49] [×] [5] [$a^{b/c}$] [=] Display: 19AA

↑ this key enters A in hex (refer to your own calculator)

Exercise 1(c)

In this exercise work out, if possible, each question both with and without a calculator.

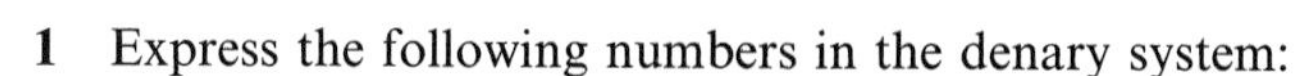

1 Express the following numbers in the denary system:

(i) 101_2	(ii) 10110_2	(iii) 1011_2
(iv) 243_5	(v) 212_3	(vi) 46_7.

2 Change the following denary numbers (a) into base 2 (b) into base 8 (c) into base 16:

(i) 9 (ii) 15 (iii) 35 (iv) 120.

3 Find the number represented by X in the following:

(i) $X_2 = 200_3$ (ii) $X_8 = 100111_2$
(iii) $666_8 = X_2$ (iv) $354_6 = X_8$.

4 Evaluate the following in the base shown:

(i) $11_2 + 111_2$ (ii) $101_2 + 111_2$
(iii) $24_8 + 36_8$ (iv) $101_2 - 11_2$
(v) $1011_2 - 111_2$ (vi) $67_8 - 46_8$
(vii) $312_4 + 213_4 - 33_4$ (viii) $101_2 \times 10_2$
(ix) $21_3 \times 2_3$ (x) $73_8 \times 4_8$

Working with fractions and decimals

2.1 Understanding fractions

The word **fraction** is used to describe a *part* of something. If you split (divide) a quantity into three equal parts, then each part is a third $(\frac{1}{3})$ of the quantity. If you split a quantity into four equal parts, then each part is a quarter $(\frac{1}{4})$ of the quantity. A fraction such as $(\frac{2}{5})$ (two-fifths) means divide the amount into five equal parts (each one is a fifth), and take two of the fifths.

The top line of a fraction is called the **numerator**.

The bottom line of a fraction is called the **denominator**.

Example 2.1

(a) Find $\frac{1}{4}$ of 84. (b) Find $\frac{2}{3}$ of 24.

Solution

(a) To find $\frac{1}{4}$ of something means to break it into 4 equal parts. This can be done by dividing by 4. Notice how the word '*of*' is treated slightly differently to how it was used in Section 1.3. However, $\frac{1}{4}$ of 84 can also mean $\frac{1}{4} \times 84$ (i.e. 'of' meaning multiply).

Hence $\frac{1}{4}$ of 84 is 21.

'of' means multiply

(b) First find $\frac{1}{3}$ of 24. This is 8 $(24 \div 3)$.

Then $\frac{2}{3}$ of 24 is $2 \times 8 = 16$.

(i) Equivalent fractions

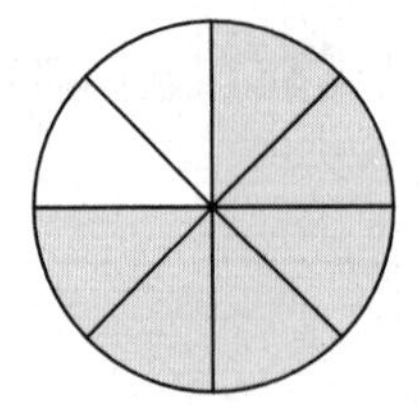

Figure 2.1

Figure 2.1 shows a circle divided into 8 equal parts. Each part represents $\frac{1}{8}$ of the complete circle. Since 6 of the parts have been shaded, the diagram represents $\frac{6}{8}$.

If you consider the circle to be only divided into 4 equal parts, then 3 have been shaded. The total shaded is hence $\frac{3}{4}$.

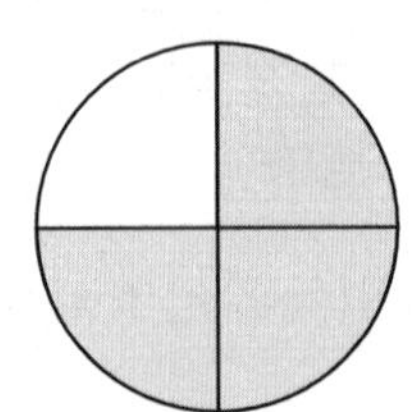

Figure 2.2

Clearly, each answer represents the same thing. It follows that $\frac{6}{8} = \frac{3}{4}$.

We say that $\frac{6}{8}$ and $\frac{3}{4}$ are **equivalent fractions**.

Division of the circle into 16 parts would give 12 shaded.

We have the continuing series of fractions:

$$\frac{3}{4} = \frac{6}{8} = \frac{12}{16} = \frac{24}{32} = \dots$$

Each fraction is obtained from the next by multiplying the top line (*numerator*) and bottom line (*denominator*) by the number 2.

If the numerator and denominator of $\frac{3}{4}$ are multiplied by 3, we get:

$$\frac{3}{4} = \frac{3 \times 3}{4 \times 3} = \frac{9}{12}$$

If we multiply by 5 we get:

$$\frac{3}{4} = \frac{3 \times 5}{4 \times 5} = \frac{15}{20}$$

Continuing this, the complete set of equivalent fractions is obtained:

$$\frac{3}{4} = \frac{6}{8} = \frac{9}{12} = \frac{12}{16} = \frac{15}{20} = \frac{18}{24} = \dots$$

$\frac{3}{4}$ is the **simplest form** of all of these fractions.

Reversing this process is called **cancelling**.

Start with the fraction $\frac{24}{36}$. Divide the numerator and denominator by 12:

So

$$\frac{24}{36} = \frac{24 \div 12}{36 \div 12} = \frac{2}{3}$$

Hence $\frac{24}{36} = \frac{2}{3}$. Since $\frac{2}{3}$ cannot be reduced any further, then we say that $\frac{24}{36} = \frac{2}{3}$, in its **lowest terms**.

This cancelling is often written $\frac{\overset{2}{\cancel{24}}}{\underset{3}{\cancel{36}}}$ (dividing by 12).

If you cannot see the full cancellation in one step, you could proceed in several steps.

Hence

$$\frac{24}{36} = \frac{\overset{(\div 2)}{\cancel{24}}\,^{12}}{\underset{(\div 2)}{\cancel{36}}\,_{18}} = \frac{\overset{(\div 2)}{\cancel{12}}\,^{6}}{\underset{(\div 2)}{\cancel{18}}\,_{9}} = \frac{\overset{(\div 3)}{\cancel{6}}\,^{2}}{\underset{(\div 3)}{\cancel{9}}\,_{3}} = \frac{2}{3}$$

(ii) Common denominator

There are many occasions when you want to change two or more fractions into fractions which have the same denominator. This will be referred to as the common denominator of the fractions. The technique is particularly useful when you are adding or comparing fractions.

Study the following examples carefully.

Example 2.2

Write the given fractions with the same denominator:
(a) $\frac{2}{5}, \frac{3}{4}$ (b) $\frac{25}{16}, \frac{17}{24}, \frac{31}{48}$.

Solution

(a) The best common denominator will be the LCM of the denominators 5 and 4 which is 20, since it is the smallest number that is a multiple of 5 and 4.

So

$$\frac{2}{5} = \frac{2 \times 4}{5 \times 4} = \frac{8}{20}$$

$$\frac{3}{4} = \frac{3 \times 5}{4 \times 5} = \frac{15}{20}.$$

(b) Here, the LCM of 16, 24 and 48 is 48.

$$\frac{25}{16} = \frac{25 \times 3}{16 \times 3} = \frac{75}{48}$$

$$\frac{17}{24} = \frac{17 \times 2}{24 \times 2} = \frac{34}{48}.$$

The fraction $\frac{31}{48}$ is left unchanged.

Example 2.3

Arrange the fractions $\frac{2}{3}$, $\frac{3}{5}$ and $\frac{5}{12}$ in increasing order of size.

Solution

The lowest common denominator is the LCM of 3, 5 and 12, which is 60.

$$\frac{2}{3} = \frac{2 \times 20}{3 \times 20} = \frac{40}{60}$$

$$\frac{3}{5} = \frac{3 \times 12}{5 \times 12} = \frac{36}{60}$$

$$\frac{5}{12} = \frac{5 \times 5}{12 \times 5} = \frac{25}{60}$$

So $\frac{25}{60} < \frac{36}{60} < \frac{40}{60}$

Hence in increasing size we have $\frac{5}{12}, \frac{3}{5}, \frac{2}{3}$.

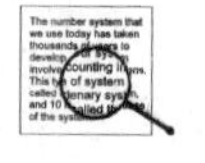

Investigation IV: Shading fractions

1 In the diagrams in Figure 2.3, shade as many regions as necessary to represent the fractions stated.

(i) $\frac{2}{9}$

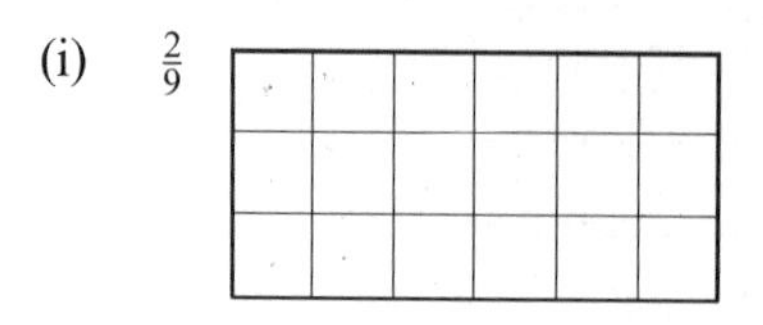

(ii) $\frac{3}{5}$

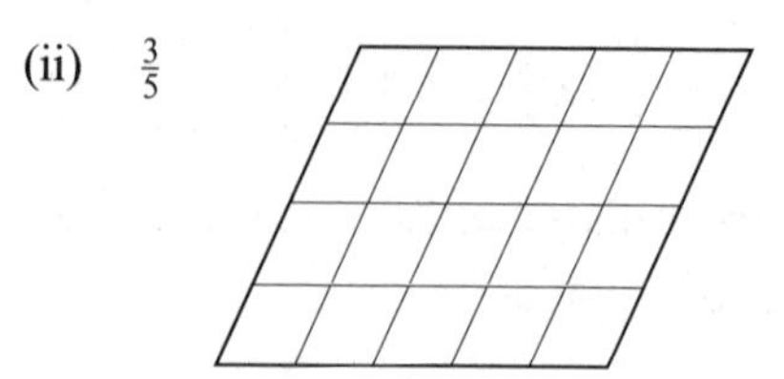

(iii) $\frac{1}{3}$

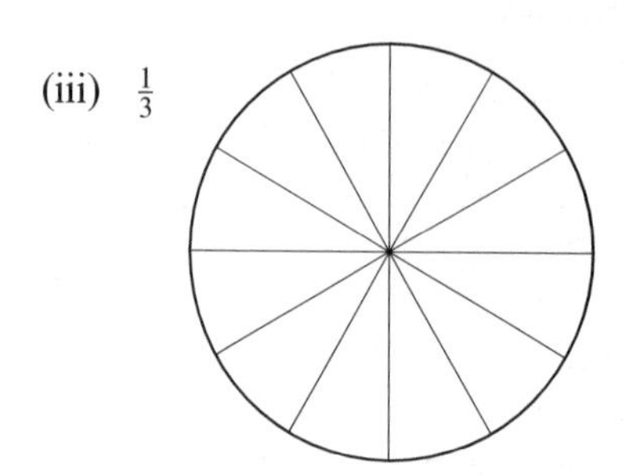

(iv) $\frac{5}{8}$

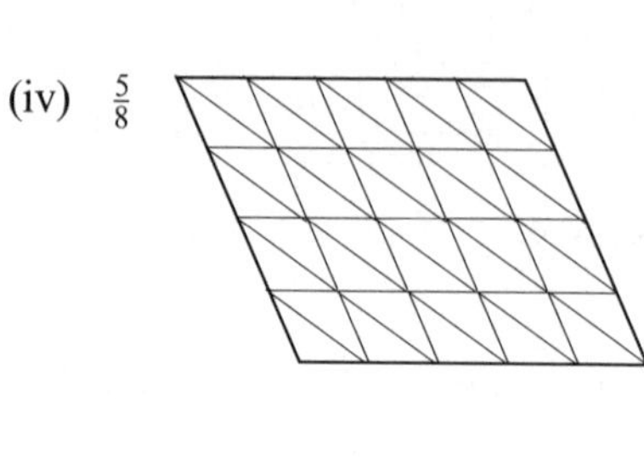

Figure 2.3

2 In the diagrams in Figure 2.4, a certain fraction has been shaded. State, in its simplest form, what that fraction is.

(i)

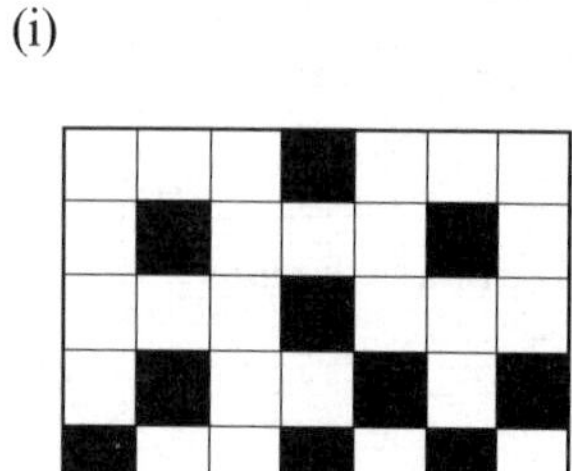

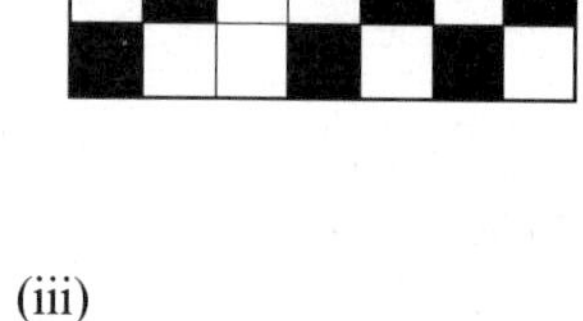

(ii)

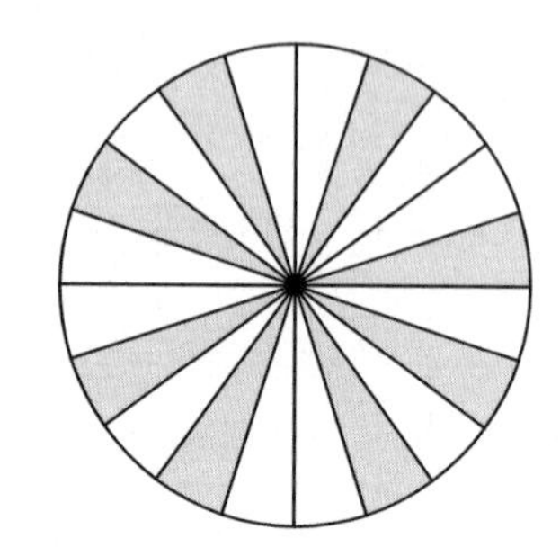

(iii)

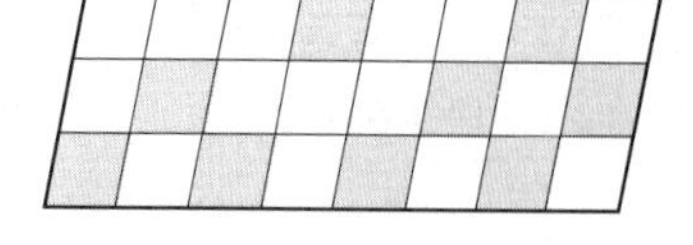

(iv)

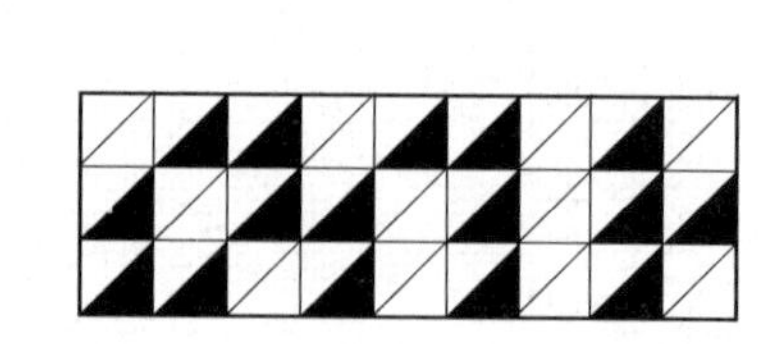

Figure 2.4

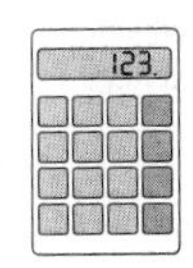

Working with fractions has always been found difficult by people, and a calculator with a fraction button $\boxed{a^{b/c}}$ will make it easier. In order to enter say $\frac{2}{3}$ on the calculator, you proceed as follows:

Display

$\boxed{2}$ $\boxed{a^{b/c}}$ $\boxed{3}$ $\boxed{2 \lrcorner 3.}$

The calculator will also reduce a fraction to its simplest form. To simplify $\frac{24}{80}$, you proceed as follows:

Display

$\boxed{24}$ $\boxed{a^{b/c}}$ $\boxed{80}$ $\boxed{=}$ $\boxed{3 \lrcorner 10.}$

So $\frac{24}{80} = \frac{3}{10}$.

(iii) Improper fractions and mixed numbers

An improper fraction has its numerator *greater* than its denominator, for example $\frac{17}{3}$. Since this means seventeen thirds, it can be split up in the following way:

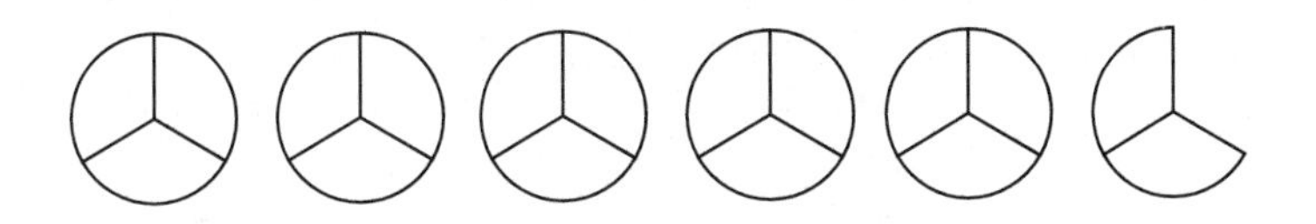

Figure 2.5

i.e. 5 and $\frac{2}{3}$ or $5\frac{2}{3}$. This type of number is called a **mixed number**.

$17 \div 3 = 5$ remainder 2.

Hence $\frac{17}{3} = 5$ and $\frac{2}{3} = 5\frac{2}{3}$

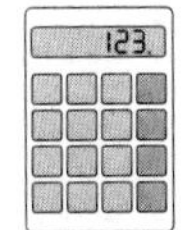

The calculator handles improper fractions very easily:

Display

$\boxed{17}$ $\boxed{a^{b/c}}$ $\boxed{3}$ $\boxed{=}$ $\boxed{5 \lrcorner 2 \lrcorner 3.}$

This shows how $5\frac{2}{3}$ would be displayed. To actually enter $5\frac{2}{3}$ to start with, you would proceed:

Display

$\boxed{5}$ $\boxed{a^{b/c}}$ $\boxed{2}$ $\boxed{a^{b/c}}$ $\boxed{3}$ $\boxed{5 \lrcorner 2 \lrcorner 3.}$

The reverse process of changing a mixed number into an improper fraction is relatively straightforward:

$$6\frac{3}{5} = 6 + \frac{3}{5} = \frac{6 \times 5}{5} + \frac{3}{5} = \frac{30 + 3}{5} = \frac{33}{5}$$

(iv) Working with fractions

Addition and subtraction

Fractions can *only* be added or subtracted if they have the same denominator.

For example

$$\frac{2}{9} + \frac{3}{9} = \frac{5}{9}.$$

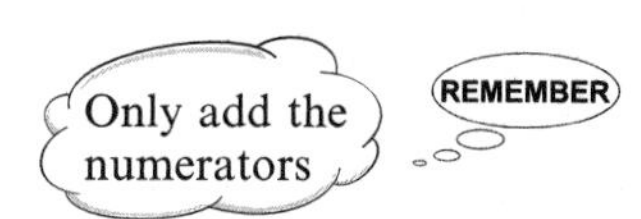

[*Note*: You do *not* change the denominator.]

Example 2.4

Find:

(a) $\frac{1}{4} + \frac{5}{8}$ (b) $\frac{2}{3} + \frac{1}{2}$ (c) $2\frac{1}{2} + 3\frac{1}{3}$ (d) $5\frac{1}{3} - 1\frac{11}{12}$.

Solution

(a) The smallest number that 4 and 8 divide into exactly is 8 (this is the common denominator).

Hence $\frac{1}{4}$ is written $\frac{2}{8}$.

So

$$\frac{1}{4} + \frac{5}{8} = \frac{2}{8} + \frac{5}{8} = \frac{7}{8}.$$

(b) The smallest number that 3 and 2 divide into exactly is 6.

Hence $\frac{2}{3}$ is written $\frac{4}{6}$, and $\frac{1}{2}$ as $\frac{3}{6}$.

So

$$\frac{2}{3} + \frac{1}{2} = \frac{4}{6} + \frac{3}{6} = \frac{7}{6} = 1\frac{1}{6}.$$

(c) It is probably easier to change each mixed fraction into an improper fraction first.

So

$$2\tfrac{1}{2} = 2 + \frac{1}{2} = \frac{2 \times 2}{2} + \frac{1}{2} = \frac{5}{2}$$

and $$3\tfrac{1}{3} = 3 + \frac{1}{3} = \frac{3 \times 3}{3} + \frac{1}{3} = \frac{10}{3}$$

The common denominator will be 6.

Hence

$$\frac{5}{2} + \frac{10}{3} = \frac{15}{6} + \frac{20}{6} = \frac{35}{6} = 5\tfrac{5}{6}.$$

(d) $$5\tfrac{1}{3} - 1\tfrac{11}{12} = \frac{16}{3} - \frac{23}{12} = \frac{64}{12} - \frac{23}{12} = \frac{41}{12} = 3\tfrac{5}{12}.$$

Using a calculator on parts (a) and (d) of Example 2.4, for example, we have as follows:

Display

(a) [1] [$a^{b/c}$] [4] [+] [5] [$a^{b/c}$] [8] [=] [7⌟8.]

(d) [5] [$a^{b/c}$] [1] [$a^{b/c}$] 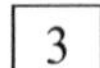  [1] [$a^{b/c}$] [11] [$a^{b/c}$]

Display

[12] [=] [3⌟5⌟12.]

Multiplication

$$\frac{1}{4} \times \frac{1}{2} \text{ means } \frac{1}{4} \text{ of } \frac{1}{2} = \frac{1}{8}$$

Hence

$$\frac{1}{4} \times \frac{1}{2} = \frac{1 \times 1}{4 \times 2} = \frac{1}{8}.$$

To multiply two (or more) fractions, you multiply the two (or more) denominators, and multiply the two (or more) numerators.

To multiply two (or more) mixed numbers, convert them to improper fractions first.

Example 2.5

Work out:

(a) $\frac{2}{5} \times \frac{5}{8}$ (b) $\frac{14}{15} \times \frac{5}{21}$.

Solution

(a) $$\frac{2}{5} \times \frac{5}{8} = \frac{2 \times 5}{5 \times 8} = \frac{10}{40} = \frac{1}{4} \qquad \left[\text{or: } \frac{\cancel{2}^{\,1}}{\cancel{5}_{\,1}} \times \frac{\cancel{5}^{\,1}}{\cancel{8}_{\,4}} = \frac{1}{4}\right].$$

(b) This example is *much easier* if you cancel before multiplying:

$$\frac{\overset{(\div 7)}{\cancel{14}}\,^{2}}{\underset{(\div 5)}{\cancel{15}}\,_{3}} \times \frac{\overset{(\div 5)}{\cancel{5}}\,^{1}}{\underset{(\div 7)}{\cancel{21}}\,_{3}} = \frac{2 \times 1}{3 \times 3} = \frac{2}{9}.$$

By calculator, part (b) of Example 2.5 would become:

Display

[14] [$a^{b/c}$] [15] [×] [5] [$a^{b/c}$] 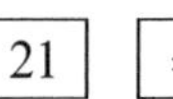 [21] [=] [2⌟9.]

Notice that the calculator *always* gives the answer in its lowest terms.

Example 2.6

Find as a single number $3\frac{1}{7} \times 4\frac{1}{2} \times 1\frac{5}{9}$.

Solution

$$3\frac{1}{7} = \frac{22}{7}, \quad 4\frac{1}{2} = \frac{9}{2}, \quad 1\frac{5}{9} = \frac{14}{9}$$

So

$$3\frac{1}{7} \times 4\frac{1}{2} \times 1\frac{5}{9} = \frac{\cancel{22}^{11}}{\cancel{7}_{1}} \times \frac{\cancel{9}^{1}}{\cancel{2}_{1}} \times \frac{\cancel{14}^{2}}{\cancel{9}_{1}} = \frac{11 \times 1 \times 2}{1 \times 1 \times 1} = 22.$$

Division

To find $\frac{1}{2} \div \frac{3}{4}$, you invert the second fraction and multiply.

So

$$\frac{1}{2} \div \frac{3}{4} = \frac{1}{2} \times \frac{4}{3} = \frac{4}{6} = \frac{2}{3}.$$

To understand this is quite difficult.

The statement $\frac{1}{2} \div \frac{3}{4}$ is asking how many $\frac{3}{4}$ make $\frac{1}{2}$. Since $\frac{1}{2}$ is less than $\frac{3}{4}$, it can only be a part of the $\frac{3}{4}$.

In fact it is $\frac{2}{3}$ of it.

Figure 2.6 shows you that $\frac{2}{3}$ of the shaded region has been double shaded. This cross-hatched region is $\frac{1}{2}$ of the complete square.

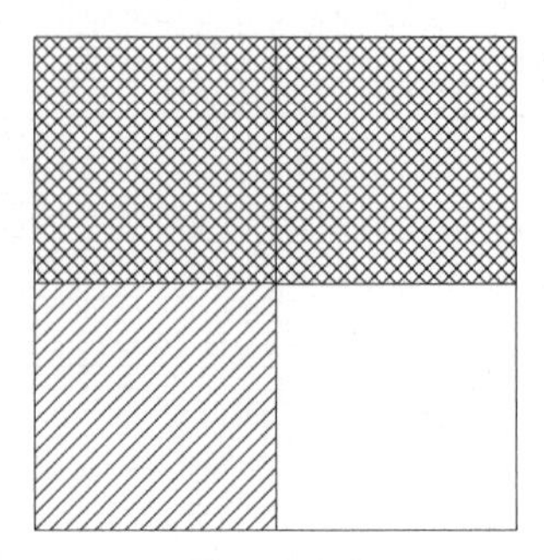
Figure 2.6

Hence

$$\frac{2}{3} \text{ of } \frac{3}{4} = \frac{1}{2} \qquad \left[\frac{2}{3} \times \frac{3}{4} = \frac{1}{2}\right]$$

or

$$\frac{1}{2} \div \frac{3}{4} = \frac{2}{3}.$$

Example 2.7

Work out:

(a) $\frac{7}{12} \div 1\frac{1}{2}$ (b) $6 \div \frac{1}{3}$ (c) $\frac{3}{5} \div 4$ (d) $3\frac{1}{7} \div 4\frac{5}{7}$.

Solution

(a) $\frac{7}{12} \div 1\frac{1}{2} = \frac{7}{12} \div \frac{3}{2} = \frac{7}{\cancel{12}_6} \times \frac{\cancel{2}^1}{3} = \frac{7 \times 1}{6 \times 3} = \frac{7}{18}$

(b) $6 \div \frac{1}{3} = \frac{6}{1} \times \frac{3}{1} = 18$ $\left[\textit{Note } 6 \text{ is written as } \frac{6}{1}\right]$

(c) $\frac{3}{5} \div 4 = \frac{3}{5} \div \frac{4}{1} = \frac{3}{5} \times \frac{1}{4} = \frac{3}{20}$

(d) $3\frac{1}{7} \div 4\frac{5}{7} = \frac{22}{7} \div \frac{33}{7} = \frac{\cancel{22}^2}{\cancel{7}_1} \times \frac{\cancel{7}^1}{\cancel{33}_3} = \frac{2}{3}.$

Example 2.8

Tina is given pocket money every week by her mother. She saves one-half of it, spends one-quarter of it on a magazine, and one-third of the remainder on sweets. She is left with 80p in her purse. How much pocket money is she given each week?

Solution

This problem requires a good knowledge of fractions. It is the sort of problem that people find quite difficult.

Look at the fractions involved (leaving out the units).

$$\left.\begin{array}{ll}\text{savings} & = \dfrac{1}{2} \\[2ex] \text{magazine} & = \dfrac{1}{4}\end{array}\right\} \quad \frac{1}{2} + \frac{1}{4} = \frac{3}{4}$$

This means she has $\frac{1}{4}$ left (the remainder).

$$\text{Now } \tfrac{1}{3} \text{ of the remainder} = \frac{1}{3} \text{ of } \frac{1}{4}$$

$$= \frac{1}{3} \times \frac{1}{4} = \frac{1}{12}$$

$$\text{Hence the amount spent} = \frac{3}{4} + \frac{1}{12} = \frac{9}{12} + \frac{1}{12} = \frac{10}{12} = \frac{5}{6}$$

At the end of all of this then, $\frac{5}{6}$ has been spent, and so $\frac{1}{6}$ has not been spent.

Therefore, $\frac{1}{6}$ of the pocket money is 80p [The amount left.]

The pocket money $= 6 \times 80\text{p} = £4.80$.

Exercise 2(a)

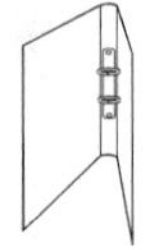

1 Reduce the following fractions to their lowest terms:

(i) $\frac{6}{16}$ (ii) $\frac{16}{24}$ (iii) $\frac{18}{48}$ (iv) $\frac{60}{240}$

(v) $\frac{45}{75}$ (vi) $\frac{72}{84}$ (vii) $\frac{44}{121}$ (viii) $\frac{120}{140}$.

2 In the following questions, arrange the fractions in order of increasing size:

(i) $\frac{3}{5}, \frac{4}{7}$ (ii) $\frac{3}{8}, \frac{5}{12}$ (iii) $\frac{1}{3}, \frac{2}{9}, \frac{4}{5}$

(iv) $\frac{3}{8}, \frac{1}{3}, \frac{2}{5}$ (v) $\frac{7}{10}, \frac{3}{4}, \frac{4}{5}$ (vi) $\frac{5}{9}, \frac{1}{3}, \frac{5}{6}$.

3 Express as mixed numbers:

(i) $\frac{5}{2}$ (ii) $\frac{10}{3}$ (iii) $\frac{18}{8}$

(iv) $\frac{44}{14}$ (v) $\frac{86}{16}$ (vi) $\frac{44}{32}$.

4 Write as improper fractions:

(i) $2\frac{1}{4}$ (ii) $4\frac{2}{3}$ (iii) $3\frac{2}{9}$

(iv) $6\frac{5}{7}$ (v) $5\frac{12}{13}$ (vi) $10\frac{7}{24}$

5 In a jar containing red, black and green sweets, one-third were red and one-quarter were black. What fraction were green?

6 Simplify the following:

(i) $\frac{2}{5}+\frac{1}{3}$ (ii) $\frac{4}{9}+\frac{5}{6}$ (iii) $1\frac{1}{2}+\frac{4}{5}$

(iv) $2\frac{3}{7}+3\frac{5}{12}$ (v) $1\frac{1}{3}+2\frac{1}{2}$ (vi) $2\frac{3}{8}+1\frac{2}{5}$

(vii) $\frac{3}{4}-\frac{1}{8}$ (viii) $5\frac{2}{3}-3\frac{4}{9}$ (ix) $2\frac{3}{5}-\frac{5}{8}$

(x) $5\frac{1}{5}-2\frac{3}{10}$ (xi) $2\frac{1}{2}-1\frac{5}{8}$ (xii) $4-1\frac{1}{3}-2\frac{1}{2}$

(xiii) $3\frac{1}{2}+\frac{1}{4}-2\frac{1}{3}$ (xiv) $4\frac{14}{15}-2\frac{2}{3}+3\frac{5}{12}$.

7 Simplify:

(i) $\frac{3}{4}\times\frac{2}{3}$ (ii) $2\times\frac{7}{8}$ (iii) $1\frac{5}{8}\times\frac{2}{39}$

(iv) $1\frac{3}{4}\times 1\frac{3}{4}$ (v) $7\frac{4}{5}\times 1\frac{1}{4}\times 5\frac{5}{13}$ (vi) $\left(\frac{5}{9}\div 7\frac{1}{7}\right)\times 1\frac{4}{5}$.

8 What is one-fifth of $6\frac{2}{3}$?

9 Evaluate three-eighths of £96.

10 Work out:

(i) $\frac{2}{3}\div\frac{4}{9}$ (ii) $\frac{1}{4}\div\frac{1}{2}$ (iii) $1\frac{1}{7}\div 1\frac{1}{3}$

(iv) $3\frac{3}{4}\div 2\frac{1}{2}$ (v) $4\frac{1}{3}\div 2\frac{4}{5}\div 3\frac{5}{7}$.

11 The Patel family spend $\frac{1}{3}$ of their income on rent, $\frac{1}{4}$ on food, and $\frac{1}{5}$ on clothes. If they are left with £39 each week, find:

(i) The total fraction spent on rent, food and clothes
(ii) The fraction which is left
(iii) Their weekly income.

12 A piece of wood 2 metres long, is cut into three equal pieces. One of these pieces is then cut into four equal pieces. What fraction of the total length will one of the small pieces be, and how long is it?

2.2 Decimal fractions

In the denary system, fractions appear in the columns to the right of the decimal point. They are called **decimal fractions** or just **decimals**.

The headings to the columns are as follows:

Tens	*Units*	.	*Tenths*	*Hundredths*	*Thousandths*
10	1	.	$\frac{1}{10}$	$\frac{1}{100}$	$\frac{1}{1000}$
8	4	.	4	2	

The number 84.42 really means $84 + \frac{4}{10} + \frac{2}{100}$.

Example 2.9

Change the following decimals into fractions:
(a) 5.12 (b) 1.205 (c) 17.08

Solution

(a) 5.12 means $5 + \frac{1}{10} + \frac{2}{100} = 5 + \frac{10}{100} + \frac{2}{100} = 5 + \frac{\cancel{12}^{3}}{\cancel{100}_{25}} = 5\frac{3}{25}$

(b) 1.205 means $1 + \frac{\cancel{205}^{41}}{\cancel{1000}_{200}} = 1\frac{41}{200}$

(c) 17.08 means $17 + \frac{\cancel{8}^{2}}{\cancel{100}_{25}} = 17\frac{2}{25}$.

A fraction can be changed into a decimal by a simple division process.

Example 2.10

Change into a decimal:

(a) $\frac{4}{5}$ (b) $\frac{3}{8}$.

Solution

(a) $\frac{4}{5}$ means $4 \div 5 = 0.8$

$$\begin{array}{r} 0.8 \\ 5\overline{)4.0} \end{array}$$

(b) $\frac{3}{8}$ means $3 \div 8 = 0.375$

$$\begin{array}{r} 0.3\ 7\ 5 \\ 8\overline{)3.0^{6}0^{4}0} \end{array}$$

Sometimes, when changing a fraction into a decimal, the answer is not exact, and the same digits are repeated indefinitely. We say that the digits **recur**, and that the decimal is a **recurring decimal**. The repeating pattern is shown by a dot over the repeating digits.

Example 2.11

Express as a decimal:

(a) $\frac{2}{9}$ (b) $5\frac{2}{7}$.

Solution

(a) $\frac{2}{9}$ means $2 \div 9$

$$\begin{array}{r} 0.2\ 2\ 2 \\ 9\overline{)2.0^{2}0^{2}0} \end{array}$$

Since the 2 repeats $\frac{2}{9} = 0.\dot{2}$

(b) $5\frac{2}{7} = \frac{37}{7}$ means $37 \div 7$

$$\begin{array}{r} 5.\ 2\ 8\ 5\ 7\ 1\ 4\ 2\ 8 \\ 7\overline{)37.^{2}0^{6}0^{4}0^{5}0^{1}0^{3}0^{2}0^{6}0} \end{array}$$

The digits 285714 keep repeating, so that $5\frac{2}{7}$ is written as $5\frac{2}{7} = 5.\dot{2}8571\dot{4}$.

Rule

All fractions can be expressed either as an exact decimal, or a recurring decimal.

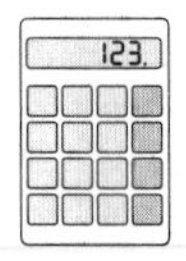

Using a calculator to express a fraction as a decimal is simply a question of carrying out a division sum the correct way round. So $\frac{2}{9}$ as a decimal is $2 \div 9$.

Investigation V: Decimal fractions

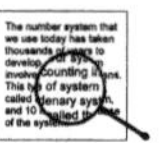

Use the calculator to change the following fractions into decimals. Write down as many decimal places as your calculator shows.

$$\frac{1}{9} = \qquad \frac{2}{9} = \qquad \frac{3}{9} = \ldots, \text{ etc.}$$

What do you notice about the answers?

Now try:

$$\frac{1}{11} = \qquad \frac{2}{11} = \qquad \frac{3}{11} = \ldots, \text{ etc.}$$

What do you notice this time?

It was not too difficult to spot what was happening

Now try:

$$\frac{1}{7} = \qquad \frac{2}{7} = \qquad \frac{3}{7} = \ldots, \text{ etc.}$$

Can you see the pattern this time?

Try some fraction sequences of your own.

2.3 Decimal places and significant figures

Decimal places (d.p.) refer to the digits to the *right* of the decimal point, and **significant figures** (sig. fig.) include all of the digits in a number to the right of the first non-zero digit. (0.0026 has two significant figures) or at the end of a number to the left of the decimal point (3600 has two significant figures). In working with numbers, too many figures are often in an answer. These are reduced by giving answers *corrected* or *rounded* to an appropriate number of decimal places or significant figures.

Rounding or correcting consists of discarding any digits beyond the specified number and adjusting the last digit retained where necessary.

The rules for rounding used in this book are:

Rules

RULE

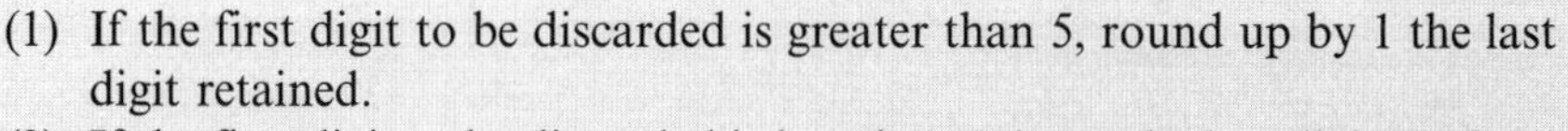

(1) If the first digit to be discarded is greater than 5, round up by 1 the last digit retained.

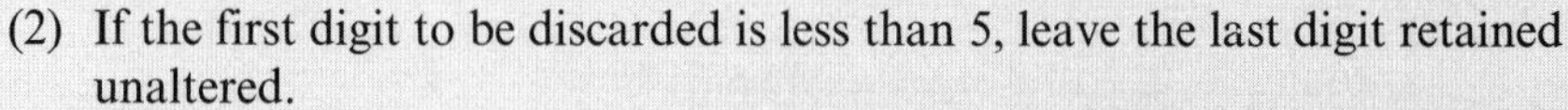

(2) If the first digit to be discarded is less than 5, leave the last digit retained unaltered.

(3) If the first digit to be discarded is 5, round up by 1 the last digit retained.

The following numbers are all rounded to 2 significant figures.

7.2607 rounds to 7.3 (rule (1))
7.2407 rounds to 7.2 (rule (2))
7.2507 rounds to 7.3 (rule (3))
827 rounds to 830 (rule (1))
999 rounds to 1000 (rule (1)).

These numbers are written to 2 decimal places:

2.7283 is 2.73 (rule (1))
2.7243 is 2.72 (rule (2))
2.0743 is 2.07 (rule (2))
1.9984 is 2.00 (rule (1))
10.845 is 10.85 (rule (3)).

It is important not to round numbers *too early* in a calculation, otherwise accuracy may be lost.

2.4 Standard index or scientific notation

(i) Positive indices

When numbers become very large or very small, **standard form** (**standard index form** or **scientific notation**) is used to make the numbers more manageable.

Since $10^6 = 1000000$, then 8.6×10^6 is said to be in standard form (a number *between* 1 and 10 (not including 10) multiplied by a power of 10).

Example 2.12

Write in standard form correct to 3 sig. fig.:
(a) 41460 (b) 3201.7.

Solution

(a) 41460 is 4.146×10^4.
Correcting 4.146 to 3 sig. fig. gives 4.15.
Hence the answer is 4.15×10^4.
(b) 3201.7 is 3.2017×10^3.
Correcting 3.2017 to 3 sig. fig. gives 3.20.
Hence the answer is 3.20×10^3.

(ii) Negative indices

The idea of a negative power is not easy to understand. Whereas 10^3 means $10 \times 10 \times 10$, 10^{-1} (that is 10 to the power negative one) does not have an obvious meaning. To make things consistent later on, we define 10^{-1} as being $1 \div 10$ or $\frac{1}{10}$. See Section 10.1 for a fuller explanation. Similarly, $\frac{1}{100}$ is written as 10^{-2} and so on.

$$
\begin{aligned}
\text{Now } 0.0024 &= 2.4 \div 1000 \\
&= 2.4 \times \frac{1}{1000} = 2.4 \times 10^{-3}.
\end{aligned}
$$

We now have a notation that can be used for *small* numbers.

Example 2.13

Write in standard form correct to 3 sig. fig.

(a) 0.00004627 (b) 0.004003.

Solution

(a)
$$
\begin{aligned}
0.00004627 &= 4.627 \div 100000 \\
&= 4.627 \times \frac{1}{100000} \\
&= 4.627 \times 10^{-5} \\
&= 4.63 \times 10^{-5} \text{ (3 sig. fig.).}
\end{aligned}
$$

(b)
$$
\begin{aligned}
0.004003 &= 4.003 \div 1000 \\
&= 4.003 \times \frac{1}{1000} \\
&= 4.003 \times 10^{-3} \\
&= 4.00 \times 10^{-3} \text{ (3 sig. fig.).}
\end{aligned}
$$

(iii) Addition and subtraction of numbers in standard form

You have to be careful not to take short cuts here if you are not using a calculator.

Example 2.14

Evaluate, giving your answers in standard form:

(a) $2.8 \times 10^4 + 1.6 \times 10^3$ (b) $2.5 \times 10^{-4} - 1.6 \times 10^{-5}$.

Solution

(a) First, express each number in full.
So $2.8 \times 10^4 = 28000$, $1.6 \times 10^3 = 1600$

Hence $2.8 \times 10^4 + 1.6 \times 10^3 = \begin{array}{r} 28000 \\ +\ 1600 \\ \hline 29600 \end{array}$

The answer in standard form $= 2.96 \times 10^4$.

(b) $2.5 \times 10^{-4} - 1.6 \times 10^{-5} = \begin{array}{r} 0.000250 \\ -0.000016 \\ \hline 0.000234 \end{array}$

Always have the same number of decimal places in each number

The answer in standard form is 2.34×10^{-4}.

Standard form can be entered directly on certain calculators. On the keyboard being used here the button is [EXP]. The following examples show you how it works, and how calculations can be carried out in standard form.

Example 2.15

Use the calculator to evaluate:

(a) $4.65 \times 10^{14} + 3.5 \times 10^{15}$
(b) $(6.85 \times 10^{-2} \times 1\frac{3}{4}) \div 5.63 \times 10^{-8}$.

Solution

(a) [4.65] [EXP] [14] [+] [3.5] [EXP] [15] [=] — **Display** [3.965^{15}]

Notice that 4.65×10^{14} is displayed as 4.65^{14}. You must always remember to insert the 10 when writing the answer, otherwise it is confused with a power. The answer to the problem is 3.965×10^{15}.

(b) [(] [6.85] [EXP] [2] [+/−] [×] [1.75] [)] [÷]

[5.63] [EXP] [8] [+/−] [=] — **Display** [2129218.5]

Notice here that 6.85×10^{-2} is displayed 6.85^{-02}, and also that the answer is not written in standard form.

A suitable answer might be 2.13×10^6 (3 sig. fig.).

Example 2.16

The distance that light travels in one year is known as a light year. The nearest star to earth apart from the Sun is Alpha Centauri which is 4 light years away. If light travels at 3×10^5 km per second, find the distance from earth to Alpha Centauri.

Solution

Astronomy is full of large numbers, and it is important to be able to use standard form properly if you are to understand the calculations involved. To find the distance that light travels in one year, we need to find the number of seconds in one year:

60	×	60	×	24	×	365
Seconds in one minute		Minutes in one hour		Hours in one day		Days in one year

$$= 31536000 \text{ seconds}$$

The distance travelled by light in this time must be:

$$31536000 \times 3 \times 10^5 \text{ km} = 9.46 \times 10^{12} \text{ km}$$

This number is 9.46 million million km. It is known as 1 light year.

The distance between the earth and Alpha Centauri is:

$$4 \times 9.46 \times 10^{12} \text{ km} = 3.784 \times 10^{13} \text{ km}.$$

2.5 Working with decimals

(i) Addition and subtraction

When adding and subtracting decimal fractions, it is important to keep decimal places in the correct column.

Example 2.17

Evaluate:

(a) $2.6 + 3.84 + 12.96$ (b) $12.8 - 6.74$ (c) $9.8 - 12.63$.

Solution

(a)

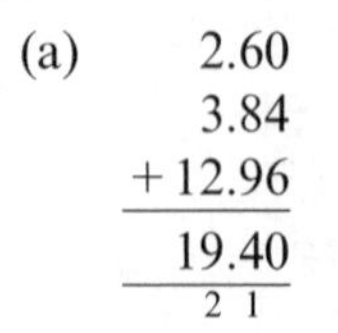

$$\begin{array}{r} 2.60 \\ 3.84 \\ +12.96 \\ \hline 19.40 \\ \hline {\scriptstyle 2\ 1} \end{array}$$

Put a zero after the 6 to give the same number of columns.

The answer is 19.40

(b)

$$\begin{array}{r} {\scriptstyle 7\ 10} \\ 12.\not{8}\,\not{0} \\ -\ 6.7\,4 \\ \hline 6.0\,6 \\ \hline \end{array}$$

Put a zero after the 8 to give the same number of columns.

Hence $12.8 - 6.74 = 6.06$.

(c) This is a much more difficult problem. The easiest way is to subtract 9.8 from 12.63 and call the answer negative.

$$\begin{array}{r} 12.63 \\ -\ 9.80 \\ \hline 2.83 \\ \hline \end{array}$$

Hence $9.8 - 12.63 = -2.83$.

(ii) Multiplication and division

Most people would use a calculator to carry out multiplication and division of decimals. Some simple examples using pencil and paper techniques are given here for completeness.

Example 2.18

Work out:

(a) 2.91×1.2 (b) 1.45×2.8 (c) $8.63 \div 1.2$ (correct to 2 d.p.)

Solution

(a) Work out 291×12 first without any decimal places.

$$291 \times 12 = 3492$$

There are 3 digits after the decimal points altogether in the two numbers, so putting 3 decimal places into the answer gives:

$$\begin{aligned} 2.91 \times 1.2 &= 3.492 \\ &= 3.49 \text{ (2 d.p.).} \end{aligned}$$

(b) Work out 145×28 first.

$$145 \times 28 = 4060$$

The problem contains 3 decimal places, hence counting 3 from the right gives 4.060.

The answer is 4.06 (2 d.p.).

Note that you would get the wrong answer if you left out the zero on the end, and worked with 406 to give an answer of 0.406!

(c) $$\frac{8.63}{1.2} = \frac{8.63 \times 10}{1.2 \times 10} = \frac{86.3}{12}$$

We have multiplied each number by 10, so that we are dividing by a *whole number*.

$$\begin{array}{r} 7.\;1\;\;9\;1\;.\,.\,. \\ 12\overline{)86.{}^{2}3{}^{11}0{}^{2}0{}^{8}0} \end{array}$$

Correct to 2 d.p. the answer is 7.19.

Exercise 2(b)

In this exercise, try and answer as many questions as you can *without* a calculator.

1 Change into fractions in simplest form:

(i) 0.8 (ii) 0.625 (iii) 2.125 (iv) 4.15.

2 Change into decimals:

(i) $\frac{3}{8}$ (ii) $\frac{7}{20}$ (iii) $\frac{3}{16}$ (iv) $\frac{19}{40}$.

3 Write, correct to (a) 2 dec. pl. (b) 3 sig. fig., the following numbers:

(i) 0.507 (ii) 0.00858 (iii) 41.125 (iv) 34.827

(v) 1985.026 (vi) 285.799 (vii) 99.99.

4 Express the following fractions as decimals correct to 3 d.p.:

(i) $\frac{9}{40}$ (ii) $\frac{11}{7}$ (iii) $\frac{8}{21}$ (iv) $\frac{5}{9}$

(v) $\frac{13}{14}$.

5 Write the following in standard index form:

(i) 0.06 (ii) 0.0058 (iii) 81.9 (iv) 6823

(v) 250000.

6 Evaluate the following:

(i) $4.32 + 2.06 + 0.63$ (ii) $2.75 - 1.086$

(iii) $16.8 - 2.09$ (iv) $2.3 - 5.8$

(v) $12.84 + 0.96 - 2.83$ (vi) $0.91 - 43 + 3.75$.

7 Express the answers to the following in standard form:

(i) $3 \times 10^2 + 6 \times 10^2$ (ii) $4.8 \times 10^3 + 2.5 \times 10^2$

(iii) $8 \times 10^{-2} + 4 \times 10^{-2}$ (iv) $2.53 \times 10^{-5} - 6.24 \times 10^{-6}$.

8 Work out exactly:

(i) 22×0.05 (ii) 0.003×1.4

(iii) $0.21 \div 0.7$ (iv) $0.081 \div 0.09$

(v) 2.06×3.1 (vi) 12.1×0.04

(vii) $13.02 \div 0.02$ (viii) $0.432 \div 0.6$

(ix) $150 \div 0.1$ (x) 25×0.12.

9 Evaluate correct to 2 d.p.

(i) $23.5 \div 1.1$ (ii) $8.1 \times 2.3 \times 0.4$

(iii) $186 \div 2.5$.

10 The speed of light is 3×10^8 m/s. Find the distance that light travels in one year (365 days). Give your answer in the form $A \times 10^n$, where A is correct to 3 significant figures. It is found that two stars are 25 light years apart. What is the distance between these two stars? Express your answer in the form $A \times 10^n$.

2.6 Squares and square roots

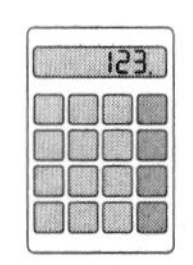

As a power, 8×8 can be written 8^2, i.e. 8 to the power two. This is more often referred to as eight squared, or the square of 8. There is usually a separate button $\boxed{x^2}$ for this on the calculator. However, on the keyboard shown in Figure 1.1, this is $\boxed{\text{INV}}\ \boxed{\sqrt{\ }}$

Now $8 \times 8 = 64$.

We say that 8 is the **square root** of 64. The symbol for square root is $\sqrt{\ }$.

Now $\sqrt{64} = 8$

similarly $\sqrt{49} = 7$.

If we try to find the $\sqrt{55}$, it must lie between 7 and 8 [Because $49 < 55 < 64$].

Display

55	√	7.416198487

Hence correct to 3 sig. fig., $\sqrt{55} = 7.416$.

If you now find 7.416^2 using the calculator

Display

7.416	INV	√	54.997056

However many decimal places you take, you will never quite get back to exactly 55. This is because 55 is *not* an *exact* decimal. It also does not recur, and is called an **irrational number**.

If you feel able to try and follow some simple algebra in the form of a flow chart, the following investigation shows an interesting way of obtaining square roots.

Investigation VI: Square roots

Follow through the flow chart alongside, and write down the first few values in the following table.

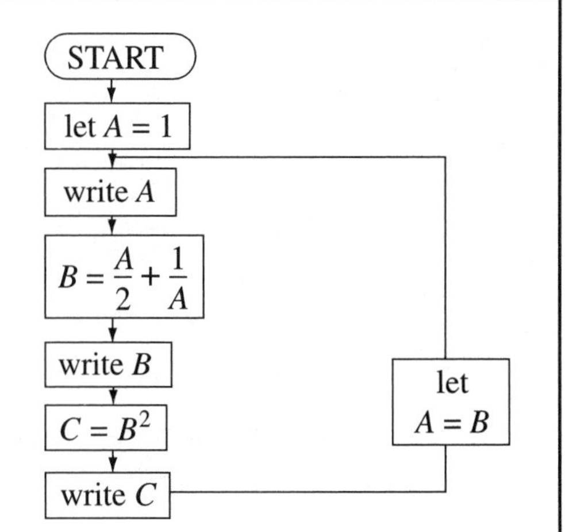

A	B	C
1		

What do you notice about the values of B and C? What do you think this flow chart calculates?

Look at the box $B = \frac{A}{2} + \frac{1}{A}$. Can you see how to alter the expression for $\frac{1}{A}$ so that it will find the square root of a *different* number?

Extension: This sort of repeating process is called an **iteration** method. See if you can invent a method of your own.

2.7 Approximation and the calculator

The ability to approximate is an essential skill in modern day mathematics, now that the calculator is used so frequently. You will often want to know whether your answer is reasonable without going through a long checking process. The major part of approximating is replacing a number that is awkward, by a number that you can easily work with. For example, 18.983×31.064 would be a difficult pencil and paper calculation, but the answer will be close to $20 \times 30 = 600$. The correct answer is in fact 589.69 (2 d.p.). The following example should show you how you can approximate in a variety of situations. The symbol $\approx$ is used to denote 'approximately equal'.

Example 2.19

Estimate to 1 significant figure, the answers to the following calculations.

(a) 2.93×14.7 (b) $8.3 \div 1.9$ (c) $(4.9 \times 3.8) \div 1.96$

(d) The cost of 83 pens at 29p each (e) $\sqrt{385}$

(f) $\left(\sqrt{(8.9)^2 + (4.6)^2} \times 18.56\right) \div 11.09.$

Solution

(a) $2.93 \times 14.7 \approx 3 \times 15 = 45.$

You could give the estimated answer as 40 or 50 to 1 sig. fig.

(b) $8.3 \div 1.9 \approx 8 \div 2 = 4.$

(c) $(4.9 \times 3.8) \div 1.96 \approx (5 \times 4) \div 2 = 10.$

(d) 83 @ 29p $\approx$ 80 @ 30p = 2400p.

£24 is the approximate cost.

(e) Choose a number near to 385 that has got an exact square root.

So $\sqrt{385} \approx \sqrt{400} = 20.$

(f) This last example would appear to be very complicated, but using the ideas above, we can get a good approximation to the answer fairly easily.

$(8.9)^2 \approx 10 \times 10 = 100, \quad (4.6)^2 \approx 5 \times 5 = 25$

So $(8.9)^2 + (4.6)^2 \approx 100 + 25 = 125$

Now $\sqrt{125} \approx 11$, and so we can approximate the sum as follows:

$$\left(\sqrt{(8.9)^2 + (4.6)^2} \times 18.56\right) \div 11.09 \approx (11 \times 20) \div 11 = 20.$$

Exercise 2(c)

Estimate to 1 significant figure the value of the following:

1 18.3×11.4

2 $129 \div 38$

3 $(7.9)^2$

4 The cost of 63 light bulbs at 42p each

5 The difference between 89^2 and 71^2

6 $1\frac{3}{4} \times (8.96 \times 1.9^2)$

7 $83 \times 59.8 \times 70.5$

8 The area of a pane of glass measuring 815 mm by 685 mm

9 $\sqrt{(11.2)^2 + (17.1)^2}$

10 $6.4 \times (293 \div 38.5)^2$.

Use a calculator to evaluate the following, giving your answers correct to 3 significant figures.

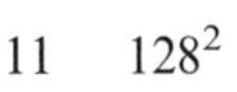

11 128^2

12 85.9×73.86

13 $86900 \div 237$

14 4.9^3

15 $1\frac{4}{5} \times 1\frac{1}{4}$

16 $(8.4 \times 7.9) \div (3.6 \times 1.2)$

17 $128 \div (18.5 \times 16.1)$

18 $\sqrt{(6.9)^2 + (7.3)^2}$

19 2.3^4

20 $128.5 - 63 \times 12.5$

21 $-18.56 \div (2.37)^2$

22 $(12.9^3 - 6.5^3) \div 8.6$

23 $4.5 \times 10^4 + 6.3 \times 10^6$

24 $(1.8 \times 10^{-5}) \div (6.35 \times 10^{-4})$

25 $\sqrt{2.8 \times 10^{-3} \times 1\frac{1}{2}}$

26 $(4.3 \times 10^{-20})^2 \div (6.8 \times 10^{11})$

2.8 The metric system

The common system of units used in everyday calculations is centred around the number 10. The Standard International (SI) system uses the **metre** as the unit of length, the **kilogram** as the unit of mass (commonly misunderstood as weight), and the **second** as the unit of time. There are a number of names associated with decimal multiples as shown in the following table:

mega means $1\,000\,000 = 10^6$
used in megawatt or megatonne

kilo (k) means $1000 = 10^3$

hecto (h) means $100 = 10^2$

deca (da) means $10 = 10^1$

micro (μ) means $\frac{1}{1\,000\,000} = 10^{-6}$
used in microsecond

milli (m) means $\frac{1}{1000} = 10^{-3}$

centi (c) means $\frac{1}{100} = 10^{-2}$

deci (d) means $\frac{1}{10} = 10^{-1}$

There are several units that are in common use that should also be known.

area: 1 **hectare** (ha) = 10 000 square metres, written 10 000 m^2
This is used in land measurements.

volume: 1 litre (l) = 1000 cubic centimetres, written 1000 cm^3
or 1000 millilitres (ml) [**Note**: 1 cm = 1 ml]
This is used mainly in liquid measurements.

1 **centilitre** (cl) = 10 cm^3
or 10 ml

It is important that you can change from one unit to another. The following example shows you how to go about this.

Example 2.20

Change:

(a) 20 m into mm
(b) 650 g into kg
(c) 200 hm into cm
(d) 10^8 μs into minutes
(e) 2 m^2 into cm^2
(f) 2000 l into m^3
(g) 6 m^3 into cl.

Solution

(a) Since 1 mm = $\frac{1}{1000}$ m, then 1 m = 1000 mm (mm means millimetre)
Hence 20 m is 20 × 1000 mm = 20 000 mm.

(b) Since 1 kg = 1000 g, then 1 g = $\frac{1}{1000}$ kg (g stands for gram)
Hence 650 g is 650 × $\frac{1}{1000}$ kg = $\frac{650}{1000}$ kg = 0.65 kg.

(c) 1 hm is 100 m = 100 × 100 cm = 10 000 cm (hm is for hectometre)
So 200 hm is 200 × 10 000 cm = 2 000 000 cm.

(d) Now 1 μs is $\frac{1}{1\,000\,000}$ s, hence 10^6 μs = 1 s

10^8 μs is $10^2 \times 10^6$ μs = $10^2 \times 1$ s
= 100 s.

But 60 seconds = 1 minute (min.)
So 10^8 μs = 1 min 40 s.

(e) Always be very careful when changing *area* or *volume* units.
2 m^2 means 2 square metres
Now 1 square metre measures 1 m by 1 m, i.e. 100 cm by 100 cm, giving an area of 100 × 100 cm^2 = 10 000 cm^2
Hence 2 m^2 = 2 × 100 × 100 cm = 20 000 cm^2.
[*Note*: You do *not* simply multiply 2 m^2 by 100 to change into cm^2.]

(f) $2000\ \text{l} = 2000 \times 1000\ \text{cm}^3$
$= 2\,000\,000\ \text{cm}^3$
$= 2\,000\,000 \times 1\ \text{cm} \times 1\text{cm} \times 1\ \text{cm}$
$= 2\,000\,000 \times \frac{1}{100}\ \text{m} \times \frac{1}{100}\ \text{m} \times \frac{1}{100}\ \text{m}$
$= 2\ \text{m}^3$.

(g) $6\ \text{m}^3$ means $6 \times 1\ \text{m} \times 1\ \text{m} \times 1\ \text{m}$
$= 6 \times 100\ \text{cm} \times 100\ \text{cm} \times 100\ \text{cm}$
$= 6\,000\,000\ \text{cm}^3$
$= 6000 \times 100\ \text{cl} = 600\,000\ \text{cl}$.

Time is a quantity that is often not handled very well. Although you may feel that you would not make a mistake in calculations with time, check that you would have answered the following question correctly.

Example 2.21

(a) How many minutes are there between 2135 on Tuesday and 0112 the next day?
(b) How many days are there from 20 March to 15 April inclusive?
(c) A bus service runs every 15 minutes. If there is a bus at 0810, how many buses are there between 0700 and 1000?

Solution

(a) When working *across* midnight, it is best to consider before and after separately.

From 2135 until midnight, is 2 h 25 min $= 2 \times 60 + 25$ minutes
$= 145$ minutes

After midnight, 1 h 12 min $= 72$ minutes

The total is $145 + 72 = 217$ minutes.

(b) March has 31 days

Now $31 - 20 = 11$.

But inclusive means you need to add 1 (check this by counting), giving 12 days.

1 April until 15 April inclusive is 15 days

The total $= 12 + 15 = 27$ days.

(c) The safest way here is to list the times of the buses. These are:
0710 0725 0740 0755 0810 0825 0840 0855 0910 0925 0940 0955
Hence there are 12 buses.

2.9 Speed, distance, time

Common units of speed are metres per second (m/s or ms^{-1}), and kilometres per hour (km/h or kmh^{-1}). Calculations involving speed are covered by three simple rules:

Rules

(1) $\text{average speed} = \dfrac{\textit{Total}\text{ distance travelled}}{\text{Time taken}}$

(2) $\text{time taken} = \dfrac{\textit{Total}\text{ distance travelled}}{\text{Average speed}}$

(3) *total* distance travelled = average speed × time taken.

When using these rules, you must be *consistent* in the units you use.

Example 2.22

(a) Find the average speed of a car that travels 45 km in 2 hours 15 minutes.

(b) Find the distance in km travelled by an aeroplane moving at 80 m/s for 5 minutes.

(c) Find the time it takes for a cyclist to make 10 laps of a 400 m cycle track cycling at an average speed of 40 km/h.

Solution

(a) You must write 2 hours 15 minutes as $2\frac{1}{4}$ hours. $\left(\frac{15}{60} = \frac{1}{4}\right)$

Then average speed $= 45 \div 2\frac{1}{4}$ km/h **(Rule 1)**

$$= 45 \div \frac{9}{4} \text{ km/h} = 45 \times \frac{4}{9} \text{ km/h}$$

$$= 20 \text{ km/h.}$$

(b) Now 5 minutes $= 5 \times 60$ s $= 300$ seconds

The distance travelled $= 80 \times 300$ metres $= 24\,000$ metres **(Rule 3)**

$= 24$ km since 1000 m $= 1$ km.

(c) 10 laps of 400 m is 4000 m. Change this to km, by dividing by 1000, to give 4 km

Time $= \frac{4}{40}$ hours $= 0.1$ hours **(Rule 2)**

Now 0.1 hours $= 0.1 \times 60$ minutes $= 6$ minutes.

Timetables require you to be aware of all of these ideas; try the examples in Exercise 2(d).

Exercise 2(d)

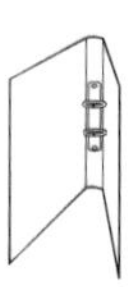

1 Change:

(i) 640 mm into cm
(ii) 86840 m into km
(iii) 6×10^5 m into km
(iv) 840 mg into g
(v) 0.65 km into mm
(vi) 4 m^2 into cm^2
(vii) 60 000 m^2 into ha
(viii) 60 000 ml into litres
(ix) 400 cl into litres
(x) 2 m^3 into cl
(xi) 64 mm^2 into cm^2
(xii) 1000 mm^3 into cl.

2 A man uses the trains between Cambridge and London (King's Cross station) and between London (Paddington station) and Cardiff, where he has to attend meetings. Parts of the relevant timetables are given below.

Cambridge (*depart*)	0530	0610	0630	0705	0735	0745
London (King's Cross) (*arrive*)	0706	0750	0801	0815	0905	0910

London (Paddington) (*depart*)	0700	0815	0845	0915
Cardiff (*arrive*)	0910	1005	1055	1110

It is necessary to allow 28 minutes between leaving the train at King's Cross and getting on the train at Paddington and to allow 23 minutes between leaving the train at Cardiff and arriving at the meeting.

(i) State the arrival time at Cardiff station of the latest train it is possible to use if the man must be at his meeting by 1130.
(ii) On another day his meeting in Cardiff begins at 1015. What is the latest time that he must leave Paddington station if he is to get to his meeting on time?
(iii) On yet a different day he has to travel from Cambridge to the meeting which begins at 1030 in Cardiff. What is the latest time that he can depart from Cambridge in order to arrive at the meeting in time?

3 I live in Rugby, which is 82 miles from London, and I have to be at the meeting in London at 10.30am.

It takes 15 minutes for me to walk from London (Euston station) to my meeting.

Table 2.1 opposite is the railway timetable for the day of my meeting. Assume that all the trains run on time.

(i) What is the departure time of the latest train that I can catch at Rugby in order to arrive at my meeting on time?
(ii) What was the average speed, to the nearest mile per hour, of that train on its journey from Rugby to London?
(iii) The meeting lasts 200 minutes.
At what time does it finish?
(iv) The saver fare from Rugby to London is £8.50.
Express this as a percentage of the return first class fare of £34.

[LEAG]

4

RICHARDSON TRAVEL LTD

STILL NO INCREASE IN FARES!!

ADULTS: £3.70 SINGLE – £6.80 RETURN

CHILDREN: £1.90 SINGLE – £3.50 RETURN

Depart
Sheffield
Pond St Bus Station

0050 0350 0550 0750 0950 1150 1350 1550 1750 1950 2150

Arrive
Manchester Airport
Bus Station

0215 0515 0715 0915 1115 1315 1515 1715 1915 2115 2315

Depart
Manchester Airport
Bus Station

0225 0530 0730 0930 1130 1330 1530 1730 1930 2130 2330

Arrive
Sheffield
Pond St Bus Station

0350 0700 0900 1100 1300 1500 1700 1900 2100 2300 0100

Birmingham and Rugby ⟶ London

Mondays to Fridays

Wolverhampton ....69 d		0641		0727		0707			0757	0741					0803		0812	0827	0816	0849					
Sandwell & Dudley ....69 d				0736					0806								0821	0836		0858					
Birmingham New Street ..... d		0721	0735	0748		0751	0804		0818	0821					0837		0844	0848	0851	0918		0921			0937
Adderley Park .... d		0725	0739			0755	0808			0825									0855			0925			
Stechford........................... d		0729	0743			0759	0812			0829									0859			0929			
Lea Hall............................. d		0731	0745			0801	0814			0831									0901			0931			
Marston Green.................. d		0734	0748			0804	0817			0834									0904			0934			
Birmingham International .. a		0737	0752	0758		0807	0821		0828	0837					0847		0853	0858	0907	0928		0937			0947
d		0737		0758		0807			0828	0837					0847		0854	0858	0907	0928		0937			0947
Hampton-in-Arden .... d		0741				0811				0841									0911			0941			
Berkswell........................... d		0745				0815				0845									0915			0945			
Tile Hill .... d		0749				0819				0849									0919			0949			
Canley............................... d		0752				0822				0852									0922			0952			
Coventry .... a		0755		0809		0825			0839	0856					0858		0903	0909	0926	0939		0956			0958
d		0758		0810		0826			0840						0858			0910		0940					0958
Rugby .... a		0810				0839									0911										1011
d		0811			0832	0841					0904				0911						0955				1011
Long Buckby .... d		0821				0851									0921										1021
Northampton .... a		0831				0901									0932										1033
d		0836				0906									0936										1036
Wolverton .... d		0850													0950										1050
Milton Keynes Central .... a		0855				0923									0955					1012	1019				1055
d	0838	0855				0927							0925	0949	0955					1012	1020		1025		1055
Bletchley .... d	0843	0900											0930		1000								1030		1100
Leighton Buzzard .... d	0850	0907											0937		1007								1037		1107
Cheddington .... d	0855												0942										1042		
Tring.................................. d	0901											0924	0948										1048		
Berkhamsted .... d	0908											0931	0955		1021								1055		1121
Hemel Hempstead .... d	0912											0936	0959		1025								1059		1125
Apsley................................ d												0939	1002										1102		
King's Langley .... d												0943	1006										1106		
Watford Junction ....59 d	0920	0930										0948	1011		1033			1006		1040	1047		1113		1135
Bushey ....59 d																									
Harrow & Wealdstone....59 d	0927											0954	1017										1119		
London Euston ....59 d	0943	0953		0924	0938	1014			0956		1010	1011	1034	1035	1054			1029		1101	1112		1136		1156

Table 2.1

Use the coach timetable above to answer these questions.

(i) What time does the 0550 from Sheffield arrive at Manchester Airport?

(ii) Mr and Mrs Davis and their 3 children have to catch an aeroplane at 1530. They must arrive at the airport at least 1½ hours before the flight time.
   (a) What is the latest coach they can catch from Sheffield?
   (b) They return from holiday 2 weeks later and catch the 1930 coach from Manchester Airport.
   What time do they arrive back at Sheffield?
   (c) How much does the return journey by coach cost them?

[MEG]

5

	YORK	NORWICH	MANCHESTER	LONDON	GLASGOW	EDINBURG	DOVER	CAMBRIDGE	BIRMINGHAM
ABERDEEN	493	771	539	798	230	198	906	724	658
BIRMINGHAM	204	251	129	177	462	460	293	159	
CAMBRIDGE	243	100	249	87	565	526	190		
DOVER	425	262	410	116	747	708			
EDINBURGH	295	573	341	600	74				
GLASGOW	336	610	343	638					
LONDON	315	179	296						
MANCHESTER	103	295							
NORWICH	287								

Table 2.2

Table 2.2 shows the distances, in kilometres, between ten major cities and towns in Great Britain.

For example, the distance between Cambridge and Norwich is 100 km and the distance between Aberdeen and York is 493 km.

(i) Use the table to find the distance from
   (a) Dover to Edinburgh
   (b) Glasgow to York.

(ii) Which of the listed towns is
   (a) furthest from London
   (b) nearest to London?

(iii) (a) Calculate the distance travelled on a return journey between Edinburgh and Norwich.

(b) Which of the single journeys Birmingham to Norwich or Manchester to Cambridge is the longer and by how many kilometres?

(iv) (a) If it takes Miss Johnson 3 hours to drive from Cambridge to Birmingham, calculate her average speed in km/h.

(b) Mr Burns drives from Glasgow to London at an average speed of 87 km/h. Calculate the time of the journey in hours and minutes.

(c) Mr Bell leaves Norwich and heads for Dover. He drives for 4 hours 20 minutes at an average speed of 57 km/h. How far will he be from Dover after this time?

3 Introducing algebra

3.1 Basic definitions

Algebra is a word that often conjures up apprehension in people. The word originates from the arabic word *al-jabr*. It is perhaps best to think of algebra as a kind of shorthand, where a letter is used to represent a number.

Algebra is also used to represent general results in a clear and concise way. Providing you *remember* that whenever a letter is used in algebra it is representing a number, you should be able to learn the rules of algebra fairly easily.

Rules

$x + y$ means *add* a number x *to* number y

$x - y$ means *subtract* number y *from* number x

xy means *multiply* x by y (notice that you do not have to include the $\times$ sign)

$\frac{x}{y}$ means x *divided* by y (i.e. $x \div y$)

$3x$ means $x + x + x$ or 3 lots of x or 3 times x

x^4 means $x \times x \times x \times x$ or x to the power 4

$2y^3$ means $2 \times y^3$ or $2 \times y \times y \times y$

3.2 Formulae and substitution

If we write down $q = p + t$, then $p + t$ is called an **algebraic expression**. This statement is also called a **formula** for q in **terms** of p and t. It is really a piece of mathematical shorthand which tells us that if we want to work out the value of q, then we add together the values of p and t. Hence if $p = 3$ and $t = 5$, then $q = 3 + 5 = 8$.

We have **substituted** the values $p = 3$ and $t = 5$ into the formula.

Example 3.1

If $y = 3t + 2q$, find y in the following cases:

(a) $t = 4, q = 1$ (b) $t = 1, q = -2$ (c) $t = \frac{2}{3}, q = -\frac{1}{2}$

Solution

(a) $3t$ means $3 \times t$ so $3 \times 4 = 12$

$2q$ means $2 \times q$ so $2 \times 1 = 2$

Hence $y = 3t + 2q = 12 + 2 = 14$.

(b) $y = 3 \times 1 + 2 \times -2$

$= 3 + -4$ (since $2 \times -2 = -4$) $= -1$.

(c) This is a more difficult example.

$$y = 3 \times \frac{2}{3} + 2 \times -\frac{1}{2}$$

$$= \frac{6}{3} + -\frac{2}{2} = 2 + -1 = 1.$$

Example 3.2

If $y = 2x^2$, find y when: (a) $x = 3$ (b) $x = -2$ (c) $x = -0.4$

Solution

It is important that you realise that $2x^2$ means $2 \times (x^2)$. In other words, you *square x first*, and *then* multiply the answer by 2.

(a) $y = 2 \times (3^2) = 2 \times 9 = 18$

(b) $y = 2 \times \left((-2)^2\right) = 2 \times 4 = 8$

Here we have used the fact that $-2 \times -2 = +4$

(c) $y = 2 \times \left((-0.4)^2\right) = 2 \times 0.16 = 0.32$.

Example 3.3

If $H = \dfrac{4t}{p}$, find H when:

(a) $t = 4, p = 2$ (b) $t = 6, p = -0.5$.

Solution

Here $H = (4 \times t) \div p$

(a) So $H = (4 \times 4) \div 2 = 16 \div 2 = 8$.

(b) Here $H = (4 \times 6) \div -0.5 = 24 \div -0.5 = -48$.

3.3 Forming expressions

It is important that you can translate a set of instructions into an algebraic expression. Simple logic is usually all that is required.

Example 3.4

If 3 oranges can be bought for N pence, how much would 5 oranges cost?

Solution

To find the cost of one orange, you would divide the total cost by 3 to give $N \div 3$ or $\frac{N}{3}$ pence. The cost of 5 oranges is 5 times the cost of one orange. That is $5 \times \frac{N}{3}$.

To simplify this, remember 5 is the same as $\frac{5}{1}$,

$$\text{hence the cost} = \frac{5}{1} \times \frac{N}{3}$$
$$= \frac{5N}{3} \text{ pence.}$$

Example 3.5

Walking at a steady speed of 3 miles per hour, how far would I walk in:
(a) 2 hours (b) x hours (c) t minutes?

Solution

(a) Distance = speed × time
$= 3 \times 2 = 6$ miles

(b) Distance $= 3 \times x = 3x$ miles.

(c) To change minutes into hours, you divide by 60,

hence t minutes $= t \div 60$ or $\frac{t}{60}$ hours.

So the distance $= 3 \times \frac{t}{60} = \frac{3t}{60} = \frac{t}{20}$ miles

A topic which requires the use of algebraic expressions is in the study of number sequences. The following investigation attempts to introduce the ideas involved. With patience, you should be able to master the techniques.

Investigation VII: Number patterns

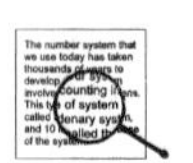

Look at the number **sequence**:

1, 2, 3, 4, 5, ... (1)

What is the 20th number if you continue this sequence? The answer is obviously 20. Now look at the sequence:

2, 4, 6, 8, 10, ... (2)

Again, it is not difficult to spot that the 20th number is 40. In fact, in sequence (1) we can write down that the Nth number in the sequence is N. In sequence 2, the Nth number is $2N$.

We say that N and $2N$ are expressions for the **general term** in each sequence respectively. Now look at this sequence:

7, 9, 11, 13, 15, ... (3)

Can you see an expression for the general term here? If you look carefully at the numbers in sequence (3), they are five more than those in sequence (2). Hence the general term (3) is $2N + 5$. If you were now asked to find the 100th number in sequence (3), we can use the expression $2N + 5$ to give $2 \times 100 + 5 = 205$.

In the sequences (1), (2) and (3), the successive numbers increase by a constant amount. This type of sequence is called an **arithmetic** sequence. Now consider the following sequence:

4, 7, 10, 13, 16, 19, ... (4)

The numbers increase by plain 3 each time, so take one from each number so that the first number becomes 3.

i.e. 3, 6, 9, 12, 15, 18,..

The Nth term here must be $3N$. We now need to add on one to get back to the original series. Hence the Nth term of sequence (4) is $3N + 1$.

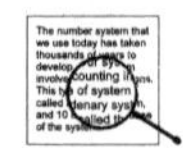

Now try the following. In each case, find an expression for the Nth term, and find the 100th number in the sequence.

(i) 8, 11, 14, 17, 20, ...

(ii) 4, 9, 14, 19, 24, ...

(iii) 2, 6, 10, 14, 18, 22, ...

(iv) 8, 19, 30, 41, 52, ...

Now invent some arithmetic sequences of your own.

3.4 Simplifying expressions

Rules

Like and unlike terms

$3x + 2x = 5x$

$6x - 2x = 4x$

$7x - 2x + 3x = 8x$

This is known as the *collection* of **like terms**

$x + x^2$ cannot be simplified, they are **unlike** terms [powers must be the same for *like* terms].

Quantities such as $4x$, $4pt$, $3y$ are called **algebraic** terms.

Indices

$x^2 \times x^3 = x \times x \times x \times x \times x = x^5$

$$x^7 \div x^4 = \frac{x \times x \times x \times \overset{1}{\cancel{x}} \times \overset{1}{\cancel{x}} \times \overset{1}{\cancel{x}} \times \overset{1}{\cancel{x}}}{\underset{1}{\cancel{x}} \times \underset{1}{\cancel{x}} \times \underset{1}{\cancel{x}} \times \underset{1}{\cancel{x}}} = x^3$$

Now $x^3 \div x^3 = x^{3-3} = x^0$. What does x^0 mean?

$x^3 \div x^3$ must give us 1.

Hence whatever x is equal to, $x^0 = 1$.

This seems a bit strange, but you will get used to it.

Brackets

$a(b + c) = ab + ac$, since $a(b + c)$ really means

$$a \times (b + c) = a \times b + a \times c$$

$a(b - c) = ab - ac$

Example 3.6

Simplify the following expressions as much as possible.

(a) $4x + 2y + 3x + 5y$ (b) $2p + 3q + 4p - 2q$

(c) $4(x + 1) + 2(x + 2)$ (d) $x^2 + x + 2x^2$

(e) $2x^2 \times x^3$ (f) $4(2t + 1) - 3(t - 2)$

(g) $12y^4 \div 3y^2$ (h) $15y^3 \div 3y^3$

Solution

(a) $4x + 2y + 3x + 5y = 7x + 7y.$ (like terms: $4x$ and $3x$; like terms: $2y$ and $5y$)

(b) $2p + 3q + 4p - 2q = 6p + q.$ (like terms: $2p$ and $4p$; like terms: $3q$ and $-2q$)

(c) $4(x + 1) + 2(x + 2)$

First remove the brackets

$$4x + 4 + 2x + 4$$

Then collect like terms, to give

$$6x + 8.$$

(d) $x^2 + x + 2x^2 = 3x^2 + x$ (like terms: x^2 and $2x^2$)

This cannot be simplified any more, $3x^2$ and x are *not* like terms.

(e) $2x^2 \times x^3$ means $2 \times x^2 \times x^3 = 2 \times x^{2+3}$

$= 2 \times x^5$

$= 2x^5.$

(f) $4(2t + 1) - 3(t - 2)$

Remove the brackets

$8t + 4 - 3t + 6 = 5t + 10.$ ($+6$ from (-3×-2))

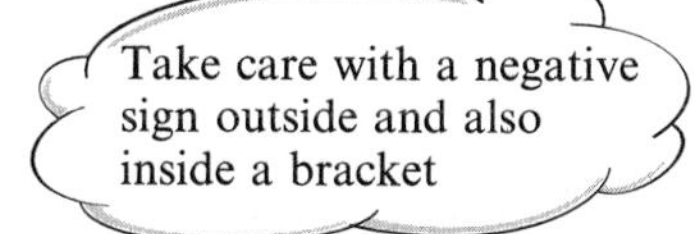

(g) $12y^4 \div 3y^2$ can be written $\dfrac{{}^{4}\not{12}y^4}{{}_{1}\not{3}y^2}$

Since $y^4 \div y^2 = y^2$, the final answer is $4y^2$.

(h) $15y^3 \div 3y^3 = \dfrac{^5\cancel{15}y^3}{_1\cancel{3}y^3}$

Now $y^3 \div y^3 = y^{3-3} = y^0 = 1$

The answer is $5y^0 = 5$.

Example 3.7

Janice earns £E per week. Nadine earns twice as much as Janice, and Paul earns £5 a week less than Nadine. How much do the three earn altogether in a year?

Solution

Twice means times 2, so Nadine earns £$2 \times E =$ £$2E$ per week.

Paul earns £5 less than Nadine, i.e. subtract £5.

Hence Paul earns £$2E$ – £5.

In one week, they earn altogether

$$£E + £2E + £2E - £5$$

which simplifies to £$5E$ – £5 = £$5(E - 1)$.

Since one year is 52 weeks, then in one year, they earn

$$£52 \times 5(E - 1) = £260(E - 1).$$

Exercise 3(a)

1 If $a = 2$, $b = -1$ and $c = \frac{1}{2}$, evaluate x in the following examples.

(i) $x = a + b$ (ii) $x = 3a + 2b$

(iii) $x = 4a - 2c$ (iv) $x = ab$

(v) $x = 3ac$ (vi) $x = ab^2$

(vii) $x = \dfrac{ab}{c}$ (viii) $x = \dfrac{4a}{b^2}$

(ix) $x = a^2 + b^2$ (x) $x = \sqrt{\dfrac{2a^2}{c}}$

2 A bag contains A pens. If B pens are removed from the bag and then a further C pens added to the bag, how many pens will be in the bag?

3 Write an expression for the number of (i) millimetres in d metres; (ii) kilograms in k grams; (iii) pence in £$(a + b)$; (iv) centimetres in $10d$ metres.

4 Write down an expression in y for the result of performing each of the following operations on y:

(i) multiply by 3 (ii) subtract 8

(iii) divide by 6 (iv) increase by 10%

5 The average speed of a cyclist is c km/h and her walking speed is w km/h. Calculate the time taken by her to cycle d kilometres. What is the total distance travelled if she walks for x hours and then cycles for y hours?

6 Simplify as much as possible the following:

(i) $7x + 4x - 2x$ (ii) $3a + 2b + 4a + b$

(iii) $5x - y + x + 4y$ (iv) $x \times 3x$

(v) $3(x + 1) + 2(x + 2)$ (vi) $3(4p + 1) - 2(p + 1)$

(vii) $2t^2 \div 2t$ (viii) $3(h + 1) - 2(h - 2)$

(ix) $4x^2 + 6x^2 - 3x^2$ (x) $4p \times 8p^2$

(xi) 4^0 (xii) $7xy^2 \div xy^2$

3.5 Equations (linear)

A linear equation in a **variable** x, is an equation which when simplified only contains a number and a term in x. When solving an equation, it is important to think of it as a mathematical balance, what you do to one side of the equation you *must* do to the other.

Hence if $x + 5 = 12$

Subtract 5 from each side: $x + 5 - 5 = 12 - 5$

So $x = 7$.

Subtracting a number from one side is the same as *adding* it to the other

You could, of course, quite easily have guessed the value of x in this case, but it is important to have a method which works however complicated the equation becomes.

REMEMBER

If $3x = 15$

Divide each side by 3: $\frac{3x}{3} = \frac{15}{3}$

So $x = 5$.

Multiplying one side by a number is the same as *dividing* the other side by that number

The following example covers all the aspects of equation solving that you need to know at this stage.

Example 3.8

Solve the following equations to find the value of the letter in each case.

(a) $2x + 1 = 15$ (b) $3x - 1 = 12$

(c) $4x + 1 = 2x + 9$ (d) $2(x + 1) = x - 3$

(e) $\frac{1}{2}y + \frac{1}{4} = \frac{1}{3}$ (f) $\frac{12}{p} = 4$

(g) $4(q + 1) - 3(q - 2) = 7$ (h) $\frac{5}{(x+1)} = \frac{3}{(x+2)}$.

Solution

(a)

$$2x + 1 = 15$$

-1 from each side: $2x + 1 - 1 = 15 - 1$

$$2x = 14$$

$\div$ each side by 2: $\frac{2x}{2} = \frac{14}{2}$

So $x = 7$

(b)

$$3x - 1 = 12$$

$+1$ to each side: $3x - 1 + 1 = 12 + 1$

$$3x = 13$$

$\div$ each side by 3: $\frac{3x}{3} = \frac{13}{3}$

Hence $x = 4\frac{1}{3}$.

(c) When x appears on both sides, it is necessary to end up with the x terms only on one side (it doesn't matter which side)

$$4x + 1 = 2x + 9$$

$-2x$ from each side: $4x + 1 - 2x = 2x + 9 - 2x$

$$2x + 1 = 9$$

-1 from each side: $2x + 1 - 1 = 9 - 1$

$$2x = 8$$

$\div$ each side by 2: $\frac{2x}{2} = \frac{8}{2}$

Hence $x = 4$.

(d) If an equation contains brackets, they should be removed

$$2(x+1) = x-3$$

Hence: $2x + 2 = x - 3$

$-x$ from each side: $2x + 2 - x = x - 3 - x$

$x + 2 = -3$

-2 from each side: $x + 2 - 2 = -3 - 2$

So $x = -5$.

(e) When an equation contains fractions, they should first be removed by multiplying each term by the LCM of the denominators

$$\tfrac{1}{2}y + \tfrac{1}{4} = \tfrac{1}{3}$$

The LCM of 2, 3 and 4 is 12

Hence: $12 \times \tfrac{1}{2}y + 12 \times \tfrac{1}{4} = 12 \times \tfrac{1}{3}$

Multiply each term by 12

Simplify to give $6y + 3 = 4$

-3 from each side: $6y + 3 - 3 = 4 - 3$

So $6y = 1$

$\div$ each side by 6: $y = \tfrac{1}{6}$.

(f) This type of equation occurs in trigonometry, see Example 7.2(b)

$$\frac{12}{p} = 4$$

$\times$ each side by p: $\not{p} \times \dfrac{12}{\not{p}} = 4 \times p$

$12 = 4p$

$\div$ each side by 4: $3 = p$

i.e. $p = 3$.

Equations of the type (i) $ax = b$, (ii) $\dfrac{x}{a} = b$ and (iii) $\dfrac{a}{x} = b$ can also be solved using the triangle rule.

Just complete the triangle (i) as shown.

(i)

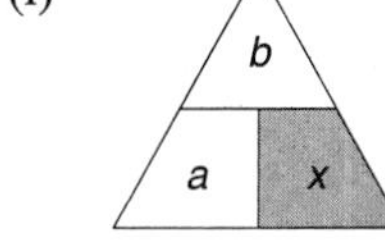

ax means $a \times x$ and so a is written alongside x. To find x (shaded) you ignore this section, and so $x = \dfrac{b}{a}$.

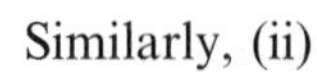

Similarly, (ii)

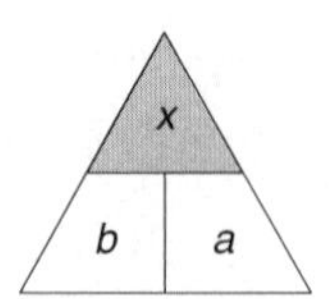

x is above a, so you put x in the top space, and a in *either* of the other two.

Hence $\quad x = b \times a = ba.$

Also (iii)

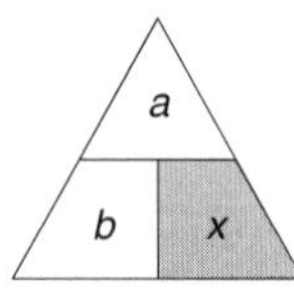

$$x = \frac{a}{b}$$

Applying this to our equation above

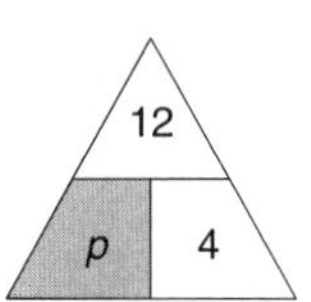

$$p = \frac{12}{4} = 3.$$

(g) Once again, the equation contains brackets that should be removed

$$4(q+1) - 3(q-2) = 7$$

REMEMBER

So: $\quad 4q + 4 - 3q + 6 = 7$

$(-3 \times -2 = +6)$

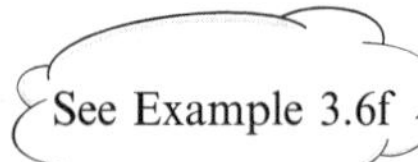

Simplify: $\quad q + 10 = 7$

-10 from each side: $\quad q + 10 - 10 = 7 - 10$

So $\quad q = -3.$

(h) At first sight, this may not appear to be a linear equation.

REMEMBER

Since $\quad \dfrac{5}{x+1} = \dfrac{3}{x+2}$

then: $\quad 5 \times (x+2) = 3 \times (x+1)$

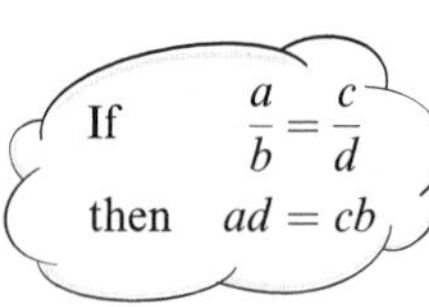

This is sometimes called **cross multiplying**.

So: $\quad 5x + 10 = 3x + 3$

-10 from each side: $\quad 5x = 3x - 7$

$-3x$ from each side: $\quad 2x = -7$

$\div$ each side by 2: $\quad x = -3\frac{1}{2}.$

Exercise 3(b)

Solve the following equations to find the unknown letter.

1 $x + 8 = 15$

2 $x - 8 = 15$

3 $2x + 8 = 15$

4 $8 - x = 4$

5 $2x - 3 = 9$

6 $2(x + 1) = 5$

7 $2p + 1 = p + 5$

8 $3(2r + 1) = 8$

9 $8 - y = 2 - 3y$

10 $\frac{1}{4}t + 5 = 6$

11 $\frac{1}{2}x + \frac{1}{3}x = \frac{1}{4}$

12 $\frac{6}{y} = 12$

13 $\frac{4}{x+1} = \frac{3}{x+2}$

14 $2(x + 1) + 3(x + 2) = 20$

15 $2(t + 1) - (t + 6) = 15$

16 $3(2x + 1) - 5(x - 2) = 12$

17 $\frac{1}{3}(x + 2) = 6$

18 $\frac{2}{2x+1} = \frac{3}{2x-4}$

19 $5(3x - 5) - 2(x - 7) = 11$

20 $\frac{7}{2x} = 3.$

Investigation VIII: Matchstick problems

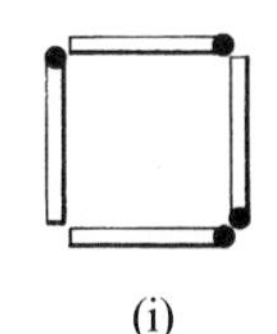

(i)

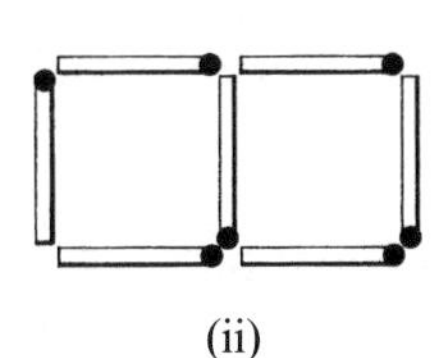

(ii)

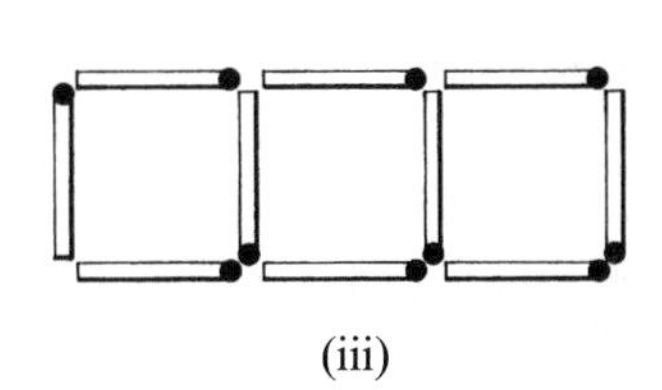

(iii)

The diagram shows a series of patterns made from matches. If you let the length of one match be 1 unit, then diagram (i) requires 4 matches, and has a perimeter of 4 units. Diagram (ii) requires 7 matches, and has a perimeter of 6 units and so on. Complete the table below.

No. of squares (n)	1	2	3	4	5	10
No. of matches (m)	4	7				
Perimeter (P)	4	6				

Can you see how the number patterns continue?

(a) How many squares can you make with 241 matches?

(b) How many squares have a perimeter of 92 units?

(c) Can you find formulae for the number of matches (m) and the perimeter (P) for n squares?

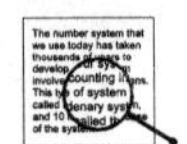

Extension:

(i) Try with different shapes such as triangles or hexagons.

(ii) Look at different types of patterns.

3.6 Problem solving with equations

Equations become most useful when they are used in problem solving. Using the techniques of algebraic expressions, we can represent a problem by an equation that can then be solved.

Example 3.9

Danny is twice as old as Alan, who in turn is 5 years older than Joanna. If their total age is 47 years, how old is each of them?

Solution

Let Joanna's age be x years.

Alan is 5 years older than this, and so his age is $x + 5$ years.

Danny is twice Alan's age, and hence his age is $2(x + 5)$ years.

The total age is 47, and so we have the equation $x + (x + 5) + 2(x + 5) = 47$.

[*Note*: The equation does not contain units (years).]

Removing the brackets, we have:

$$x + x + 5 + 2x + 10 = 47$$

Hence $$4x + 15 = 47$$

$$4x = 32$$

$$x = 8$$

So, Joanna is 8 years old, Alan is 13 years old, and Danny is 26 years old.

Example 3.10

Asif works in the local supermarket. His pay is £2.50 an hour, and he is paid overtime at time-and-a-half. His normal working week is 35 hours. How many hours of overtime did he work if in a certain week he earned £117.50?

Solution

Let the number of hours of overtime he worked be N.

He is paid $1\frac{1}{2} \times £2.50 = £3.75$ for this, hence his overtime pay is $£3.75 \times N$.

Hence

$$2.5 \times 35 + 3.75 \times N = 117.5$$
$$87.5 + 3.75N = 117.5$$

So
$$3.75N = 117.5 - 87.5 = 30$$
$$N = 30 \div 3.75 = 8$$

Asif worked 8 hours of overtime.

Exercise 3(c)

1 Tim thought of a number, multiplied it by $3\frac{1}{2}$, and added 16 to the result. If his final answer was 30, what number did he think of?

2 A mother and daughter together took £420 each week from the profits of their shop. If the mother takes twice as much as the daughter, how much do they each earn per week

3 The total length of the three sides of a triangle is 45.5 cm and two of the sides are respectively $2\frac{1}{2}$ times and $3\frac{1}{2}$ times the third side. By calling the length of the shortest side x cm form an equation in x and calculate the length of each side of the triangle.

4 In one week, Sarah worked 36 hours at the standard rate, 9 hours of overtime at $1\frac{1}{2}$ times the standard rate and 3 hours at treble pay. If her total pay was £140.40, what is the standard rate?

5 A consignment of magazines was made up into 45 parcels each weighing either 5 kg or 1 kg. Calculate the number of 5 kg parcels in the consignment if the total weight was 153 kg.

6 Of the 18 coins in my pocket, I had some pennies, twice as many twenty-pence pieces and the rest fifty-pence pieces. If I had £4.64 in total, how many of each type of coin did I have?

3.7 Rearranging formulae (changing the subject)

If we write down a formula such as $x = at + 4$, then x is the **subject** of that formula. We are now going to look at the process of rearranging the formula so that either a or t might become the subject. This process is often referred to as **changing the subject** of the formula. The methods involved are identical to those used in solving an equation. We will build up the method in simple stages.

(i) If $x = at$

$\div$ each side by t: $\frac{x}{t} = a$

or $a = \frac{x}{t}$

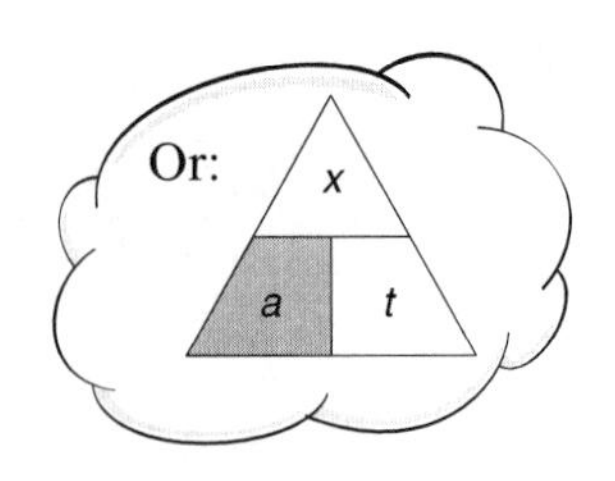

We have now made a the subject.

(ii) If $x = y - q$

$+q$ to each side: $x + q = y$

or $y = x + q$

We have now made y the subject.

(iii) If $x = at + 4$

-4 from each side: $x - 4 = at$

$\div$ each side by a: $\frac{x-4}{a} = t$

or $t = \frac{x-4}{a}$

We have now made t the subject.

Example 3.11

The speed of a train v m/s is given by the formula $v = u + \frac{1}{4}t$, where u m/s is the speed when timing starts (i.e. when $t = 0$), and t seconds is the time elapsed since timing began.

(a) Find v if $u = 10$, $t = 20$.

(b) Find a formula for u in terms of v and t, and use this formula to find u, if $v = 25$ and $t = 16$.

(c) Find a formula for t in terms of v and u, and use this formula to find t if $v = 30$ and $u = 8$.

Solution

(a) No rearranging is needed here, so $v = 10 + \frac{1}{4} \times 20 = 10 + 5 = 15$.

(b) $v = u + \frac{1}{4}t$

$-\frac{1}{4}t$ from each side: $v - \frac{1}{4}t = u$

or $u = v - \frac{1}{4}t$

Hence $u = 25 - \frac{1}{4} \times 16 = 25 - 4 = 21$.

(c) $$v = u + \tfrac{1}{4}t$$

$-u$ from each side: $$v - u = \tfrac{1}{4}t$$

$\times$ each side by 4: $$4v - 4u = t$$

or $$t = 4v - 4u$$

Hence $$t = 4 \times 30 - 4 \times 8 = 120 - 32 = 88.$$

Exercise 3(d)

1 For each of the following formulae, rearrange to make x the subject:

(i) $y = xH$ (ii) $y = x + H$ (iii) $y = H - x$

(iv) $y = \dfrac{xt}{H}$ (v) $y = ax + b$ (vi) $q = 4t - x$

(vii) $I = \dfrac{PTx}{100}$ (viii) $v = u + 2ax$ (ix) $v^2 = u^2 + 2ax$

(x) $y = \dfrac{H}{x}$.

2 The formula for the surface area (S) of a cylinder is approximately given by $S = 6r^2 + 6rh$.

Rearrange this formula to find an expression for h. Find h if $r = 2$, and $S = 100$.

3 In the following number series, find the next number, the 10th number in the series, and the nth number.

(i) 1, 3, 5, 7, 9, . . .

(ii) 3, 7, 11, 15, 19, . . .

(iii) 9, 19, 29, 39, . . .

(iv) −3, 1, 5, 9, . . .

(v) 40, 37, 34, 31, 28, . . .

4 Mr Jones drives his car from Coventry to Brighton via Oxford, Heathrow and Gatwick.

From Coventry to Oxford he travels a distance of 52 miles.

From Oxford to Heathrow he travels a distance of 38 miles.

He travels x miles from Heathrow to Gatwick and he travels $(x - 12)$ miles from Gatwick to Brighton.

(i) Write down an expression, in terms of x, for the total distance that Mr Jones travels.

(ii) The total distance travelled by Mr Jones is 166 miles. Write down an equation in x and solve it.

[LEAG]

4 Ratios and percentages

From now on, it will be assumed that all calculations are being carried out on a calculator. However, any button not so far described will be explained in full.

4.1 Percentages

The use of percentages is very common practice in everyday life. The percentage sign % means per hundred, or out of 100. It follows that a percentage is really a fraction, and so 8% means 8 out of 100, or $\frac{8}{100}$.

(i) Simplifying

Since $8\% = \frac{8}{100}$, this can be simplified to $\frac{2}{25}$.

Also, $\frac{8}{100}$ means $8 \div 100 = 0.08$.

Hence, 8% can be written as $\frac{2}{25}$ or 0.08.

Now

$$8\tfrac{1}{3}\% \text{ means } \frac{8\frac{1}{3}}{100}$$

The number on the top is not easy to manipulate, so multiply top and bottom by 3.

$$\frac{8\frac{1}{3} \times 3}{100 \times 3} = \frac{25}{300} = \frac{1}{12}$$

$$8\tfrac{1}{3}\% = \frac{1}{12}$$

Hence

$$8.15\% \text{ means } \frac{8.15}{100} = \frac{8.15 \times 100}{100 \times 100} = \frac{815}{10000}$$

$$= \frac{163}{2000} \quad \text{(by cancelling).}$$

This would not be very useful in this form, and in this case it is much more sensible to express the percentage as a decimal, so

$$8.15 \div 100 = 0.0815$$

Hence $8.15\% = 0.0815$.

These examples lead us to:

> **Rule**
>
> To change a percentage into a fraction or a decimal, divide by 100.

(ii) Fractions or decimals into percentages

The fraction $\frac{1}{5}$ can be written $\dfrac{1 \times 20}{5 \times 20} = \dfrac{20}{100}$

So we have written $\frac{1}{5}$ as 20 out of 100, i.e. 20%.

The decimal 0.28 can be written as $\frac{28}{100}$, i.e. as 28%.

In each case, if the fraction $\frac{1}{5}$, or decimal 0.28 is multiplied by 100, you get the percentage.

So $\frac{1}{5} \times 100 = 20$, and $0.28 \times 100 = 28$.

> 
> **Rule**
>
> To change a fraction or decimal into a percentage, multiply by 100.

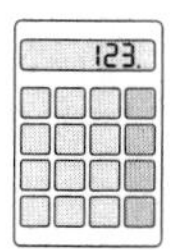

If the calculator has a % button (in fact [INV] [=] on the keyboard illustrated in section 1.1), then it will change a fraction into a percentage for you.

Hence $\frac{3}{8}$ as a percentage would be

				Display
[3]	[÷]	[8]	[%]	37.5

Therefore $\frac{3}{8} = 37.5\%$.

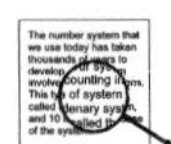

Investigation IX: Percentages worth knowing

We have seen in the preceding sections that fractions and percentages can often be interchanged. So that, if you are asked to find 20% of something, since $20\% = \frac{1}{5}$ it may be possible by mental calculation to find one-fifth. Copy and complete the following table of percentage/fraction/decimal equivalents, and try to learn some of them.

Percentage	*Fraction*	*Decimal*	*Percentage*	*Fraction*	*Decimal*
5	$\frac{5}{100} = \frac{1}{20}$	0.05	$12\frac{1}{2}$		
10			$33\frac{1}{3}$		
15				$\frac{1}{16}$	
20				$\frac{1}{32}$	
25					0.625
40					0.3125
50					0.6875

(iii) Percentage of a quantity

Since 8% of £84 means 'eight hundredths' of £84, then we must find

$$\frac{8}{100} \times £84 = \frac{£8 \times 84}{100} = \frac{£672}{100} = £6.72$$

Hence 8% of £84 is £6.72.

or since $8\% = 0.08$,

$$0.08 \times £84 = £6.72.$$

18.5% of 1 m 25 cm, would mean $\frac{18.5}{100} \times 1.25$ m.

i.e. $$\frac{18.5 \times 1.25}{100}\text{ m} = \frac{23.125}{100}\text{ m} = 0.23125\text{ m}$$

or $$18.5\% = 0.185$$

$$0.185 \times 1.25\text{ m} = 0.23125\text{ m}.$$

(iv) One quantity as a percentage of another

To express £7 as a percentage *of* £8.40, you would proceed as follows:

$$\text{£7 as a fraction of £8.40 is } \frac{7}{8.40} = 0.833$$

To change a fraction to a % we multiply by 100 so

$$0.833 \times 100 = 83.3\%.$$

> **Rule**
>
> When comparing quantities using percentages, they must both be measured in the same units.

If you want to express 80 cm as a percentage of 2 m, then you must change either 80 cm into 0.8 m, or 2 m into 200 cm.

Using the latter, we get that

$$\text{80 cm as a fraction of 200 cm is } \frac{80}{200} = 0.4$$

To change into a percentage, multiply 0.4 by 100.

Hence $0.4 \times 100 = 40\%$.

Using the calculator, this last example would be

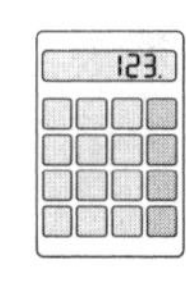

Display

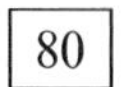

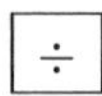

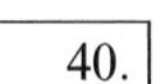

(v) Percentage changes

Example 4.1

The cost of a rail ticket is increased from £8.50 by 20%. Find the new cost of the ticket.

Solution

20% = 0.2, so 20% of £8.50 is $0.2 \times$ £8.50 = £1.70.

The new ticket costs £8.50 + £1.70 = £10.20.

An alternative and often easier way of doing this is to say that:

an increase of 20% gives a total of 100% + 20% = 120% of the original cost

Now 120% is $\frac{120}{100} = 1.2$

1.2 is known as a multiplier

REMEMBER

The new cost = £8.50 × 1.2 = £10.20.

Example 4.2

The length of a skirt is reduced by 8% from 27 cm. Find the new length.

Solution

8% = 0.08, so 8% of 27 cm is 0.08 × 27 cm = 2.16 cm.

The new length = 27 cm − 2.16 cm = 24.84 cm.

Alternatively, a decrease of 8% gives a total of 100% − 8% = 92% of the original length.

REMEMBER

Now 92% is $\frac{92}{100} = 0.92$

The multiplier is 0.92 = 1 − 0.08

Hence the new length = 27 × 0.92 = 24.84 cm.

Example 4.3

The sides of a rectangle are altered. The length is increased by 10%, and the width is decreased by 10%. Find the percentage change in the area of the rectangle, stating whether it is an increase or a decrease.

Solution

The length becomes 100% + 10% = 110% of the old length.

The width becomes 100% − 10% = 90% of the old width.

But 110% = 1.1, and 90% = 0.9.

Hence the area is multiplied by 1.1 × 0.9 = 0.99

Hence the area has decreased by 0.01 × the original area, which is the same as 1% of the original area.

Hence the area of the rectangle decreases by 1%.

(vi) Profit and loss

Example 4.4

Jane is able to buy sweaters for her boutique for £18.50. She sells them to make a profit of 30% on the cost. How much do the sweaters sell for? How much profit is made?

Solution

Since we want to know the profit, it is better to find 30% of £18.50 first

$$\text{So } \frac{30}{100} \times £18.50 = £18.50 \times 0.3 = £5.55.$$

The profit = £5.55.

The selling price = £18.50 + £5.55 = £24.05.

Example 4.5

Jason has just sold his car for £4200, making a loss of 16% on his original outlay. How much did he pay for the car?

Solution

This type of problem is often solved incorrectly. You must not find 16% of £4200 and add it back on.

A loss of 16% gives 100% – 16% = 84% of the original cost.

$$\text{So} \quad £4200 = 84\% \text{ of the original cost}$$

$$\frac{£4200}{84} = 1\% \text{ of the original cost}$$

$$\frac{£4200}{84} \times 100 = 100\% \text{ of the original cost} = £5000.$$

Jason paid £5000 for the car.

A formula which is often used when calculating percentage profits or losses is:

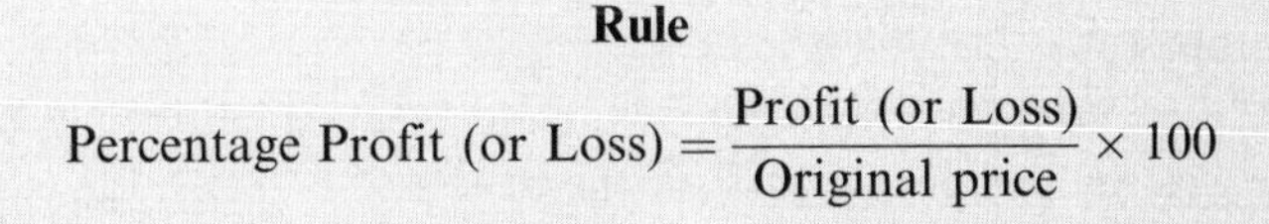

Rule

$$\text{Percentage Profit (or Loss)} = \frac{\text{Profit (or Loss)}}{\text{Original price}} \times 100$$

Example 4.6

The Patel family bought a house in 1994 for £80 000. When they sold it 18 months later, they only received £70 400 after selling costs had been deducted. Find the percentage loss on the sale.

Solution

The actual loss = £80 000 − £70 400 = £9600.

$$\text{So the \% loss} = \frac{9600}{80\,000} \times 100$$

$$= 12\%.$$

Exercise 4(a)

1 Express as percentages:

(i) $\frac{1}{4}$ (ii) $\frac{3}{8}$ (iii) $2\frac{1}{3}$ (iv) $\frac{4}{5}$

(v) 0.28 (vi) 1.65 (vii) $\frac{1}{9}$.

2 Change the following percentages into fractions (expressed in lowest terms):

(i) 50% (ii) 29% (iii) 18% (iv) $60\frac{2}{3}\%$

(v) 120% (vi) 64%.

3 Evaluate:

(i) 15% of £98.20 (ii) 45% of 540 mm

(iii) 12% of 3600 bricks (iv) $2\frac{1}{4}\%$ of $1\frac{1}{2}$ million

(v) $12\frac{1}{2}\%$ of £64.80 (vi) 140% of 2300 m.

4 If 48 employees at a factory were absent and this was 4% of the people employed, calculate the total number employed at the factory.

5 On a production line, 23 out of 4570 machine parts are found to be defective. Express the number of defectives as a percentage of the total, giving your answer to 2 significant figures.

6 If I spend £171 a month on household expenses, what percentage is this of a monthly salary of £760?

7 Jan sold a car for £3360 at a 30% loss. How much did she pay for the car?

8 A bar of chocolate can be bought by a shopkeeper in boxes of 12 for £2.20. The bar is sold for 28p, find the percentage profit made on a box.

9 John made a 36% loss on an investment and was £30.24 out of pocket. How much was his investment?

10 Mr White sold his car for £2800, which he had originally paid £4100 for. What was his percentage loss on the deal?

4.2 Interest

In many financial calculations of investment or debt repayment, interest is either gained or paid out. There are two types of interest, namely **simple interest** and **compound interest**. Simple interest is usually only used over *short* periods of time, or in certain HP (hire purchase) agreements.

(i) Simple interest

In investment, the sum of money invested is called the principal (P). The interest rate $R\%$ is usually given per year, or *per annum*. The period in *years* for which the money is invested is denoted by T. The interest gained is denoted by I.

The quantities are related by the formula:

Rule

$$\text{Interest } I = \frac{PRT}{100}$$

Example 4.7

If £1250 is invested for 3 years at an interest rate of $7\frac{1}{2}\%$ per annum simple interest, find the value of the investment after this time.

Solution

Here $P = £1250$, $T = 3$, $R = 7\frac{1}{2}$

$$\text{So} \quad I = \frac{£1250 \times 3 \times 7\frac{1}{2}}{100} = £281.25$$

$$\begin{aligned}\text{The value of the investment} &= P + I = £1250 + £281.25\\ &= £1531.25.\end{aligned}$$

Example 4.8

Jodi has borrowed £2500 from a friend. She agrees to repay the money in 4 months with interest calculated at 18% per annum calculated as simple interest. How much does she have to repay?

Solution

In using the formula $I = \dfrac{PRT}{100}$, T is measured in years, hence 4 months is taken as $T = \frac{1}{3}$, $P = £2500$ and $R = 18$

$$\text{So} \quad I = \frac{\pounds 2500 \times 18 \times \frac{1}{3}}{100}$$
$$= \pounds 150$$

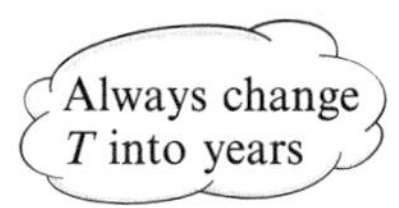

The total amount to repay is £2500 + £150 = £2650.

Example 4.9

Diana borrowed £2000 from a finance company in order to buy a car. She had to repay £100 per month for 36 months. If the interest had been calculated as simple interest, find the interest rate charged.

Solution

The total amount paid is 36 × £100 = £3600

Hence the interest I = £3600 − £2000 = £1600.

$$\text{Using} \quad I = \frac{PRT}{100}, \quad 1600 = \frac{2000 \times R \times 3}{100} = 60R,$$

$$R = \frac{1600}{60} = 26.7$$

Hence the interest rate is 26.7%.

(ii) Compound interest

In most financial transactions, interest is added to capital, or a debt, after a certain period of time and then interest is calculated on this total amount. This is known as *compounding* the interest, or just **compound interest**.

Suppose £200 is invested at 6% per annum compound interest for 5 years. We set the calculation out in the following way:

Year 1: Interest on £200 = £200 × 0.06 = £12
Value of the investment after 1 year = £200 + £12 = £212

Year 2: Interest on £212 = £212 × 0.06 = £12.72
Total after 2 years = £212 + £12.72 = £224.72

Year 3: Interest on £224.72 = £224.72 × 0.06 = £13.48
Total after 3 years = £224.72 + £13.48 = £238.20

Year 4: Interest on £238.20 = £238.20 × 0.06 = £14.29
Total after 4 years = £238.20 + £14.29 = £252.49

Year 5: Interest on £252.49 = £252.49 × 0.06 = £15.15
Total after 5 years = £252.49 + £15.15 = £267.64.

This method is of course quite lengthy. It can be done using the formula:

Rule

$$\text{Total value of the investment} = P\left(1 + \frac{r}{100}\right)^n$$

where r is the interest rate over the time period it is calculated, n is the number of times the interest is calculated, and P is the amount invested. In this last example, we have

$r = 6,\ n = 5,\ P = £200$

So $1 + \dfrac{r}{100} = 1 + 0.06 = 1.06$

Hence: total $= 200 \times (1.06)^5 = £267.64.$

To evaluate this using the calculator, the key sequence would be

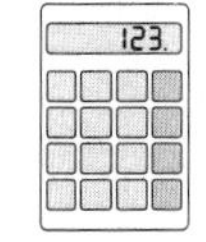

						Display
200	×	1.06	x^y	5	=	267.64512

Notice the answer would be £267.65 to the nearest penny, but £267.64 if we ignore fractions of a penny.

Investigation X: Credit cards

There are a wide variety of credit cards available today, with banks, shops and finance houses. They all tend to charge different interest rates which are very often quoted on a monthly basis. In section 4.2(ii), it was seen how the formula

$$P\left(1 + \frac{r}{100}\right)^n$$

could be used to calculate compound interest. In order to compare interest rates, the annual percentage rate (APR) is used. This is found from a similar formula:

$$\text{APR} = \left[\left(1 + \frac{r}{100}\right)^n - 1\right] \times 100\%$$

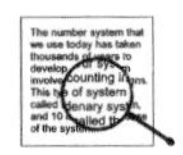

Here n is the number of times interest is calculated, and r is the percentage rate charged each time.

Find as many different types of card as you can, and calculate the APR for them. Compare your answers.

Extension: Look at other types of finance and try to find the APR for these. You may find the results quite surprising.

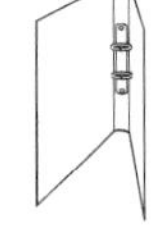

Exercise 4(b)

1 Find the simple interest on:

(i) £240 for 3 years at 7% per annum;
(ii) £725 for 2 years 6 months at 12% per annum;
(iii) £1820 for 3 months at 24% per annum.

2 What sum of money must be invested at 13% simple interest per annum in order to obtain interest of £12 per week?

3 Calculate the rate per cent per annum if £13.50 interest is paid on an investment of £50 after 2 years. Find the time taken for £450 to amount to £632.25 at this rate of interest.

4 Find the value of the following investments using compound interest:

(i) £1800 for 3 years at 10% per annum
(ii) £8000 for 10 years at 7% per annum
(iii) £725 at 6.4% per annum for 5 years.

5 A car is thought to decrease in value by 15% each year from new. Estimate the value of a car originally bought for £6000 after 4 years.

6 Roy invests £1000 on 1 January 1996 and every 1 January until 1 January 2006. Interest at 8% per annum is calculated at 31 December each year on the total value of his investment (including any earned interest). What will be the value of his investment on 31 December 2006?

4.3 Ratio and proportion

(i) Ratio

A **ratio** is really a type of fraction. For instance, if in a class of 27 children, there are 12 girls and 15 boys, the ratio of girls to boys may be written 12/15 or more usually 12 : 15 (read as 12 to 15). A ratio may be cancelled in the same way as a fraction, although it is important to remember that the numbers in a ratio must be consistent (see Example 4.10). The ratio of our children in the class could be simplified on dividing each number by 3 to give the ratio 4 : 5. It is often useful to reduce one of the numbers in a ratio to 1, and it will be asked for in the form 1 : n.

Example 4.10

(a) Express the ratio 20 cm to 15 m in the form $1 : n$.
(b) Find the ratio of 24p to £36.

Solution

(a) Units must be consistent, and so we want the ratio of 20 cm to 1500 cm (a ratio of lengths)

i.e. $20 : 1500$
$\div 20$ $1 : 75$

(b) 24p to £36 is really 24p to 3600p to be consistent

i.e. $24 : 3600$ (a ratio of money)
$\div 24$ $1 : 150$.

(ii) Proportion

Proportion is another name for fraction, and it is important that consistent units are used. It is sometimes more useful to represent a proportion as a percentage.

Example 4.11

In a manufacturing process, light bulbs are being made. In a batch of 30 000, 150 are rejected. What proportions of the bulbs in the batch are satisfactory?

Solution

The number *not* rejected is $30\,000 - 150 = 29\,850$.

The proportion *not* rejected is $\dfrac{29850}{30\,000} = \dfrac{199}{200}$

As a percentage, this is $\dfrac{199}{200} \times 100 = 99.5\%$.

A quantity can be increased or decreased in a given ratio by being multiplied by a suitable fraction. If a price increases in the ratio $3 : 2$, then it is multiplied by the fraction $\frac{3}{2}$.

Example 4.12

(a) Increase £4.50 in the ratio 7 : 5

(b) Decrease 60 kilometres in the ratio 5 : 8

Solution

(a) $£4.50 \times \frac{7}{5} = \frac{£31.50}{5} = £6.30$

(b) $60 \text{ km} \times \frac{5}{8} \text{ km} = \frac{300}{8} \text{ km} = 37.5 \text{ km}.$

A quantity can be divided in a given ratio using the idea of **proportional parts**.

Example 4.13

Jean aged 10 and Paul aged 12 share £55 in the ratio of their ages. How much does Paul receive?

Solution

The required ratio is $10 : 12 = 5 : 6$

Hence Jean would receive 5 parts, and Paul 6 parts of the money.

It is best to think of this as $5 + 6 = 11$ parts being equal to £55. So one part is $£55 \div 11 = £5$.

Now Paul receives 6 parts, i.e. $6 \times £5 = £30$.

Example 4.14

A piece of wood is cut into 3 pieces in the ratio 5:4:3. If the length of the shortest piece is 2 cm, how long was the piece of wood?

Solution

The smallest piece is 3 parts long

So 3 parts are 2 cm long

Hence 1 part is $\frac{2}{3}$ cm long

The piece of wood is $5 + 4 + 3 = 12$ parts

Hence its length is $12 \times \frac{2}{3}$ cm $= 8$ cm.

Exercise 4(c)

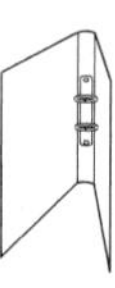

1 Express the following ratios in their simplest forms:

(i) 40 cm : 3 m | (ii) 60 g : 5 kg

(iii) 550 ml : 1.20 litre | (iv) $3^2 : 6^2$

(v) 13.65 km : 22.1 km | (vi) 24 kg : 280 kg

(vii) 4p : £3 | (viii) £4.20 : £7.70

(ix) 96 mm : 123 mm | (x} 12 minutes : $1\frac{1}{2}$ hours

(xi) 30° : 45° : 135°.

2 Express the following ratios in the form

(a) $1 : k$ (b) $k : 1$

(i) 22.2°C : 37.5°C | (ii) £7.40 : £12.80

(iii) $2^4 : 4^3$ | (iv) 18 minutes : 3 hours

(v) 1.54 g : 1.12 g | (vi) 37 ml : 3.5 litres.

3 Increase £3.50 in the ratio 9 : 7.

4 Increase 4.40 kg in the ratio 15 : 11.

5 Decrease 56 in the ratio 5 : 8.

6 Decrease 51.3 m in the ratio 2 : 3.

7 If $2 : x = 5 : 8$, find x.

8 Divide £128 in the ratio 2 : 3 : 7.

9 The sides of a triangle are in the ratio 3 : 4 : 5 and the perimeter is 105 cm. Find the lengths of the sides.

10 A sum of money is divided between three men, A, B, and C, in the ratio 7 : 5 : 1. If B has £2.40 more than C, how much does A receive?

11 The ratio of the heights of three peaks in the French Alps to each other is 2 : 5 : 8. The smallest mountain is 1200 metres in height. Find the height of the other two mountains.

12 The circumference of two circles are in the ratio 1 : 5. The smaller circle has a radius of 7 cm. What is the radius of the larger circle?

13 The radius of a circle was doubled. Express the area of the original circle as a ratio of the new one.

14 A block of wood 8 cm by 5 cm by 2.5 cm weighs 350g. Find the weight of a block of the same material with dimensions 6 cm by 3.5 cm by 1 cm.

15 At the beginning of the school year, Alan and Bill had their heights measured. The former was 145 cm and the latter 155 cm. In the year Alan grew 7 cm and Bill grew 8 cm. Which ratio of growth to height is the greater?

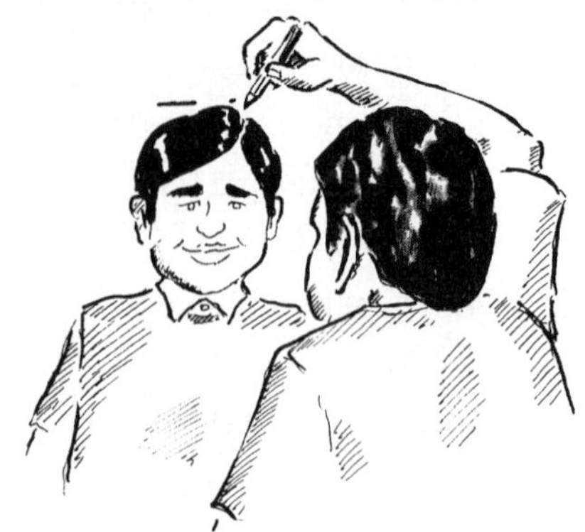

16 At a test centre one day 35 people took the driving test. If 7 failed, what proportion passed the test?

17 In a group of 30 students, 2 did not like any sport, and 12 did not play sport regularly. What proportion of the students who played sport played regularly?

4.4 Changing units and currency conversion

(i) Units

It is important that you can change from one unit to another, for example m/s into km/h. The ideas used are similar to those of ratio and proportion.

Example 4.15

(a) Change 10 m/s into km/h.

(b) Change 8 g/cm^3 into kg/m^3

(c) Change 100 km/h into cm/s.

Solution

(a) *Method 1*

10 m/s means 10 m in 1 second

i.e. $10 \times 60 \times 60$ m in 1 hour

$= 36\,000$ m in 1 hour

$\div 1000$: $= 36$ km in 1 hour

Hence the speed $= 36$ km/h

Method 2

$$10 \text{ m/s is } \frac{10 \text{ m}}{1 \text{ s}} = \frac{10 \times \frac{1}{1000} \text{ km}}{1 \times \frac{1}{3600} \text{ h}}$$

$$= \frac{\frac{1}{100}}{\frac{1}{3600}} \text{ km/h} = \frac{3600}{100} \text{ km/h} = 36 \text{ km/h}$$

So 10 m/s is the same as 36 km/h.

(b) *Method 1*

8 g/cm^3 means each 1 cm^{3}weighs 8 g

Hence 1 m^3 weighs $8 \times 100 \times 100 \times 100 = 8\,000\,000$ g

$\div 1000$: $= 8000$ kg for each 1 m^3.

Method 2

$$8 \text{ g/cm}^3 \text{ is } \frac{8 \text{ g}}{1 \text{ cm}^3} = \frac{8 \times \frac{1}{1000} \text{ kg}}{1 \times \frac{1}{100} \times \frac{1}{100} \times \frac{1}{100} \text{ m}^3}$$

$$= \frac{\frac{8}{1000}}{\frac{1}{1\,000\,000}} \text{ kg/m}^3 = \frac{8\,000\,000}{1000} = 8000 \text{ kg/m}^3$$

Hence 8 g/cm^3 is the same as 8000 kg/m^3.

(c) *Method 1*

100 km/h means 100 km in 1 hour which is the same as 10 000 000 cm in 1 hour

But 1 hour $= 60 \times 60 = 3600$ seconds

Hence the distance in 1 second $= 10\,000\,000 \div 3600$ cm $= 2778$ cm

So 100 km/h is the same as 2778 cm/s.

Method 2

$$100 \text{ km/h is } \frac{100 \text{ km}}{1\text{h}} = \frac{10\,000\,000 \text{ cm}}{60 \times 60 \text{ s}} = 2778 \text{ cm/s.}$$

(ii) Currency

Changing from one currency to another also involves the ideas of ratio. We will use the figures in Table 4.1 for currency conversion.

Exchange rate for £1 sterling	
Austria (schillings)	15.30
Belgium (francs)	44.90
Denmark (kroner)	8.48
France (francs)	7.43
Germany (Deutschemarks)	2.19
Greece (drachmas)	359.80
Italy (lire)	2344.00
Netherlands (guilders)	2.45
Spain (pesetas)	184.00
United States (dollars)	1.49

Table 4.1

Example 4.16

(a) Change 200 schillings into £ sterling
(b) Change £500 into pesetas
(c) Change 2000 dollars into French francs.

Solution

(a) Since £1 = 15.3 schillings

÷ each side by 15.3: $\frac{\pounds 1}{15.3} = 1$ schilling

× each side by 200: $\frac{\pounds 1 \times 200}{15.3} = 200$ schillings

Hence 200 schillings = £13.07.

The idea of finding what 1 unit changes to, is called the **unitary method**.

(b) £1 = 184 pesetas

× each side by 500: £500 = 500 × 184 pesetas = 92 000 pesetas.

(c) You need to change into £ sterling along the way:

So £1 = \$1.49

$\div 1.49$: $\quad \dfrac{£1}{1.49} = \$1$

$\times 2000$: So \$2000 $= 2000 \times \dfrac{£1}{1.49} = £1342.28$

Now £1342.28 = 1342.28×7.43 francs = 9973.14 francs.

(iii) Scale

Map scales also involve the ideas of ratio.

Example 4.17

A map is drawn to a scale of 1 : 20 000.

(a) Find the distance on the map between two points that are 12 km apart.

(b) The distance on the map between two villages as the 'crow flies' is 25 cm. Find the real distance between the villages.

Solution

1 : 20 000 means that 1 cm on the map represents 20 000 cm

but 1 km = 100 000 cm.

Hence 1 cm represents 0.2 km or 5 cm represents 1 km

(a) 12 km is represented by 12×5 cm = 60 cm

(b) 25 cm represents 25×0.2 km = 5 km

Exercise 4(d)

1 Make the following changes of units.

(i) 20 km/h into m/s

(ii) 8 g/cm^3 into kg/m^3

(iii) 40 m/s into km/h

(iv) 250 g/litre into g/cm^3

(v) 20 mm/s into m/min

(vi) 100 mph into ft/s

(vii) 4×10^6 m/h into cm/s.

2 Using Table 4.1 on p. 92, make the following currency changes:

(i) £50 into guilders

(ii) 10 000 lire into £ sterling

(iii) 2000 drachmas into kroner

(iv) 4000 Belgian francs into French francs

(v) 2 000 000 lire into dollars

3 A map is drawn to a scale of 1:50 000. Find:

(i) the distance on the map between two places that are 10 km apart

(ii) the real distance between two places that are 8.5 cm apart on the map.

4 A map is drawn to a scale of 1:40 000. Find:

(i) the distance between two villages that are 18.5 cm apart on the map

(ii) the time it takes a car travelling at 70 km/h to travel a distance on a motorway that measures 33 cm on the map.

Exercise 4(e): Miscellaneous problems

1 When my house was built in 1981 the total cost was made up of the following:

materials	£12 000
wages	£24 000
office expenses	£4 000

(i) Find the total cost of the house.

(ii) In 1982 an identical house was built.

(a) By then the cost of materials was 20% more than in 1981. Find the new cost of materials.

(b) The cost of wages was 15% more than in 1981. Find the new cost of wages.

(c) The office expenses were 5% less than in 1981. Find the new cost of office expenses.

(iii) (a) Calculate the total cost of building this identical house in 1982.

(b) Calculate the overall percentage increase in cost from 1981 to 1982.

2 A man is paid £4.40 per hour for a basic 40-hour week. Any overtime is paid at 'time-and-a-half'.

(i) (a) How much will the man earn for working 40 hours?

(b) How much will the man earn for each hour of overtime?

(ii) (a) How much will the man earn if one week he works 47 hours?

(b) If in another week he earns £209, how many hours of overtime did he work?

(iii) The man finds that his deductions are 25% of his gross pay. Find his net pay (i.e. what he receives after deductions) when working a basic 40-hour week.

3 Alan and Bob win a total of £210. Bob receives twice as much as Alan.

(i) How much does Bob receive?

(ii) Bob spends his money on a two-night stay in London. His expenses are listed below:

Return train fare	£21.80
Two nights at hotel	£24.60 per night plus 17.5% VAT
Meals	£11.60 plus 10% service charge
Bus/tube fares	£10.00
Theatre tickets	£8.75
Football tickets	£4.40

(a) How much is the service charge on the meals?

(b) How much is the hotel bill for the two nights, including VAT?

(c) How much money did Bob still have left from his share of the win on his return from London?

(iii) Alan invests his share for two years at a simple interest rate of 10% per year. How much interest would he receive in the two years?

$$\left(\text{Simple interest} = \frac{P \times R \times T}{100}\right)$$

4 A trader buys 300 kg of strawberries for £210. He begins to sell them at the rate of £1.20 per kilogram. When 180 kg have been sold, the strawberries begin to go soft and he reduces the price to 60p per kilogram. He sells another 60 kg at this price, but the rest go bad and have to be thrown away. Calculate the percentage profit on his outlay.

The next day he buys another 300 kg for £210 and begins to sell them at £1.50 per kilogram. When 200 kg have been sold at this price, he decides to reduce the price. Find the price per kilogram that he must charge in order to make a profit of 55% on this day's outlay, assuming that all the strawberries are sold and none are thrown away.

5 A book publisher produced a book which sold at £1.20 per copy. The publisher agreed to pay the author $7\frac{1}{2}$% of the selling price of the first 2000 copies sold, 10% of the selling price of the next 3000 copies sold and 11% of the selling price of the remainder.

(i) If 5425 copies were sold, calculate the amount the author received.

(ii) If production costs together with the author's fees were £5400 find, correct to 3 significant figures, the percentage profit made by the publisher.

(iii) From a paperback edition which sold at 40p per copy the author received 8% of the selling price of each copy sold. If he received £137.12 from this edition, calculate the number of paperback copies sold.

6 A shopkeeper buys in various articles of children's clothing and fixes the market price of each article at 50% above what he has paid for it. In a sale, customers are given a 12% discount off the marked price.

(i) The shopkeeper has paid £14.00 for a raincoat.

Find (a) the marked price; and

(b) the sale price of the raincoat.

(ii) The sale price of a T-shirt is £1.98.

Find (a) the cash discount allowed off the marked price; and

(b) how much the shopkeeper has paid for the T-shirt.

(iii) In buying a blazer in the sale a customer is given a discount of £2.16. What cash profit does the shopkeeper make on this transaction?

(iv) A pair of jeans is sold in the sale. Express the actual profit from this transaction as a percentage of the sale price, giving your answer correct to the nearest integer.

5 Angle and shape

5.1 Definitions

The need to measure distances and directions accurately gave rise to the development of geometry in Egypt and Greece round about the sixth century BC. In the third century BC Euclid formalised many of the known ideas and discovered many new theorems. He wrote a series of books known as *The Elements* which remained the main source of geometrical studies for the next 2000 years, a quite remarkable achievement.

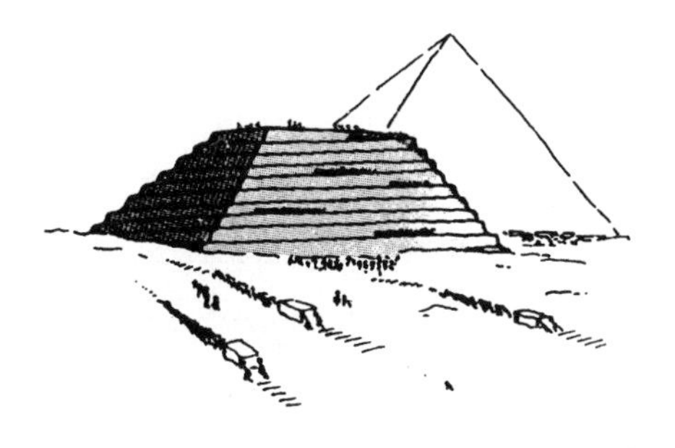

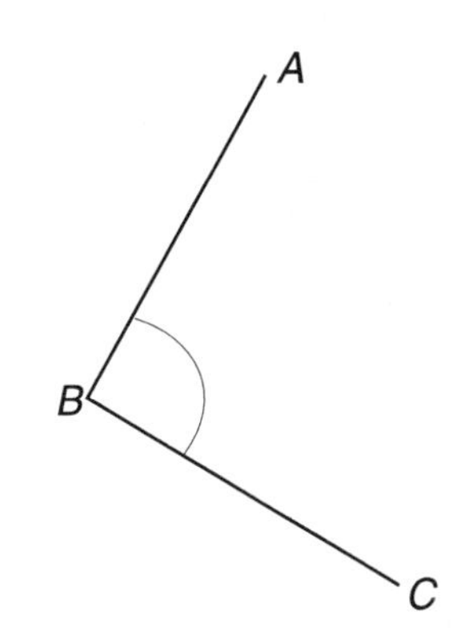

Figure 5.1

The vertex of the angle B is the middle of the letters

A **point** in geometry marks a position in space. It has *no* size (i.e. no dimensions). A **line** is the join between 2 points. The *shortest* line between two points is called a **straight** line. A line has one **dimension**, its length. A **plane** is a flat surface, which has *two* dimensions, length and width.

If two lines AB and BC meet at a point B, then the difference in the direction of the lines is known as the **angle** between the two lines. Alternatively, you can think of the amount of turn required to move BC to the position of BA. Angles are usually measured in *degrees* (°); for example, 20° or 145°. The symbol for angle used in this book is $\angle$. The angle in Figure 5.1 would be denoted by $\angle ABC$, or $\angle CBA$.

A complete *turn* consists of 360°, as shown in Figure 5.2, and the angle between two lines can be measured using a *protractor*.

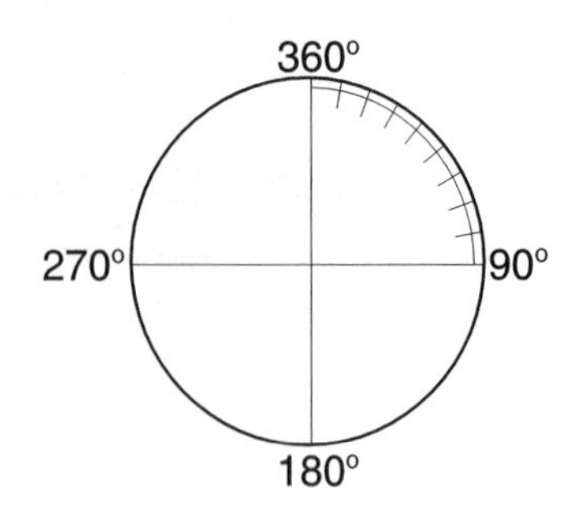

Figure 5.2

Types of angle

Angles between 0° and 90° are called **acute**

Angles between 90° and 180° are called **obtuse**

Angles greater than 180° are called **reflex**

The angle 90° is referred to as a **right angle**

An angle of 180° is clearly a straight line

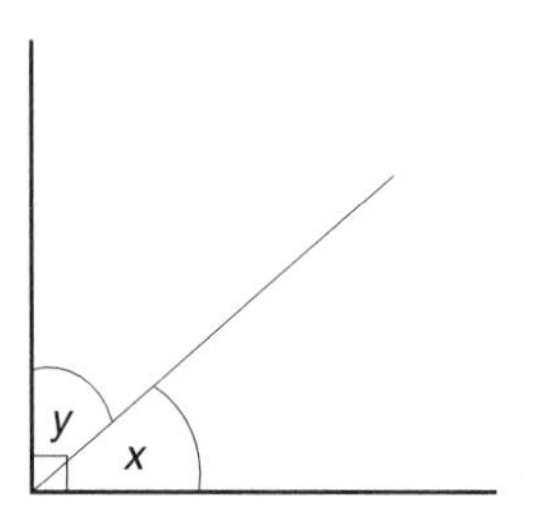

Figure 5.3

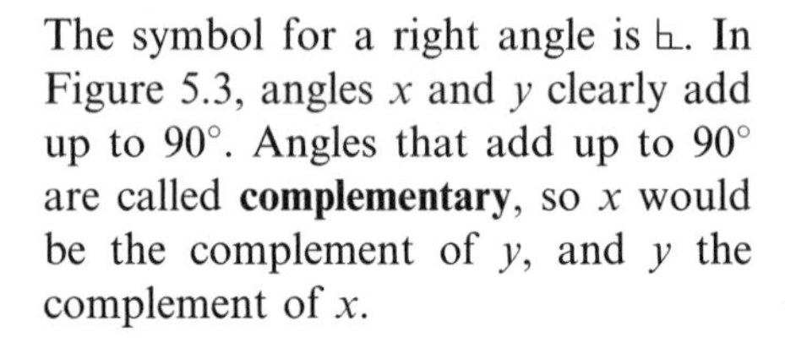

The symbol for a right angle is ∟. In Figure 5.3, angles x and y clearly add up to 90°. Angles that add up to 90° are called **complementary**, so x would be the complement of y, and y the complement of x.

In Figure 5.4, AOB is a straight line, and so angles a, b and c add up to 180°. Angles which add up to 180° are called **supplementary**.

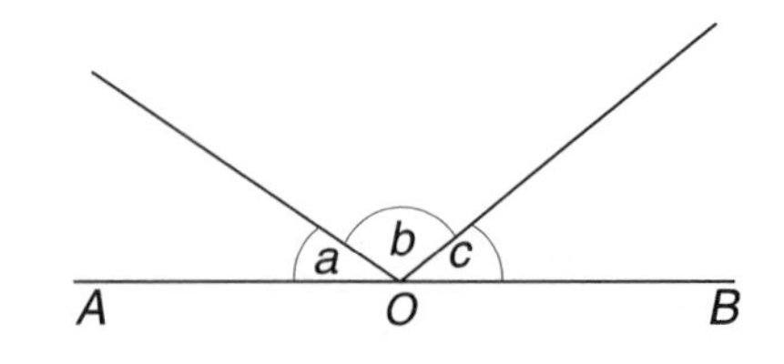

Figure 5.4

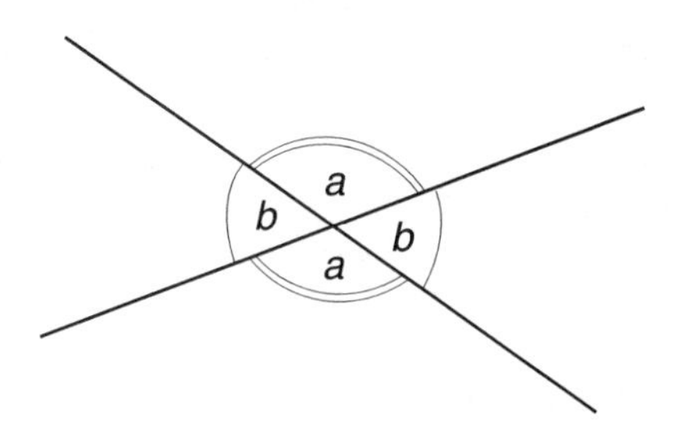

Figure 5.5

When two lines cross, they form two pairs of equal angles which are called **vertically opposite** angles (see Figure 5.5).

If $a = b = 90^\circ$, then the lines meet at 90°, and we say they are **perpendicular**. The symbol $\perp$ is used, so referring to Figure 5.6 $AB \perp CD$.

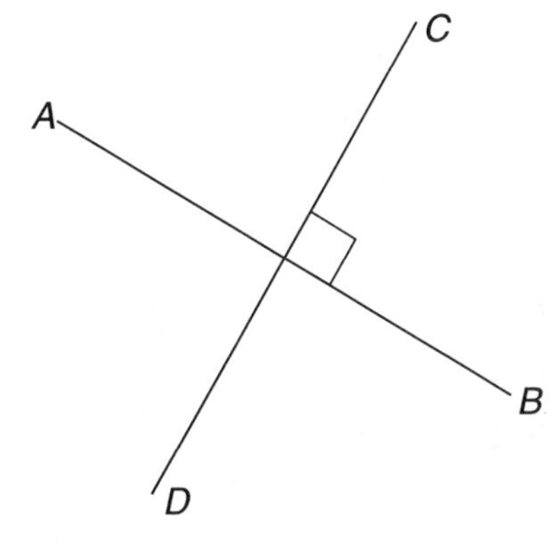

Figure 5.6

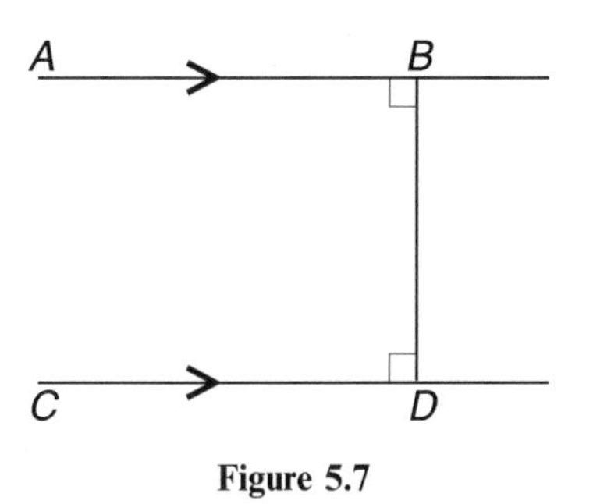

Figure 5.7

Straight lines which have the same direction are said to be **parallel**. The distance between parallel lines is the same at any point along their length (i.e. it is constant). Lines which are parallel are marked with arrows. The shortest distance between two parallel lines will be along a line that is perpendicular to both parallel lines. The symbol for parallel is $\|$. So in Figure 5.7 $AB\|CD$. The shortest distance between the lines is the length BD.

A line that crosses a series of parallel lines is called a **transversal**. These situations give rise to Z shapes (or S shapes) and F shapes. In Figure 5.8 the Z shape contains two equal angles a called **alternate** angles. Also the shape gives two equal angles c also called **alternate** angles. The F shape produces two equal angles b called **corresponding** angles. Angles between parallel lines, such as b and e will add up to 180° and are therefore **supplementary**, and are called **interior** angles.

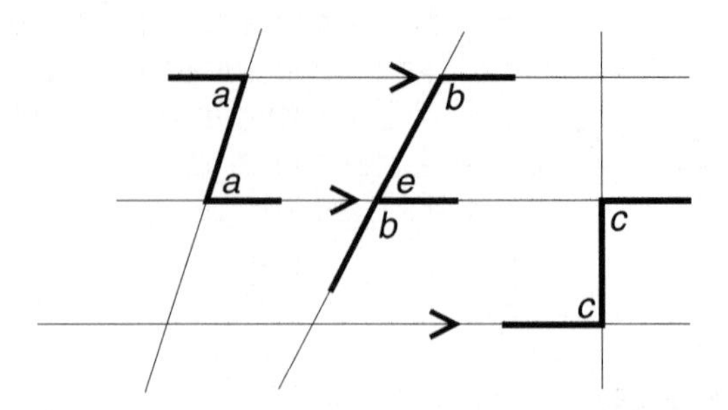

Figure 5.8

Example 5.1

Calculate the angles marked with a letter, giving reasons, in the diagrams in Figure 5.9:

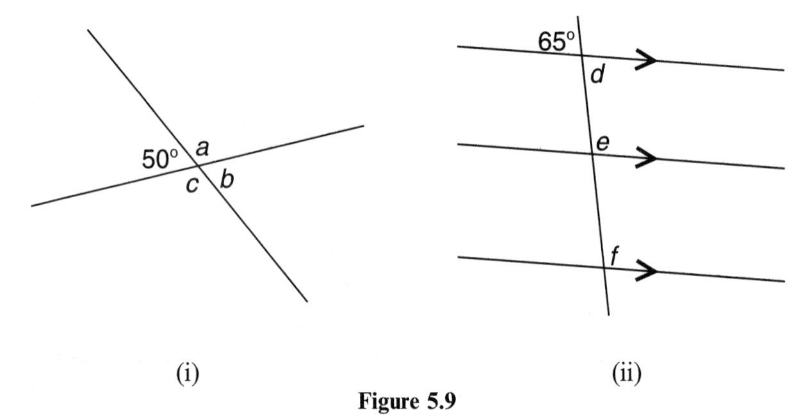

Figure 5.9

Solution

(i) $\angle b = 50°$ (vertically opposite)

$\angle a = 180° - 50°$ (supplementary angles on a straight line)

$= 130°$

$\angle c = 130°$ (vertically opposite).

(ii) $\angle d = 65°$ (vertically opposite)

$\angle d + \angle e = 180°$ (supplementary)

So $\angle e = 180° - 65° = 115°$

$\angle f = \angle e$ (corresponding)

$= 115°$.

Example 5.2

Find the size of $\angle x$ in Figure 5.10.

Solution

Sometimes, it is not possible to find an angle straight away, without adding to the diagram. A third parallel line has been added through B (see Figure 5.11).

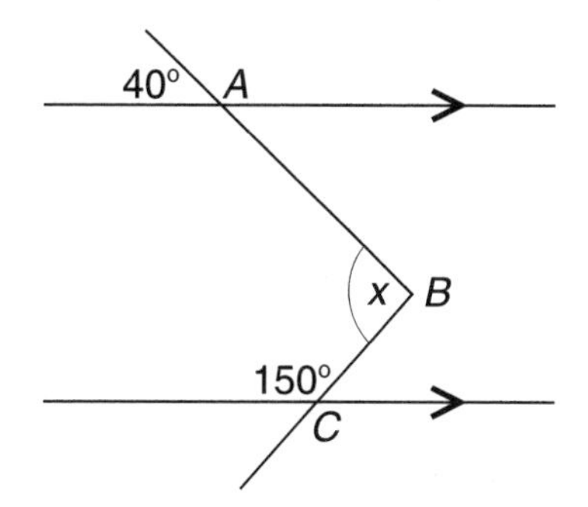

Figure 5.10

It can be seen that x has been split into two parts, one part corresponds to 40°, and the other is between two parallel lines and is supplementary with 150° ($30° + 150° = 180°$)

Hence $x = 40° + 30° = 70°$.

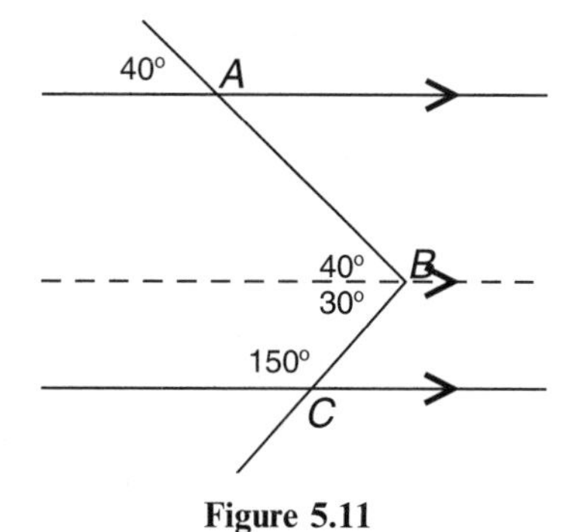

Figure 5.11

Exercise 5(a)

Calculate the size of the angles marked with a letter in the following diagrams.

1

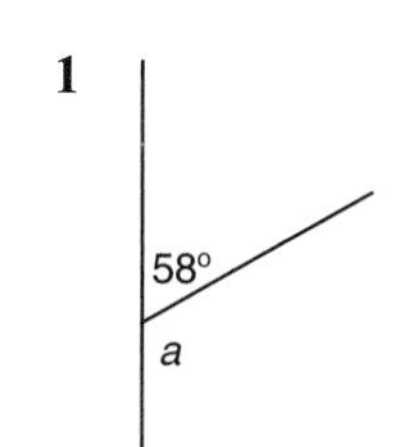

2

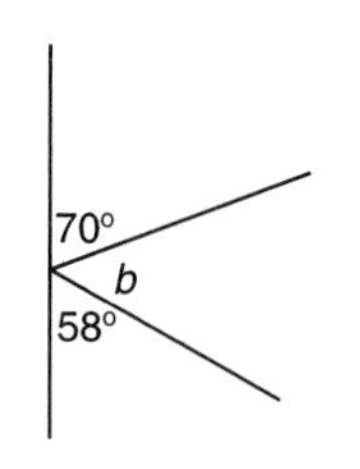

3

Figure 5.12

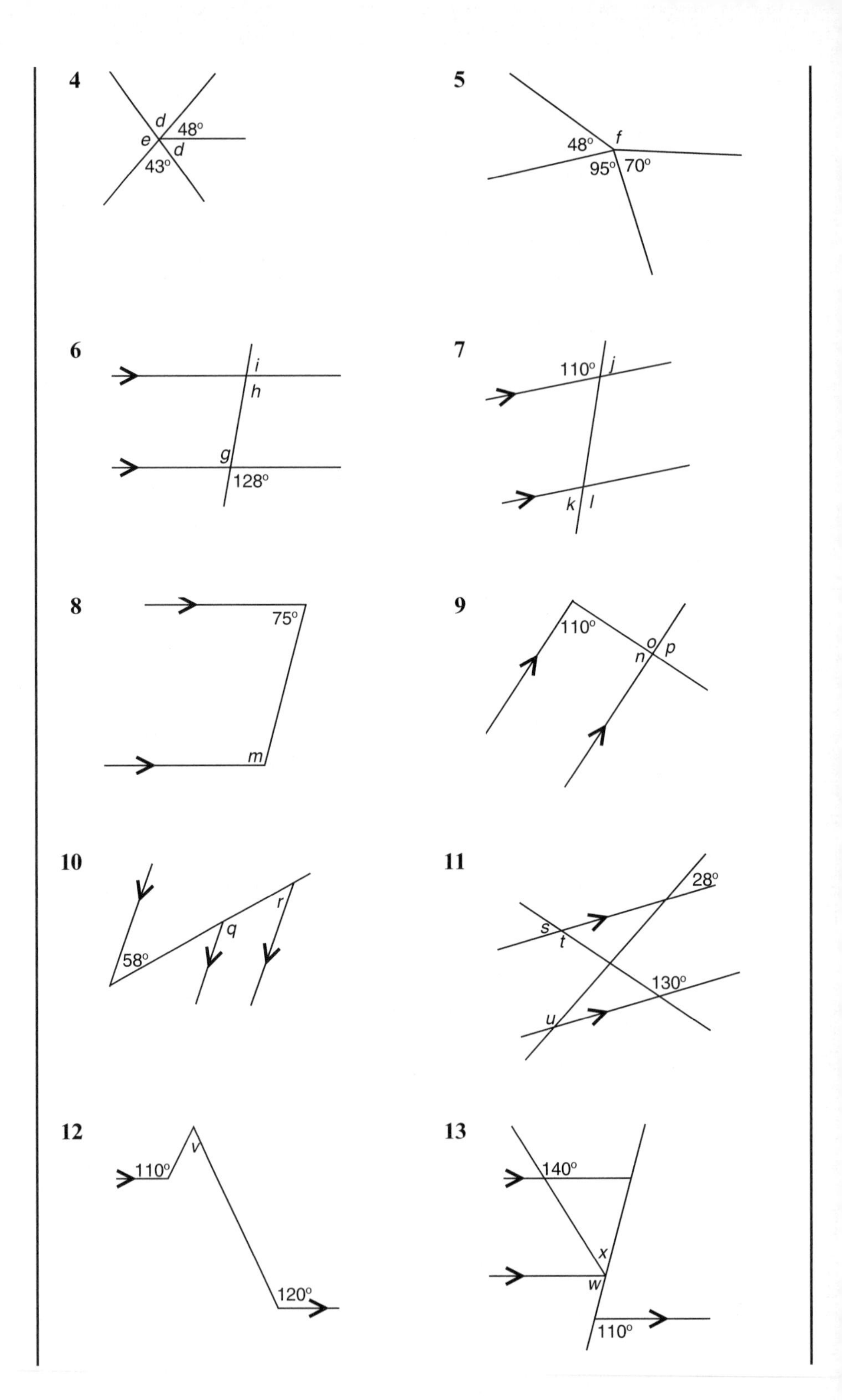
4
d
48°
e
d
43°
5
48°
f
95°
70°
6
i
h
g
128°
7
110°
j
k
l
8
75°
m
9
110°
o
p
n
10
r
q
58°
11
28°
s
t
130°
u
12
v
110°
120°
13
140°
x
w
110°

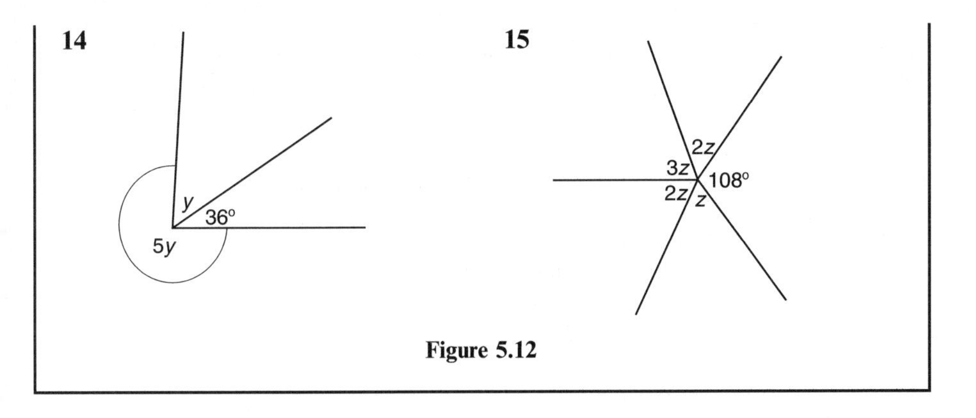

Figure 5.12

5.2 Polygons

Geometry is concerned with shapes formed by lines and planes. A plane shape that has several straight sides forming its boundary is called a **polygon**. The point at which two sides of a polygon meet is called a **vertex** of the polygon.

Figure 5.13 shows a five-sided polygon. If a side is extended, it produces an **exterior** angle. The angles inside the polygon are called **interior** angles. If all sides are the same length and all angles are equal, then the polygon is called **regular**. Table 5.1 gives a few special names associated with polygons.

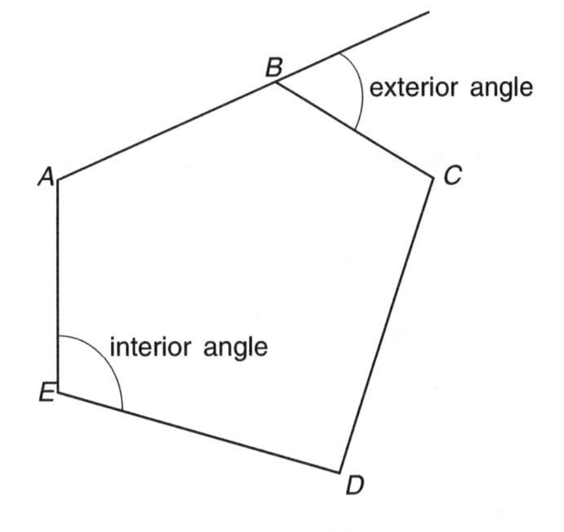

Figure 5.13

No. of sides	*Name*
3	triangle
4	quadrilateral
5	pentagon
6	hexagon
7	heptagon
8	octagon
9	nonagon
10	decagon

Table 5.1

There are certain properties of the angles of a polygon which are useful in solving problems.

(i) **The exterior angles of a polygon always add up to 360°**

This can be seen in Figure 5.14. If all of the exterior angles are moved to a fixed point, it can be seen that they form a complete turn.

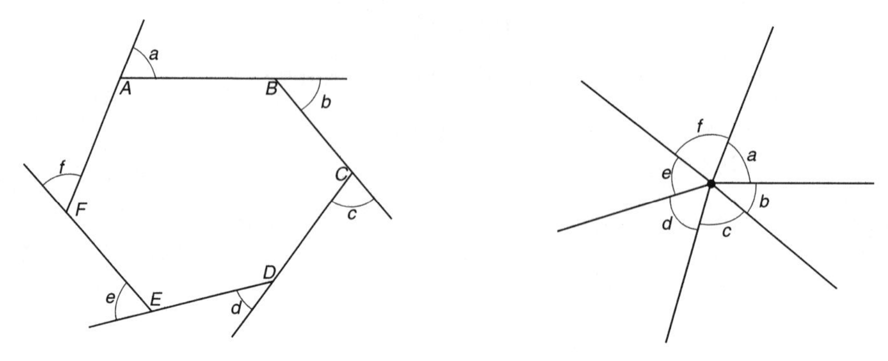

Figure 5.14

(ii) **Exterior angle + interior angle = 180°**

(iii) For a regular polygon with n sides,

each exterior angle = 360° ÷ n.

(iv) **The interior angles add up to $(n - 2) \times 180°$**, where n is the number of sides.

In Figure 5.15 you can see that a 7-sided polygon can be split into 5 triangles (i.e. $7 - 2$). The total of all the angles must be $5 \times 180° = 900°$.

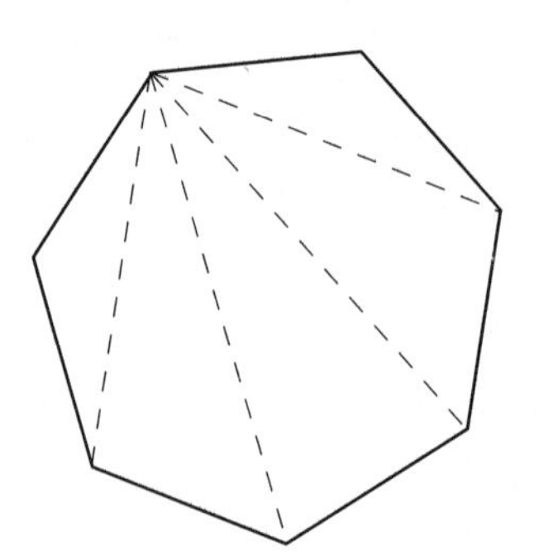

Figure 5.15

If any interior angle of a polygon is more than 180°, the polygon is called **concave**. If all interior angles are less than 180°, the polygon is called **convex** (see Figure 5.16).

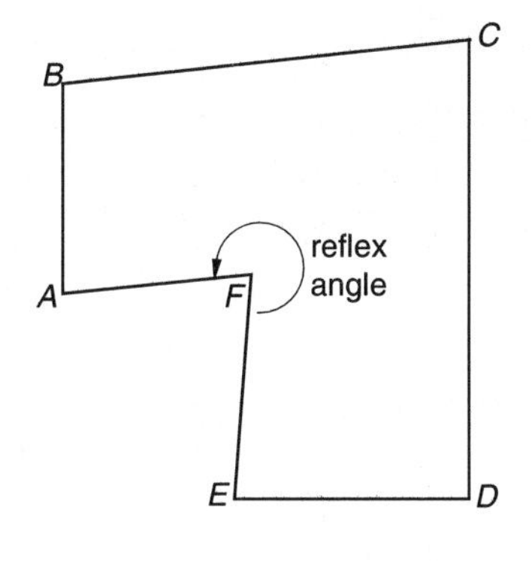

Figure 5.16

Investigation XI: Polygons

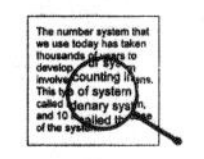

Using the result of the previous section, copy and complete the following table. Look for any patterns that occur and see if you could predict how the table continues. Assume all the polygons are regular.

No. of sides	*Name*	*Exterior angle*	*Interior angle*	*Sum of interior angles*
3	triangle		60°	
4	square	90°		360°
5	pentagon			
6	hexagon			
7				
8	octagon			
9				
10				
12				

Example 5.3

A regular polygon has 12 sides. For this polygon, find:

(a) the exterior angle
(b) the interior angle
(c) the sum of the interior angles.

Solution

(a) The exterior angle $= \dfrac{360°}{n}$

$$= \frac{360°}{12} = 30°.$$

(b) The interior angle $= 180° -$ the exterior angle

$$= 180° - 30° = 150°.$$

(c) This can be done in two ways.

(i) Since one interior angle is 150°, then the sum of all 12 angles is $12 \times 150° = 1800°$.

(ii) Using the formula, sum $= (n - 2) \times 180°$

$$= (12 - 2) \times 180°$$
$$= 1800°.$$

Example 5.4

The interior angle of a regular polygon is 172°, find the number of sides.

Solution

Since interior angle + exterior angle = 180°

the exterior angle $= 180° - 172°$

$$= 8°.$$

Always work with exterior angles

The exterior angles add up to 360°,

and so the number of sides $n = 360° \div 8 = 45$.

5.3 Symmetry of polygons

The study of polygons can lead to many types of pattern. Polygons often possess **reflective** or **rotational** symmetry. Consider the regular polygon shown in Figure 5.17. Each dotted line behaves like a mirror, in that the shape is identical on either side of these lines. These lines are called **mirror lines**, or **lines of symmetry**. If the shape is rotated about the centre O in the direction of the arrows, there will be 5 positions, where the pentagon does not appear to have been moved. We say that the **order** of rotational symmetry is 5, or the shape has *5-fold* rotational symmetry. The point O is called the **centre of rotation**.

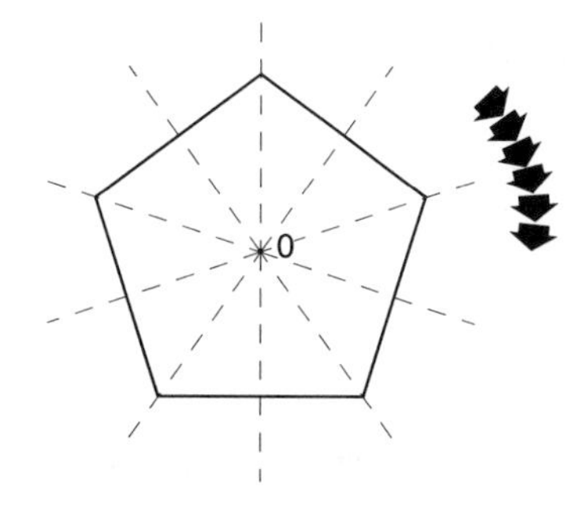

Figure 5.17

There is one particular aspect of rotation which often causes a problem, and that is where a shape is rotated about a point that is not part of the shape. The following Investigation will help you to understand the ideas involved. The ideas are extended later in the work on transformation.

Investigation XII: Centre of rotation

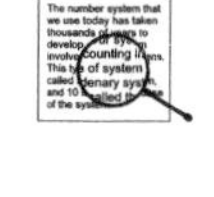

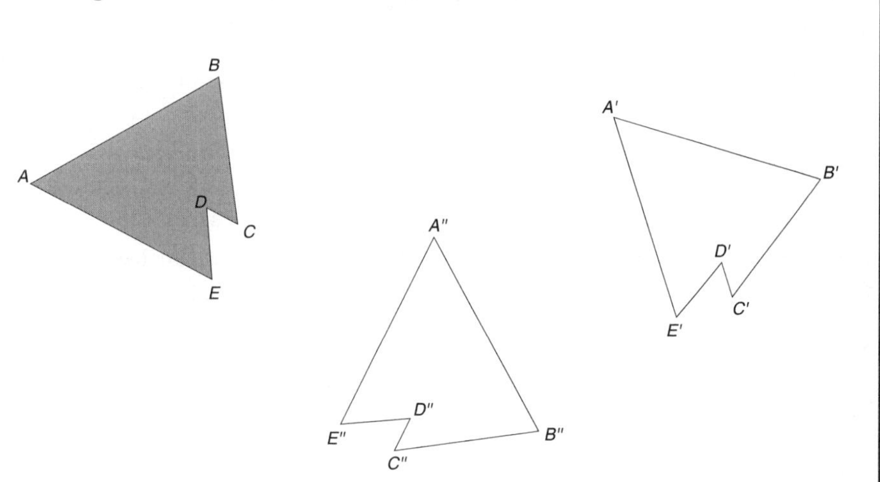

(Trace this diagram on to a sheet of plain paper).

When a shape is rotated about a point O, it is usually fairly easy to see where the centre of rotation O is, if O is a point in the shape. If, however, you look at the shapes in the above diagram, then the centres of rotation of the shaded shape to the other two positions are not very obvious. The following method will help:

(i) Join AA' (or any corresponding points). Find the centre of this line, and draw a line (L) at right angles to AA' through this point.

(ii) Join BB' (or any other pair of corresponding points). Find the centre of this line, and draw a line (M) at right angles to BB' through this point.

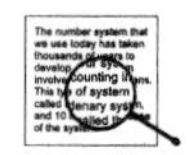

(iii) Extend L and M until they meet. This is the centre of rotation. Check this using tracing paper. Now try to find the centre of rotation for $A''B''C''D''E''$. Can you see any situation where this method will not work?

5.4 Properties of three and four sided shapes

(i) Triangles

Although the triangle is the simplest type of polygon, there are a number of facts and properties that should be known.

(i) The length of two sides of a triangle always add up to *more* than the length of the other side.

(ii) The interior angles of a triangle add up to 180°.

(iii) The longest side of a triangle is always *opposite* the largest angle.

(iv) A triangle with all sides equal in length, has all angles equal to 60° and is called **equilateral**.

(v) A triangle with two sides of the same length, has two angles equal and is called **isosceles**.

(vi) A triangle with all sides different in length is called **scalene**.

(vii) A triangle with all angles less than 90° is called an **acute-angled** triangle.

(viii) A triangle with one angle greater than 90° is called an **obtuse-angled** triangle.

(ii) Quadrilaterals

(a) The rectangle

All angles are 90°.

The diagonals are equal in length and bisect each other (meet at their midpoints).

There are 2 lines of symmetry.

There is rotational symmetry of order 2 about the centre of the rectangle.

Figure 5.18

(b) The square (special case of a rectangle)

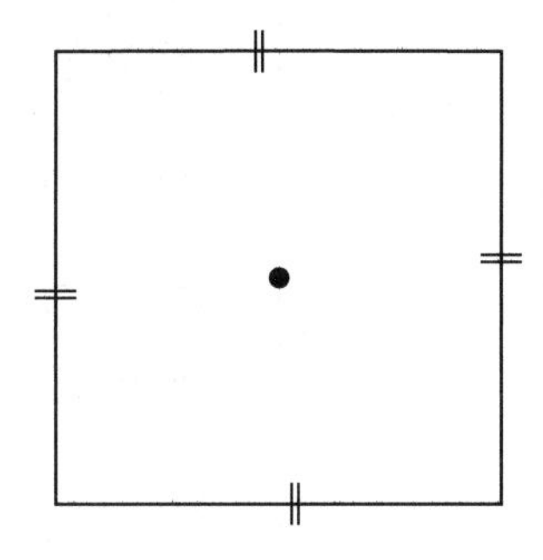

Figure 5.19

All sides are equal.

All angles equal 90°.

Diagonals are equal in length and bisect each other at right angles.

There are 4 lines of symmetry.

There is rotational symmetry of order 4 about the centre of the square.

(c) The parallelogram

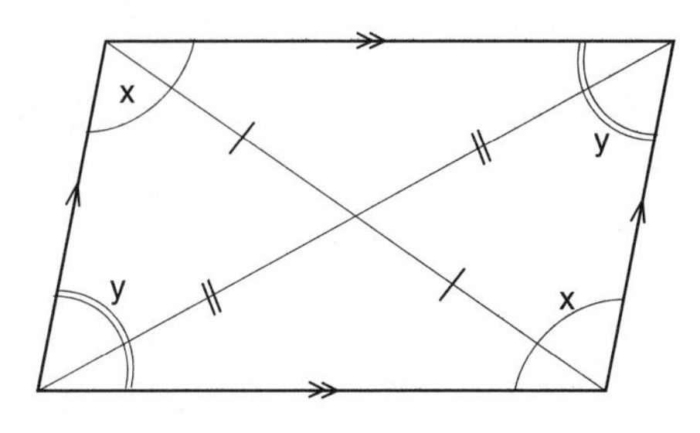

Figure 5.20

Opposite sides are parallel and equal in length.

Opposite angles are equal.

The diagonals bisect each other.

There is rotational symmetry of order 2 about the intersection of the diagonals.

There is no line symmetry.

(d) The rhombus

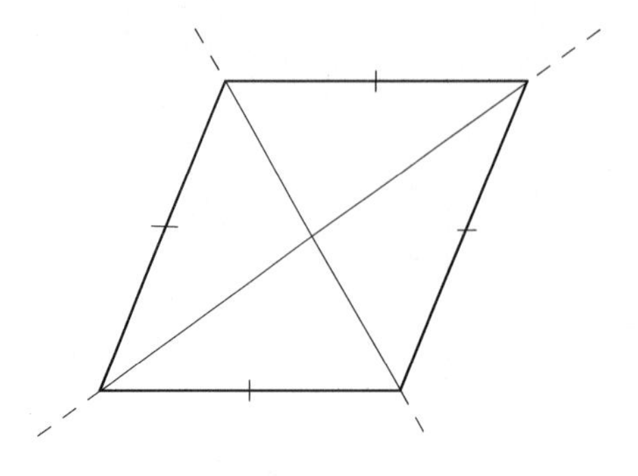

Figure 5.21

A rhombus is a parallelogram with all sides equal in length.

The diagonals bisect each other at right angles.

There are 2 lines of symmetry.

There is rotational symmetry of order 2 about the centre of the rhombus.

(e) The kite

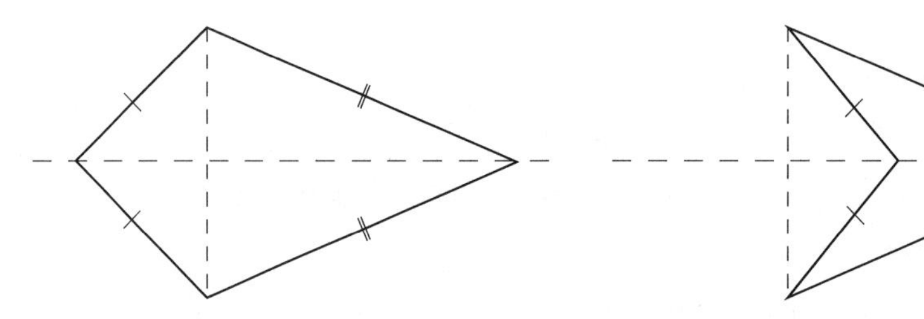

Figure 5.22

Two pairs of adjacent sides are equal in length.

The diagonals are perpendicular.

There is 1 line of symmetry.

(f) The trapezium

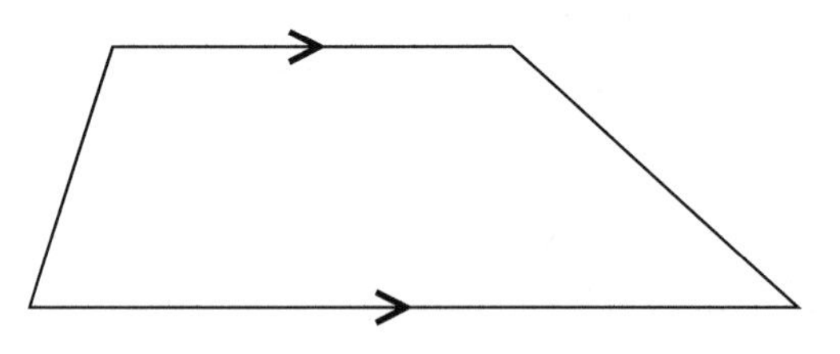

Figure 5.23

Two opposite sides are parallel.

In general there is no symmetry.

Example 5.5

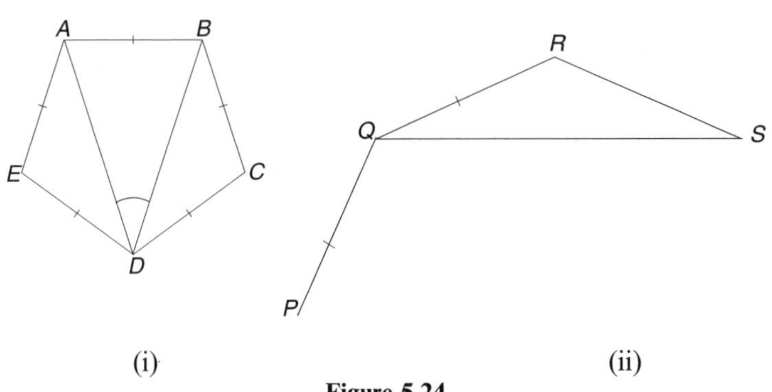

Figure 5.24

(a) Calculate $\angle ADB$ in a regular pentagon $ABCDE$ (Figure 5.24(i)).

(b) In Figure 5.24(ii) P, Q, R, S are adjacent vertices of a regular decagon. Calculate the size of $\angle PQR$ and $\angle RQS$.

Solution

(a) Each interior angle of a regular pentagon is 108°.

In the isosceles triangle AED

$$\angle EAD = \angle EDA = 36° \qquad (36° + 36° + 108° = 180°)$$

Similarly in triangle BCD, $\angle CDB = 36°$.

Therefore $\angle ADB = 108° - (36° + 36°) = 36°$.

(b) The exterior angle of a regular decagon $= 360° \div 10 = 36°$

Hence each interior angle of a regular decagon is 144° (180° − 36°)

Therefore $\angle QRS = 144°$

Triangle QRS is isosceles, since $RQ = RS$.

Hence $\angle RQS + \angle RSQ = 180° - 144° = 36°$

Therefore $\angle RQS = 18°$.

Example 5.6

Find angles a, b, c and d in Figure 5.25.

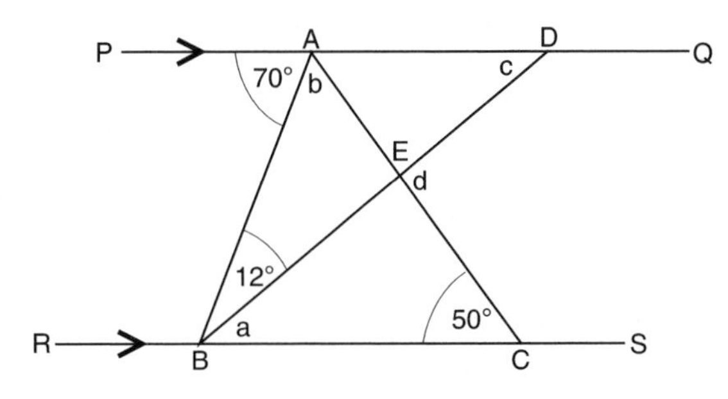

Figure 5.25

Solution

Now $\angle ABC = \angle PAB$ (alternate angles)

i.e. $\quad 12° + a = 70°$

Hence $\quad a = 58°$.

$\angle ABC + \angle BAC + \angle ACB = 180°$ (angles of a triangle)

i.e. $\quad 70° + b + 50° = 180°$

Hence $\quad b = 60°$.

$\angle ADB = \angle CBD$ (alternate angles)

So $\quad c = a = 58°$.

$\angle AEB + \angle EAB + \angle EBA = 180°$ (angles of a triangle)

i.e. $\quad \angle AEB + b + 12° = 180°$

Hence $\quad \angle AEB = 180° - 12° - 60° = 108°$

Also $\quad \angle DEC = \angle AEB$ (vertically opposite)

So $\quad d = 108°$.

Exercise 5(b)

1 Find the size of the angles marked x, y, z in Figure 5.26.

(i)

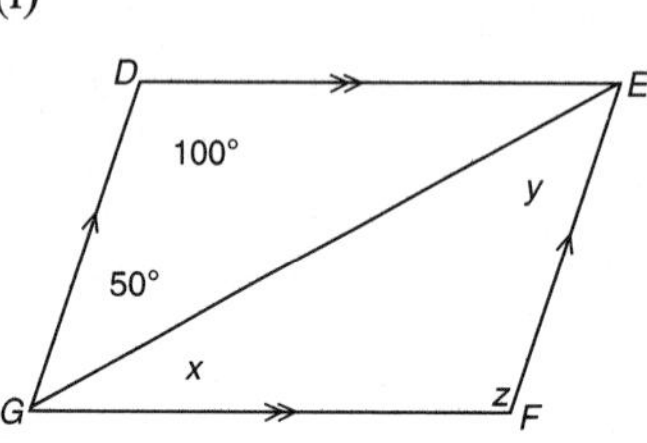

(ii)

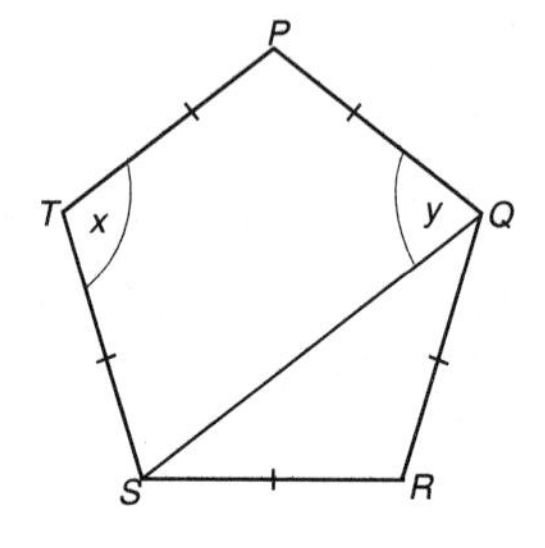

(iii)

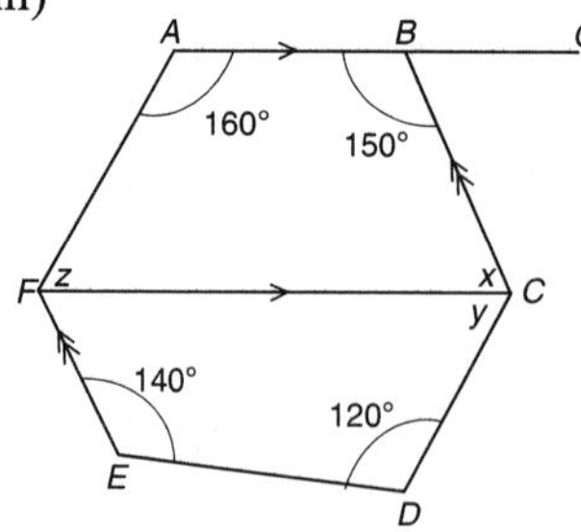

(iv)

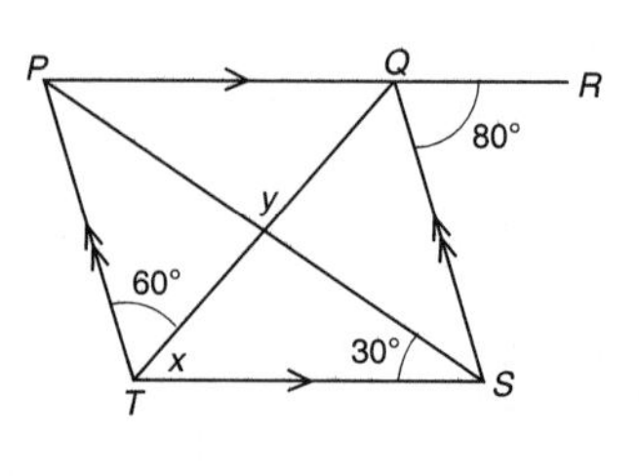

Figure 5.26

2 The interior angle of a regular polygon is three times the exterior angle. How many sides has the polygon?

[*Hint*: let the exterior angle be $x°$.]

3 Find angles a, b and c in Figure 5.27. O is the centre of rotation of the regular hexagon in (i)

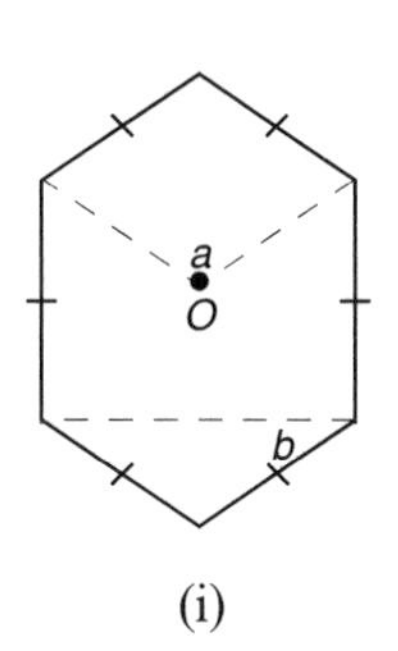

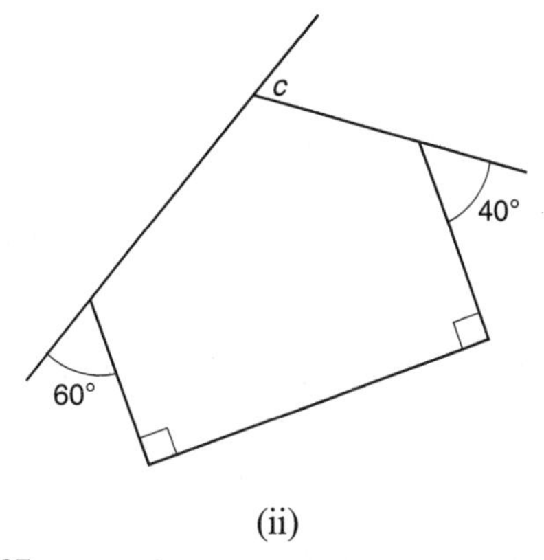

(i) (ii)

Figure 5.27

4 The interior angles of a polygon add up to 4320°. How many sides does the polygon have?

5 A hexagon has interior angles $x°$, $2x°$, $x + 50°$, 130°, 140° and 150°. Calculate the value of x, and hence the size of the angles of the hexagon.

5.4 Similarity and congruence

(i) Similarity

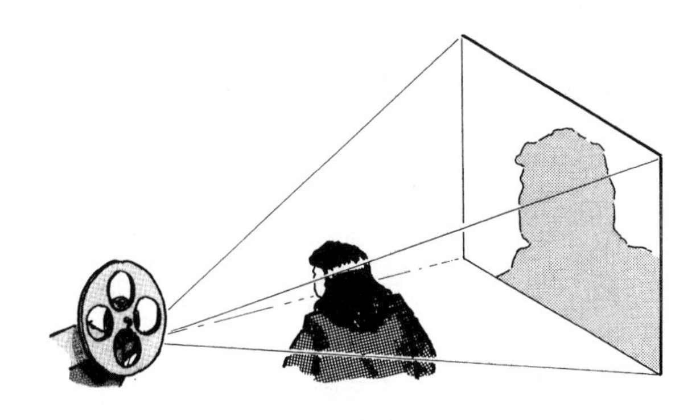

If a shape is projected on to a screen which is parallel to the shape, the resulting image is mathematically **similar** to the original shape. It will have been *enlarged*. The size of the enlargement is measured by a quantity called the **scale factor**. This is equal to the ratio obtained by dividing a length in the image by the corresponding length in the object (the original shape). It is also true that the shape is similar to the image, and is obtained by multiplying by a *fractional* scale factor (reciprocal of the original scale factor).

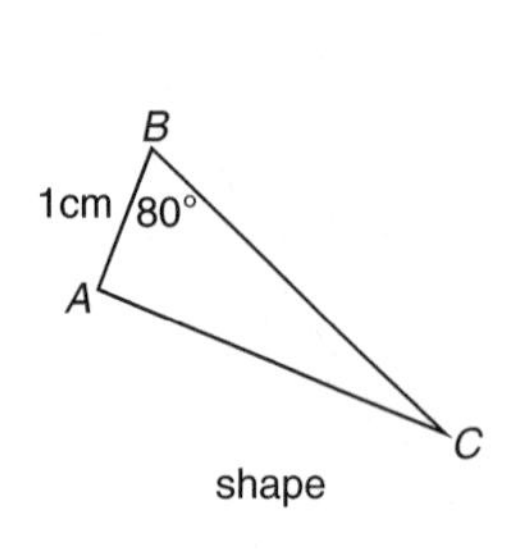

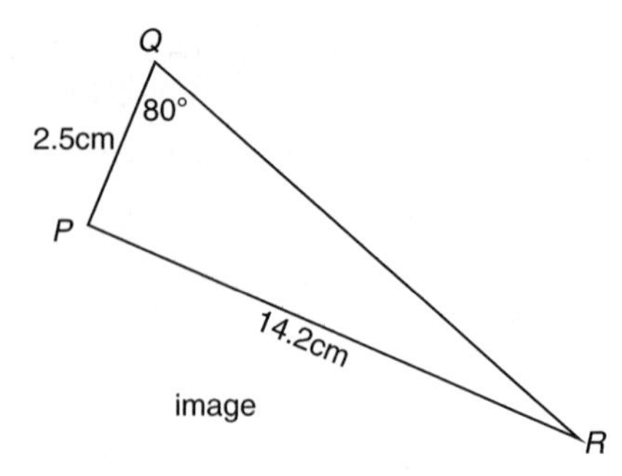

Figure 5.28

In Figure 5.28, the angles in triangle ABC are the same as those in triangle PQR. Hence triangle PQR is an enlargement of triangle ABC.

$$\frac{PQ}{AB} = \frac{2.5}{1} = 2.5$$

So the scale factor is 2.5.

Hence lengths in the shape are multiplied by 2.5 to find lengths in the image. Since AC is the side in the same position as PR, then $AC \times 2.5 = 14.2$.

$$\text{So} \quad AC = \frac{14.2}{2.5} = 5.68.$$

Example 5.7

In Figure 5.29, XY is parallel to BC, $AX = 3$ cm, $AY = 2$ cm, $XB = 9$ cm and $BC = 8$ cm. Show that $\triangle AXY$ is similar to $\triangle ABC$ and calculate the lengths of XY and YC.

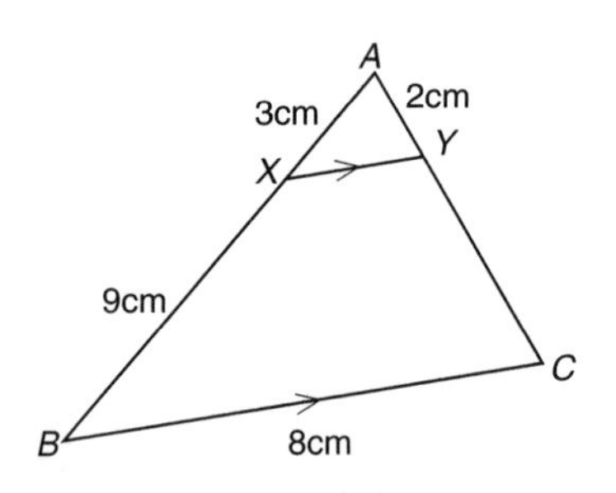

Figure 5.29

Solution

Since $XY \| BC$, $\angle AXY = \angle ABC$ (corresponding angles)

$\angle AYX = \angle ACB$ (corresponding angles)

$\angle XAY = \angle BAC$ (same angle)

Hence the angles of each triangle are the same, and so the triangles are similar. If we split the diagram up (always a useful technique), we get:

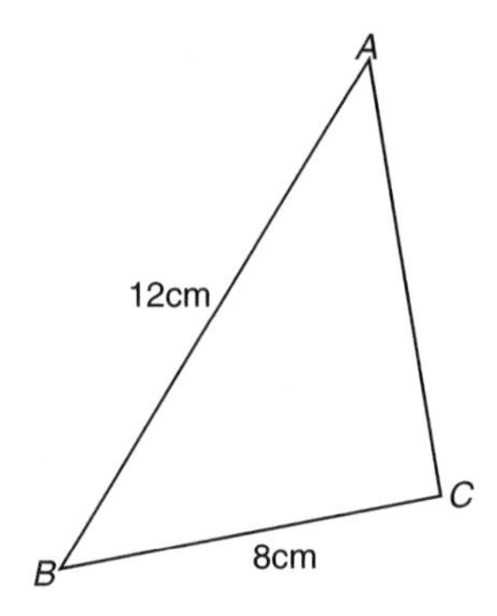

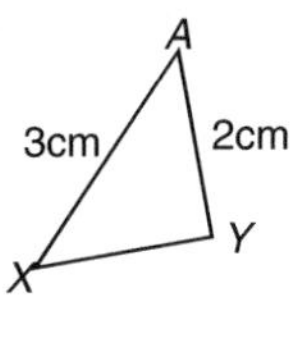

Figure 5.30

Since $\frac{AB}{AX} = \frac{12}{3} = 4$, triangle ABC is an enlargement of triangle AXY by a scale factor 4

Hence $XY \times 4 = BC = 8$ cm, so $XY = 2$ cm

Also, $AC = AY \times 4 = 2 \text{ cm} \times 4 = 8$ cm

Hence $YC = AC - AY = 8 \text{ cm} - 2 \text{ cm} = 6$ cm

(ii) Congruence

Two figures are said to be **congruent**, when they are exactly the same size, as well as the same shape. The symbol for congruent is $\equiv$ which means identical or equal in all respects. When referring to triangles, there are four different conditions for triangles to be congruent. They are illustrated in Figure 5.31.

(i) Three sides equal (S-S-S)
(ii) Two sides and the included angle equal (S-A-S) (included means between the sides).
(iii) Two angles and a corresponding side equal (A-A-S) or (A-S-A).
(iv) A right angle, the hypotenuse, and one other side equal (R-H-S)

(i)

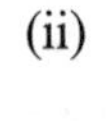

(ii)

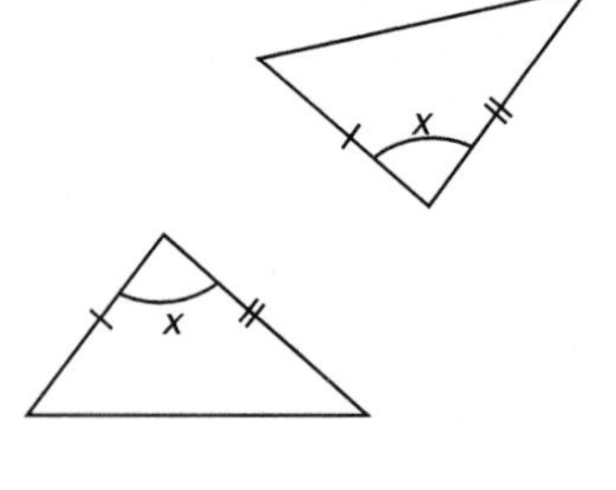

(iii)

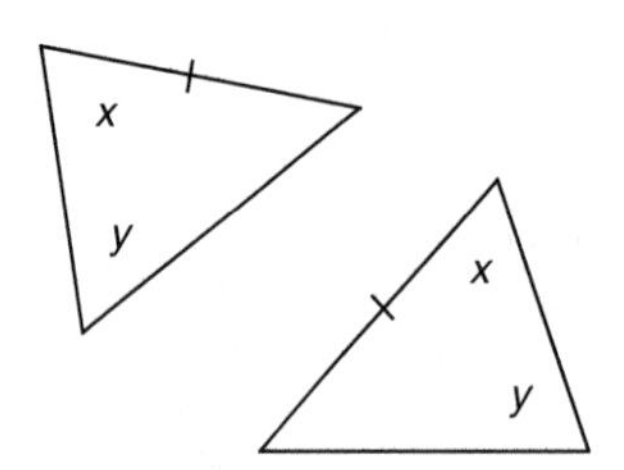

(iv)

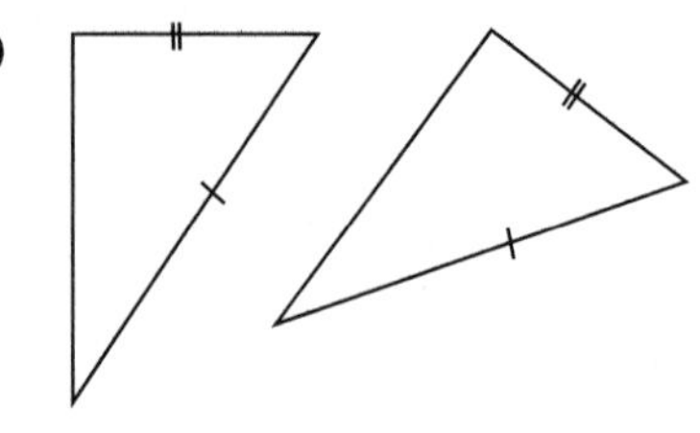

Figure 5.31

Example 5.8

$PQRS$ is a parallelogram. Show that $\triangle PSR$ is congruent to $\triangle RQP$. [*Note*: The corresponding letters must be in the same position.]

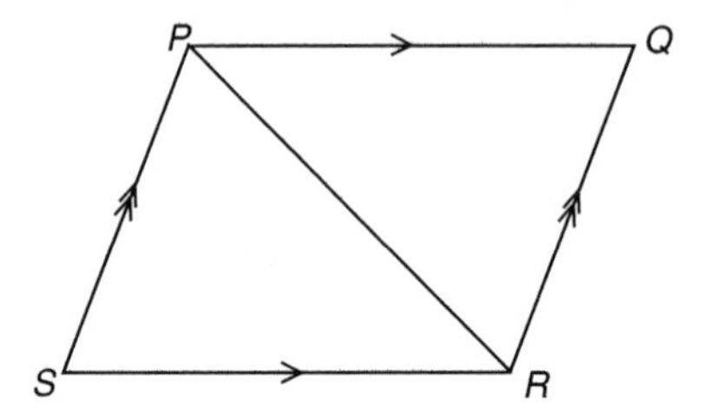

Figure 5.32

Solution

$\angle SPR = \angle PRQ$ (alternate angles)

$\angle PRS = \angle RPQ$ (alternate angles)

PR is common to both triangles

Hence $\triangle PSR \equiv RQP$ (A-A-S)

Exercise 5(c)

1 Find p and q in the following diagrams:

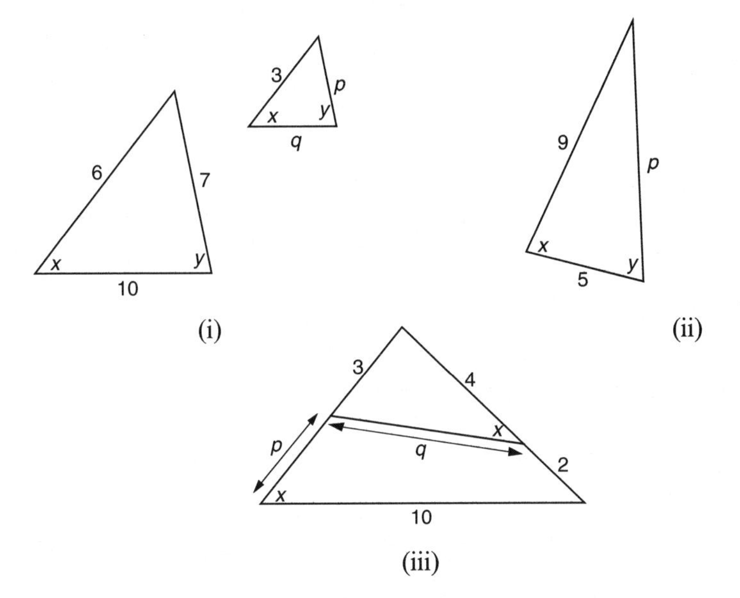

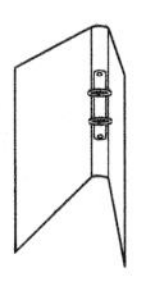

Figure 5.33

2 Triangle ABC has a right angle at A. N lies on BC, and AN is perpendicular to BC. Prove that $\triangle ANC$ and $\triangle BAC$ are similar. If $CN = 6$ cm and $AC = 8$ cm find AN and AB.

3 The following conditions apply to triangles ABC and XYZ. Sketch each pair of triangles, state whether they are congruent, and if so give the reason.

(i) $AC = XZ$, $CB = ZY$, $\angle C = \angle Z$
(ii) $AC = XY$, $\angle B = \angle Y$, $\angle A = \angle X$
(iii) $AB = YZ$, $\angle A = \angle X$, $\angle C = \angle Z$.

5.6 Tessellations

In Figure 5.34, there are a number of repeating patterns, each of which can be extended as far as you like in any direction. Such patterns are called **tessellations**. A shape tessellates if it covers the plane without any gaps.

If the basic shape (referred to as the motif) is a regular polygon, the pattern is called a **regular tessellation**. If you consider any vertex of a regular tessellation, since the angles there must add up to 360°, it follows that the internal angles of a polygon in a regular tessellation must divide exactly into 360°. If you try a few possible angles it can be seen that the only possible regular tessellations are with equilateral triangles, regular hexagons and squares.

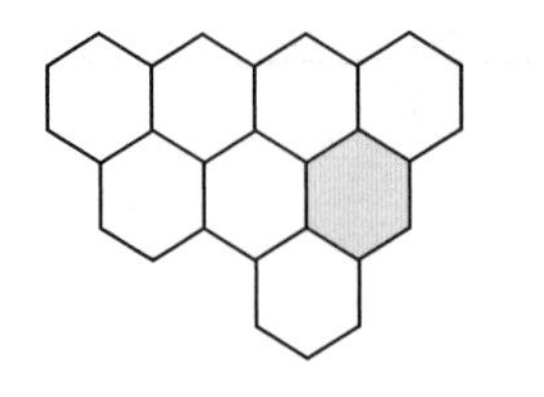
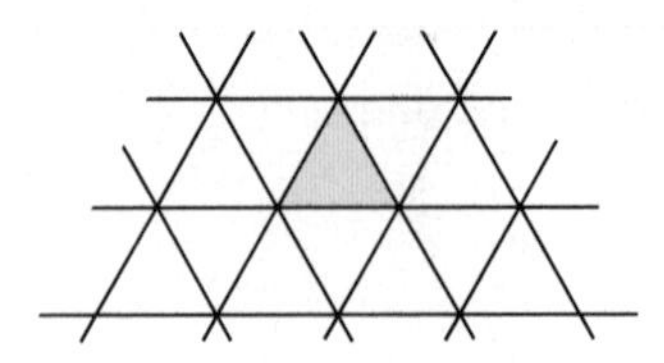
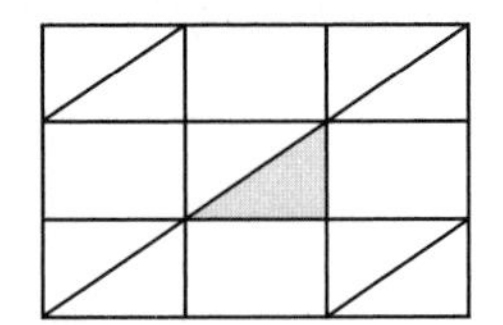

Figure 5.34

In Figure 5.35 a number of tessellations can be seen which are based on the regular tessellations.

(i) combines together 5 squares
(ii) modifies the equilateral triangle
(iii) combines together 2 hexagons.

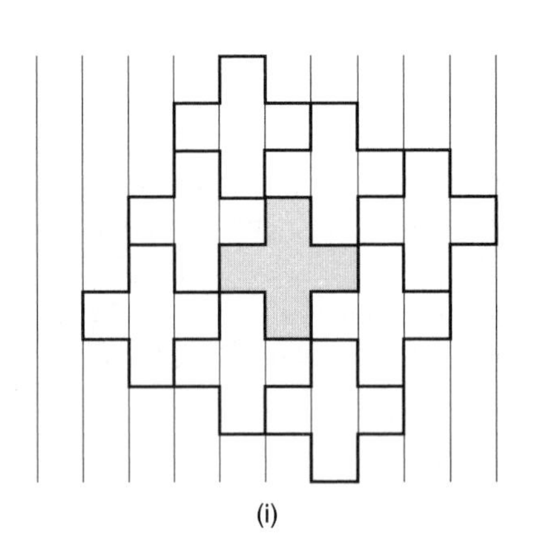

(i)

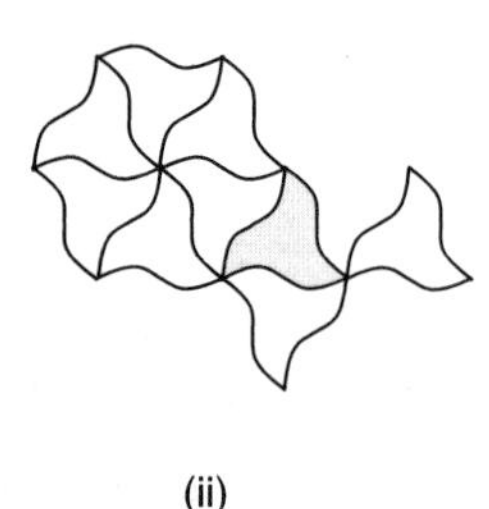

(ii)

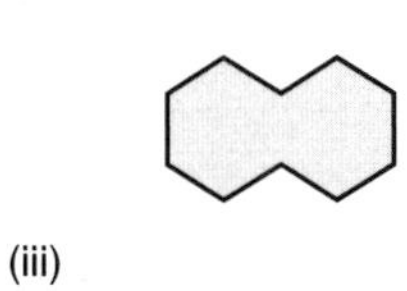

(iii)

Figure 5.35

The tessellations shown in Figure 5.36 are examples of **semi-regular tessellations**, based on two or more different regular polygons.

It can be seen that a basic motif, shown shaded, can be found to tessellate the pattern. At each vertex of a semi-regular tessellation there must be the same combination of polygons in the same order. For this reason, there are only eight semi-regular tessellations. See if you can find them.

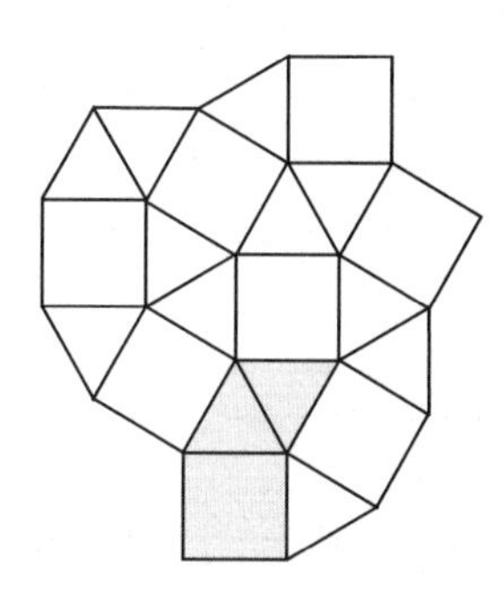

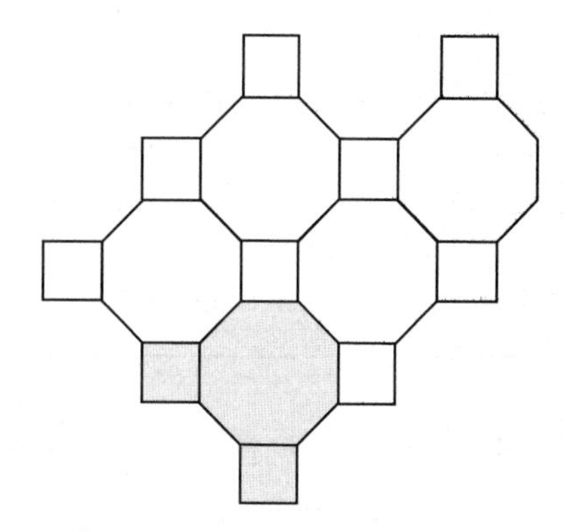

Figure 5.36

Investigation XIII: Pentominoes

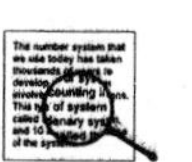

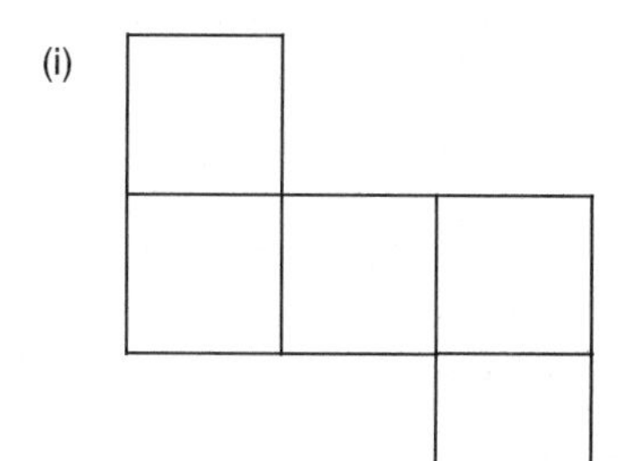

(i)

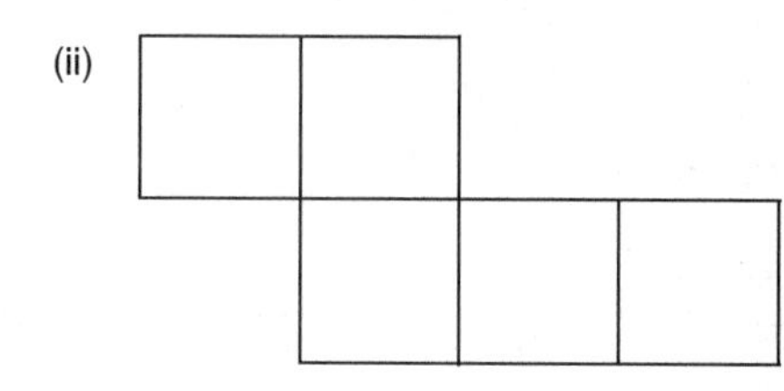

(ii)

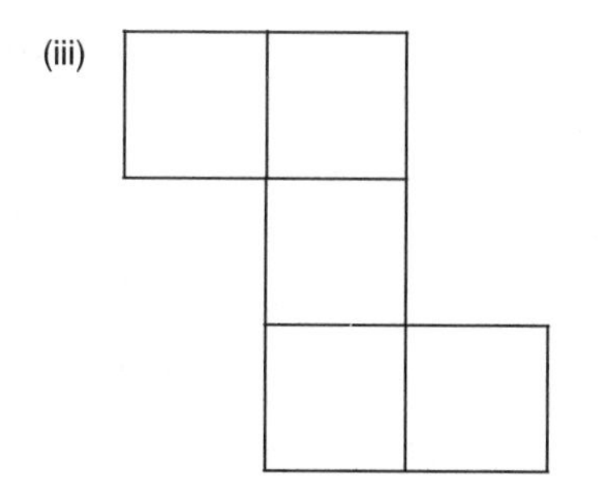

(iii)

Figure 5.37

A pentomino is the name given to an arrangement of five squares such as those shown (all squares must be joined along at least one side). The arrangements (i) and (ii) are different, but (i) and (iii) are regarded as the same. There are altogether 12 such pentominoes, can you find them? If you try and fit these pentominoes into a rectangle that measures 12 by 3 squares, it can only be done in a limited number of ways. How many ways can you find?

Extension

(a) See if the pentominoes can be fitted into other shaped rectangles.

(b) (i) Try and make up your own shapes with 6 different squares.

(ii) If your shapes were nets for constructing a solid, can they all be made into cubes?

(c) Use shapes other than squares.

Exercise 5(d)

1 Figure 5.38 shows a pattern of tiles made of 5 squares, and 4 regular figures each with 8 sides.

(i) Calculate the value of x.

(ii) Calculate the sum of the interior angles of one regular eight-sided figure.

(iii) Calculate angle UPS.

(iv) Calculate angle APU.

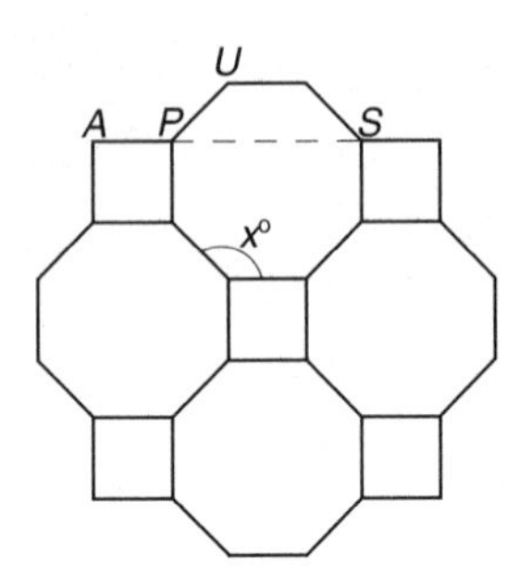

Figure 5.38

2 In Figure 5.39, E, F, G and H are the mid-points of the sides of the 5 cm square $ABCD$.

(i) What is the order of rotational symmetry of the diagram (ignore the letters in the diagram).

(ii) Explain why the central shape $JKLM$ in the diagram is a square.

(iii) The triangles EJB, FKC, GLD and HMA are each rotated through 180° about the points E, F, G and H respectively. Sketch the resulting shape.

(iv) Deduce the area of $JKLM$.

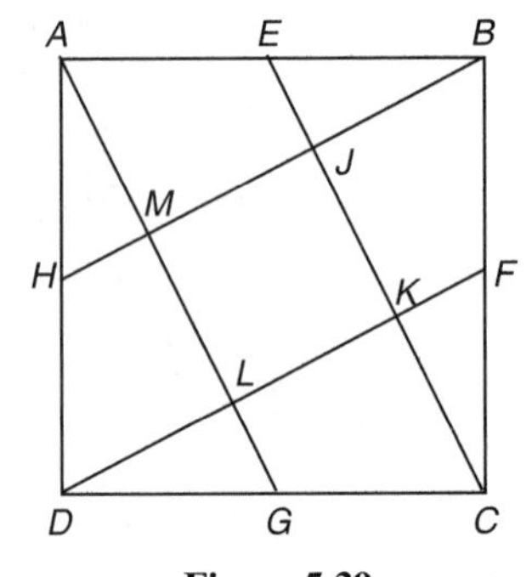

Figure 5.39

[MEG]

3 In Figure 5.40, $PQRST$ is one half of a regular octagon.

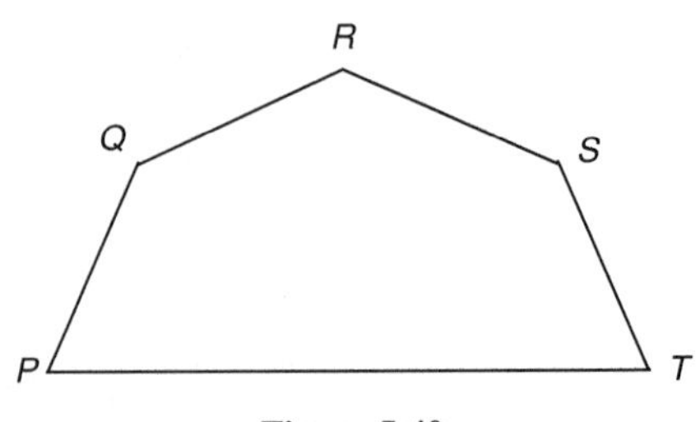

Figure 5.40

Calculate:

(i) $\angle PQR$

(ii) $\angle QPT$

(iii) $\angle PTQ$

(iv) the sum of the angles of the pentagon.

6 Length, area and volume

6.1 Perimeter of plane figures

(i) Polygons

A plane shape (one drawn on a flat surface) which is composed of straight sides is called a **polygon** (see Section 5.2). Examples are shown in Figure 6.1.

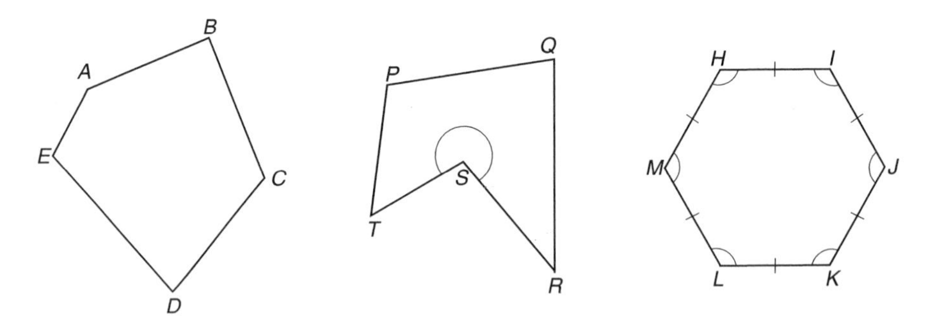

Figure 6.1

(i) A **convex** polygon has all its interior angles less than 180°.
(ii) A **concave** polygon has at least one interior angle greater than 180°.
(iii) A **regular** polygon has all sides the same length, and all angles equal in size.

The distance round the outside of a polygon is called its **perimeter**.

For a regular polygon, if the polygon has n sides of length l,

the perimeter $= nl$.

(ii) Triangles

A triangle with sides of lengths a, b and c has a perimeter of $a + b + c$.

Sometimes we need to find half the perimeter, called the semi-perimeter, which is denoted by s.

So $s = \frac{1}{2}(a + b + c)$ (see Section 6.2(iii)).

(iii) Rectangles or parallelograms

If the sides of the rectangle or parallelogram are a and b,

the perimeter $= 2a + 2b$ (see Figure 6.2).

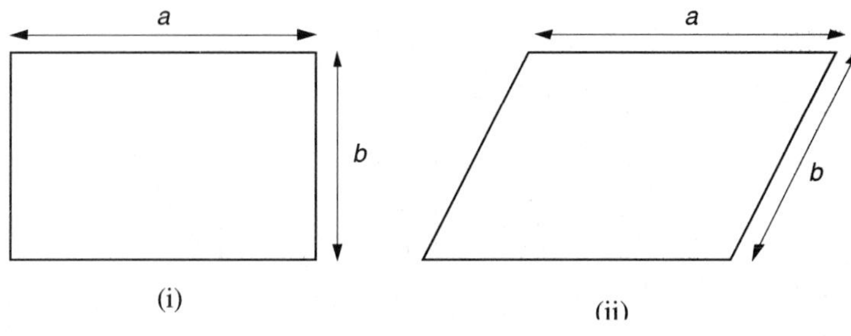

Figure 6.2

(iv) Circles

The perimeter of a circle is called its **circumference**. For any circle, the circumference divided by the diameter gives a constant value of 3.142, correct to 3 decimal places. This constant is denoted by π (pi). [*Note*: π is an irrational number, because the decimal does not terminate, and it does not repeat.]

Most scientific calculators have a button $\boxed{\pi}$ which gives more decimal places if required.

If C is the circumference and d the diameter, then it can be shown that

$$\frac{C}{d} = \pi$$

i.e. $C = \pi d$

Since the diameter $d = 2 \times$ the radius

$= 2r$

the formula can be written $C = 2\pi r$.

i.e. Circumference $C = \pi d$

or $= 2\pi r$.

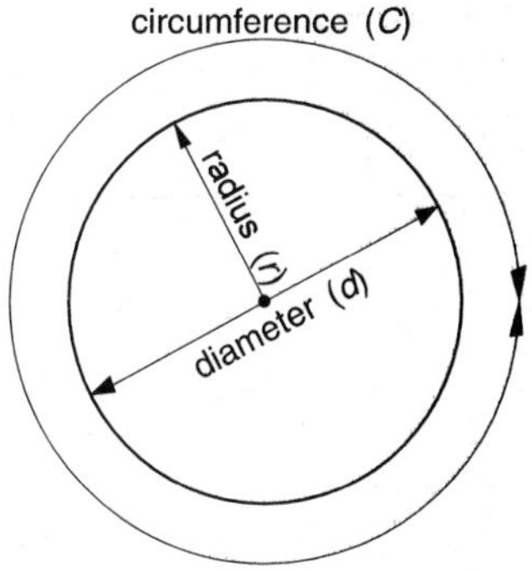

circumference (C)
radius (r)
diameter (d)

Figure 6.3

The following investigation is an exercise in trying to find the value of π by accurate drawing.

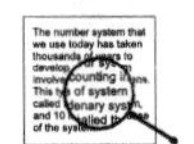

Investigation XIV: Finding the value of π

The formula for the circumference of a circle $C = \pi d$, has been known for a long time. The ancient mathematicians used the following method to calculate π.

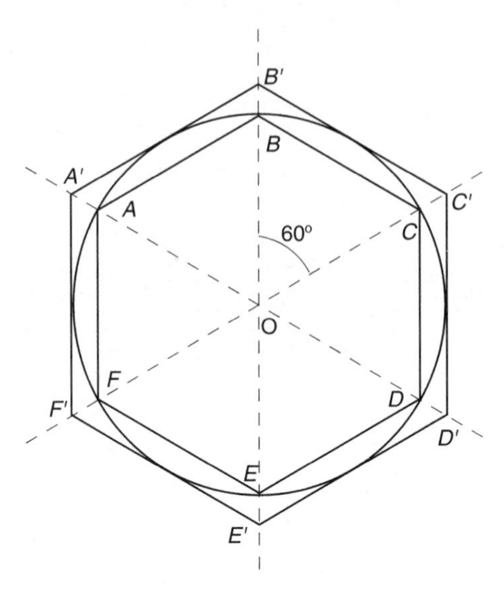

Draw a circle centre O of radius 4 cm. Measure angles at 60° intervals at O, and hence drawn the hexagon $ABCDEF$. Extend the radii and draw the hexagon $A'B'C'D'E'F'$ outside the circle, just touching the circle, with sides parallel to $ABCDEF$. An accurate measure of the perimeter of the two hexagons can be found. These are 24 cm and 27.7 cm. (Check to see how close you get.) Hence the circumference of the circle lies between 24 cm and 27.7 cm. Since circumference $= \pi \times$ diameter, and the diameter is 8, then $24 \div 8 = 3$ and $27.7 \div 8 = 3.46$.

Hence π lies between 3 and 3.46. You should now repeat the process several times, increasing the number of sides of the polygon, so you can approximate π to an increasingly high degree of accuracy.

Example 6.1

The circumference of a circle is 12 m, find its radius.

Solution

$$\text{Since} \quad C = \pi d, \quad d = \frac{C}{\pi}$$

$$\text{So} \quad d = \frac{12}{\pi} = 3.82.$$

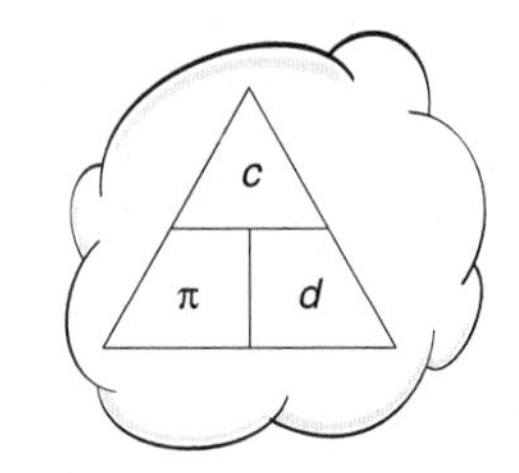

On a calculator

				Display
12	÷	π	=	3.819786

$r = \frac{1}{2}d = 3.82 \div 2 = 1.91$

The radius is 1.91 m.

Example 6.2

The length of the longest side of a parallelogram is 3 times the length of the shortest side. If the perimeter of the parallelogram is 42 m, find the lengths of the sides.

Solution

If one side is 3 times the other, then let the sides be x m and $3x$ m.

The perimeter $= x + 3x + x + 3x = 8x$

$$\text{So} \quad 8x = 42$$
$$x = \frac{42}{8} = 5.25$$

The sides are of length 5.25 m and 15.75 m.

6.2 Area

Area is a measure of the *amount* of a surface enclosed between its boundaries.

The surface does not have to be flat, as we can talk about the surface area of a solid, for example, the surface area of a sphere. Area is measured in cm^2, m^2, km^2, etc. A practical unit for large areas such as building or agricultural land is the hectare (ha), which is 10 000 m^2.

When converting area units, it is important to bear the following point in mind. Although 1 m is the same as 100 cm, it does not follow that 1 m^2 is the same as 100 cm^2. A simple square measuring 1 m by 1 m has an area of $1 \times 1 = 1$ m^2. However, its area is also 100 cm by 100 cm which gives 100 cm $\times$ 100 cm = 10 000 cm^2, or 10^4 cm^2.

$$\text{Also} \quad 1\ cm^2 = 100 mm^2$$
$$1 km^2 = 10^6\ m^2$$

(i) Rectangle

The area of a rectangle having sides of length a and b is given by the formula:

$$A = ab$$

This is easily seen when the sides are whole numbers. In Figure 6.4(i) a rectangle, measuring 3 cm by 6 cm, contains 18 one centimetre squares. So its area is 18 cm^2. If the sides are not whole numbers, it is not quite so obvious.

Referring to Figure 6.4 (ii), the rectangle measures 4.5 cm by 2.5 cm.

Its area is 4.5 cm × 2.5 cm = 11.25 cm^2.

You should convince yourself that the rectangle does contain 11.25 squares.

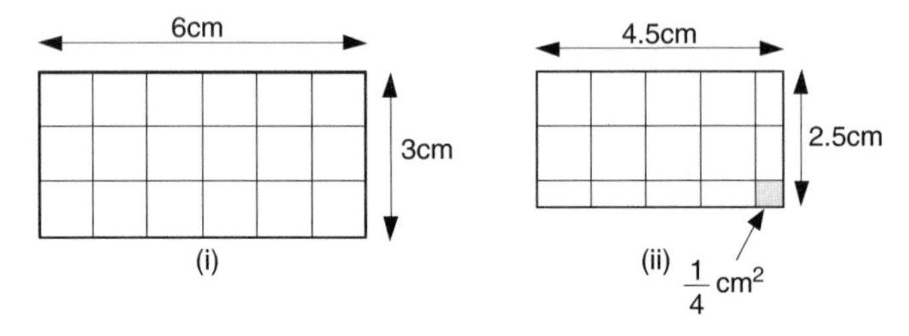

Figure 6.4

(ii) Parallelogram

The area of a parallelogram is equal to the product of the length of one of the sides and the perpendicular distance between the sides of that length.

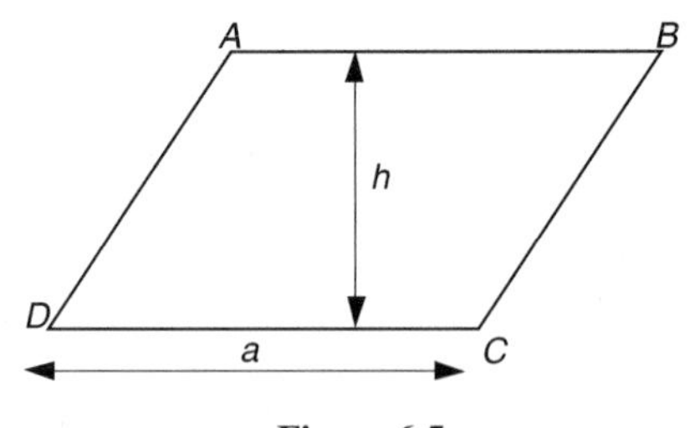

Figure 6.5

Referring to Figure 6.5

$$A = ah$$

You may prefer

$$A = \text{base} \times \text{height}$$

The justification of this is easily shown by removing a triangle from one end of the parallelogram (see Figure 6.6) and putting it at the other end to form a rectangle.

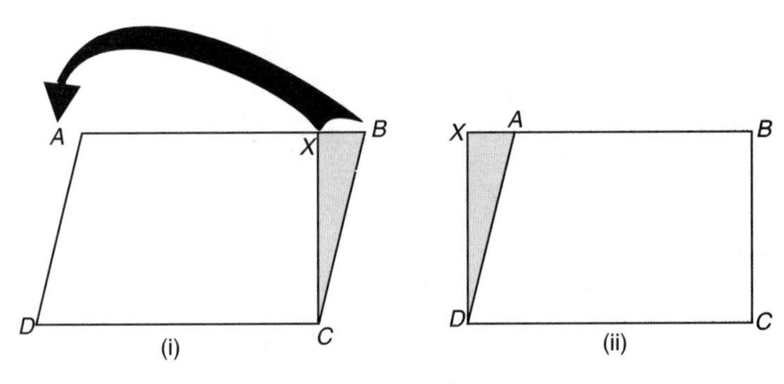

Figure 6.6

(iii) Triangle

If a parallelogram is cut in half by a diagonal, it produces two identical (congruent) triangles. Hence the area of a triangle is one half the area of the parallelogram, which is one half of the base × the height.

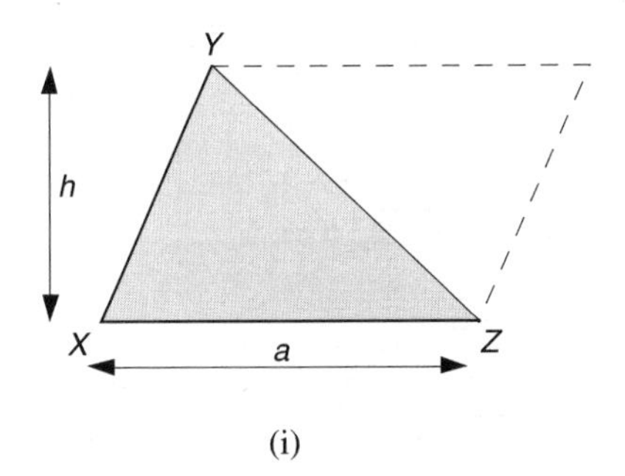

Referring to Figure 6.7(i)

$$A = \tfrac{1}{2} a \times h$$

or $A = \tfrac{1}{2}\text{base} \times \text{height}$

If the triangle is obtuse-angled, the height is found as in Figure 6.7(ii)

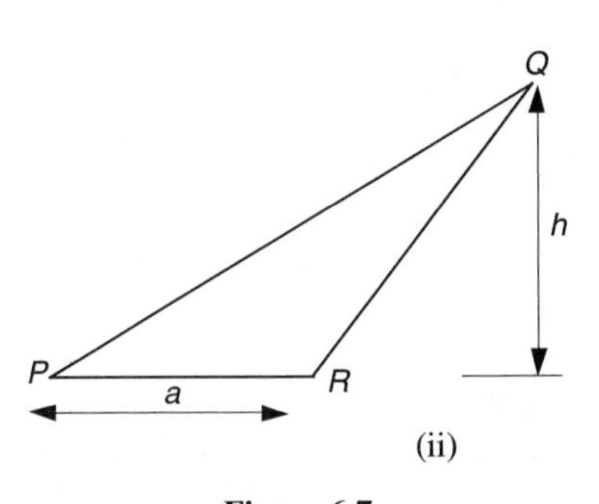

Figure 6.7

A very useful formula, attributed to Hero, gives the area (A) of a triangle, if the three sides a, b and c are known. First find the semi-perimeter

$$s = \frac{1}{2}(a + b + c),$$

then the area

$$A = \sqrt{s(s-a)(s-b)(s-c)}.$$

(iv) Trapezium

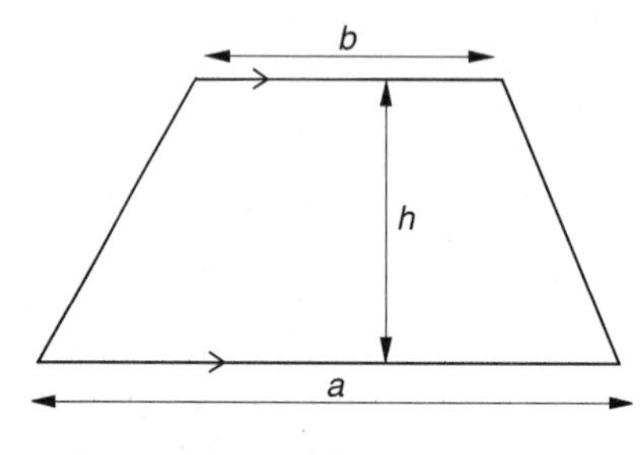

Figure 6.8

A trapezium is a quadrilateral with two sides parallel. The area A is found from the formula

$$A = \frac{1}{2}h(a + b)$$

where a and b are the lengths of the parallel sides, and h is the distance between them. (Try and prove this formula.)

Example 6.3

Figure 6.9 shows a triangular field PQR, with $PR = 200$ m, $PQ = 80$ m, and $QR = 180$ m. Find the area of the field in hectares.

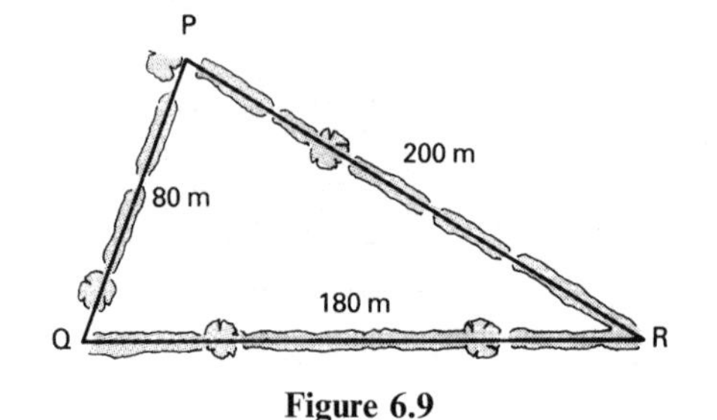

Figure 6.9

Solution

Since all three sides are known we can use the formula:

$$\text{area} = \sqrt{s(s-a)(s-b)(s-c)}$$

(where a, b, c are the sides of the triangle).

$$\text{Now } s = \frac{1}{2}(a + b + c) = \frac{1}{2}(80 + 200 + 180)$$
$$= 230$$

$$\text{The area} = \sqrt{230(230-200)(230-180)(230-80)}\ \text{m}^2$$
$$= \sqrt{230 \times 30 \times 50 \times 150}\ \text{m}^2$$
$$= 7193.7\ \text{m}^2$$

Now 1 hectare $= 10\,000\ \text{m}^2$

Hence the area $= 0.7$ ha (approximately).

(v) Irregular areas

A very good approximation to the area of an irregular shape can be found by covering the region with a square grid and counting squares.

Figure 6.10 shows a region covered by a grid of squares measuring $\frac{1}{2}$ cm by $\frac{1}{2}$ cm. The area of one square is $\frac{1}{2} \times \frac{1}{2} = 0.25\ \text{cm}^2$.

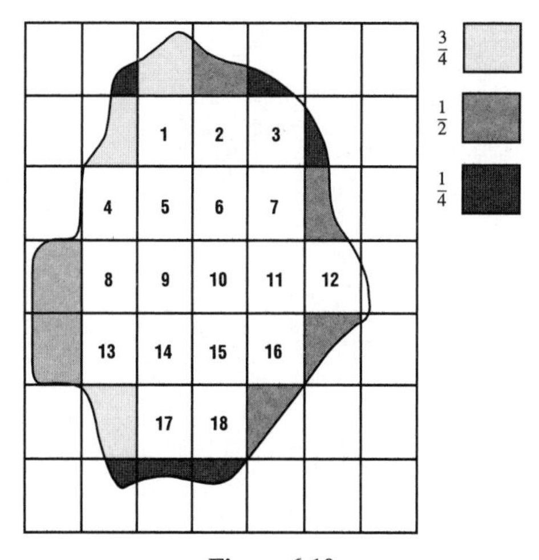

Figure 6.10

A count of the squares gives

$$\begin{aligned}
18 \times \text{full squares} &= 18 \\
5 \times \tfrac{3}{4}\ \text{squares} &= 3\tfrac{3}{4} \\
4 \times \tfrac{1}{2}\ \text{squares} &= 2 \\
6 \times \tfrac{1}{4}\ \text{squares} &= 1\tfrac{1}{2} \\
\hline
& \quad 25\tfrac{1}{4}\ \text{squares}
\end{aligned}$$

$$\begin{aligned}
\text{The area of the shape} &= 25\tfrac{1}{4} \times 0.25\ \text{cm}^2 \\
&= 6.3125\ \text{cm}^2
\end{aligned}$$

Example 6.4

A rectangle measures 8.4 cm by 3.6 cm. If the longest side is increased by 10%, and the shortest side is decreased by 10%, find the percentage change in the area of the rectangle.

Solution

Initial area $= 8.4 \times 3.6 = 30.24\ \text{cm}^2$

$8.4\ \text{cm} + 10\%$ of $8.4\ \text{cm} = 8.4 \times \dfrac{110}{100} = 9.24\ \text{cm}$

$3.6 - 10\%$ of $3.6\ \text{cm} = 3.6\ \text{cm} \times \dfrac{90}{100} = 3.24\ \text{cm}$

Note: the area has decreased

The new area $= 9.24\ \text{cm} \times 3.24\ \text{cm} = 29.9376\ \text{cm}^2$

The decrease in area $= 30.24\ \text{cm}^2 - 29.9376\ \text{cm}^2 = 0.3024\ \text{cm}^2$

$\%$ decrease $= \dfrac{0.3024}{30.24} \times 100 = 1\%$.

(vi) Circle

The area of a circle of radius r, is given by the formula

area $\quad A = \pi r^2$

(a) **Sector**

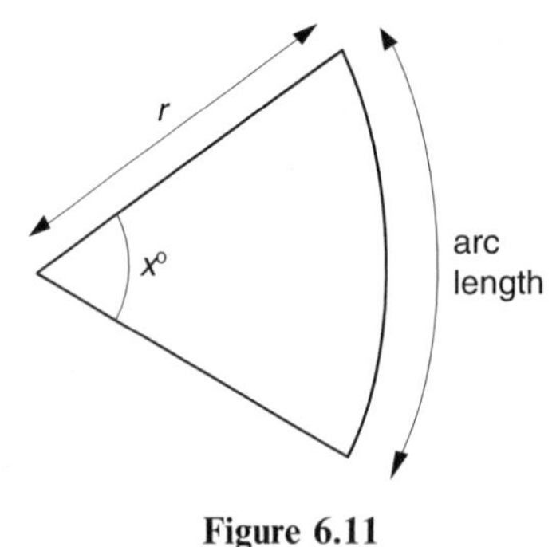

Figure 6.11

A sector is part of a circle bounded by two radii and an arc, enclosing an angle $x°$ (see Figure 6.11).

$$\text{Arc length} = 2\pi r \times \frac{x}{360}$$

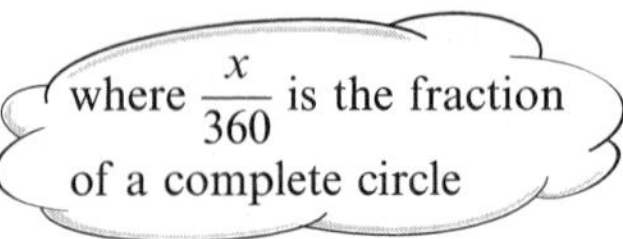

$$\text{Sector area} = \pi r^2 \times \frac{x}{360}$$

(b) Segment

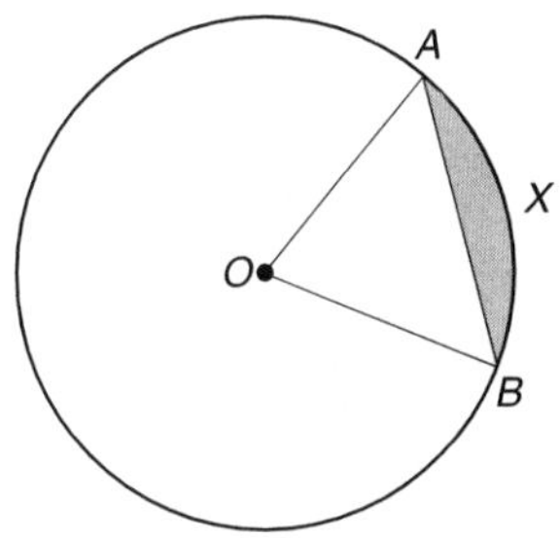

Figure 6.12

A chord AB divides a circle into two segments. The smaller segment (shown shaded in Figure 6.12) is called the minor segment, and the larger is called the major segment.

The area of the smaller segment $AXB =$ the area of the sector $OAXB -$ area of triangle AOB.

Example 6.5

The area of a semi-circle is 40 cm^2. Find its radius.

Solution

A semi-circle is exactly half of a circle, hence the full circle would have an area $= 2 \times 40$ cm$^2 = 80$ cm^2.

$$\text{So} \quad \pi r^2 = 80$$

$$r^2 = \frac{80}{\pi}$$

$$r = \sqrt{\frac{80}{\pi}} = 5.05.$$

The radius would be 5.05 cm.

Exercise 6(a)

1 The measurements of a rectangle were given as 2.2 cm by 2.6 cm, correct to the nearest mm. Give the range of possible values of

(i) the perimeter

(ii) the area of the rectangle.

2

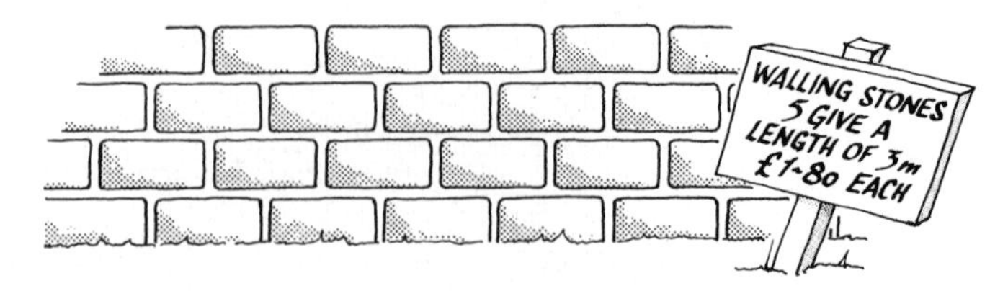

Figure 6.13

(i) What is the length of a walling stone in centimetres?
(ii) A wall is to be built five walling stones high and approximately 5 m long. How many walling stones are needed?
(iii) What is the total cost of building the wall?

3 A ship has circular portholes of diameter 90 cm. Each one is surrounded by a sealing ring.

(i) Find the area of one porthole.
(ii) Find the length of sealing strip needed for the ship, if it has 30 portholes.

4 Find the area of an equilateral triangle with sides of length 30 cm.

5 A tyre of a car has a diameter of 550 mm. If the wheel rolls a distance of 8 km without slipping, find how many complete revolutions the wheel makes to travel this distance.

6 The total land area of Belgium is 3 million hectares. Of this total, 1.4 million hectares are cultivated, and 600 000 hectares are forest. Find the area of the remaining land in km^2, giving your answer in standard form.

7 Figure 6.14 shows an outline of Sussex drawn on a grid with squares measuring 10 miles by 10 miles. Estimate the area of Sussex.

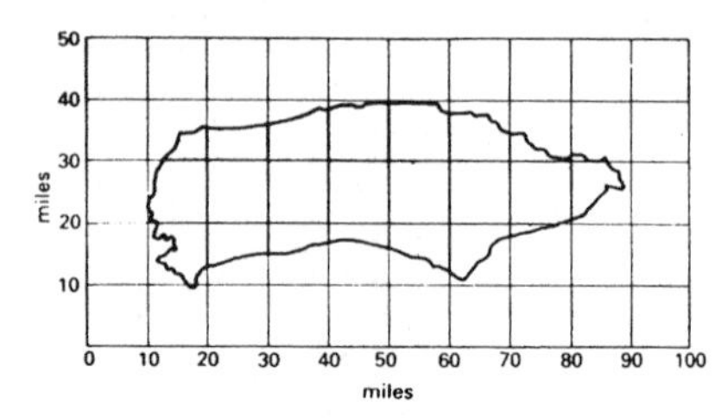

Figure 6.14

8 Calculate the circumference and area of a circle with diameter 1.2 m. If the circumference is increased by 60 cm, what is the new diameter?

9 If the diagonals of a rhombus are 15 cm and 8 cm, what is its area?

10 The parallel sides of a trapezium are of length d cm and $3d$ cm. The distance between them is 15 cm. If the area of the trapezium is 120 cm^2, find d.

11 The top of a wooden table is in the form of a semi-circle of radius 1.5 cm. Find (i) the area of the table (ii) the perimeter of the table.

12 A field is in the shape of a trapezium, with parallel sides of length 180 m and 95 m, distance 84 m apart. Find the area of the field in hectares.

6.3 Nets and surface area

The total area of the surfaces of any solid is referred to as the **surface area** of the solid. In many cases, the easiest way of finding the surface area is to look at the net which is needed to construct the three-dimensional object. (Tabs for joining the shape together are left out.) Listed below are a number of solids with the appropriate net, and how to calculate the surface area.

(i) Cuboid

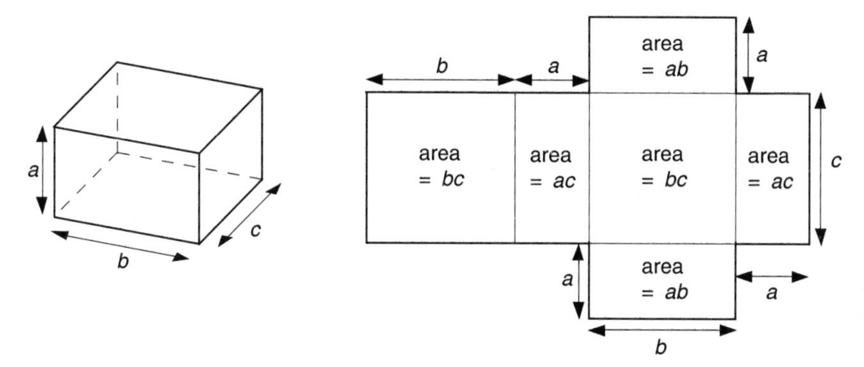

Figure 6.15

Surface area $= 2ac + 2bc + 2ab$

(ii) Cylinder

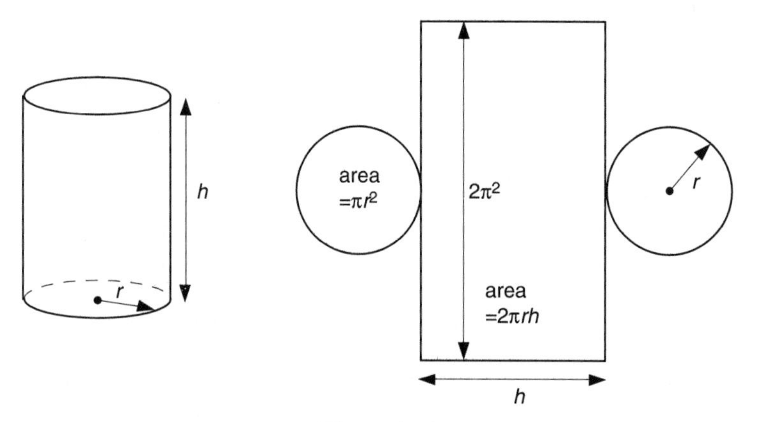

Figure 6.16

$$\text{Surface area} = \underset{\text{2 ends}}{\underset{\uparrow}{2\pi r^2}} + \underset{\text{curved surface}}{\underset{\uparrow}{2\pi rh}}$$

(iii) Square-based pyramid

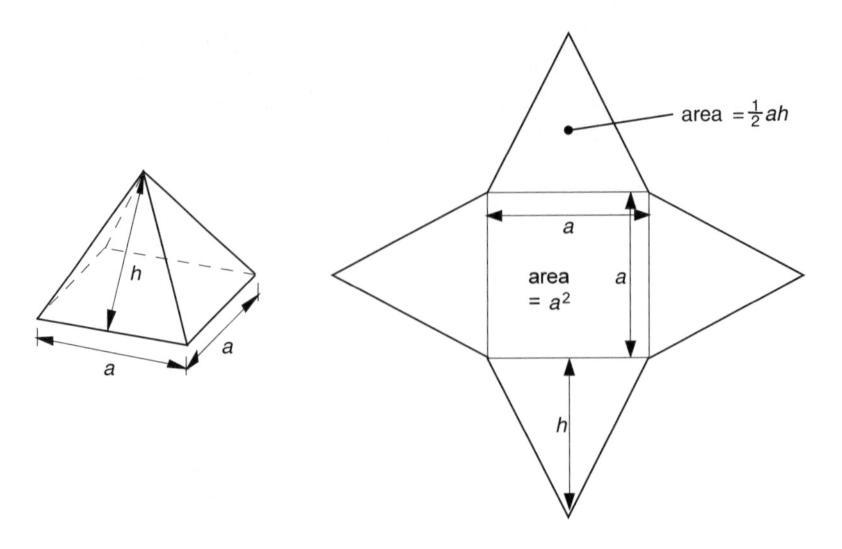

Figure 6.17

Surface area $= a^2 + 2ah$

(iv) Cone

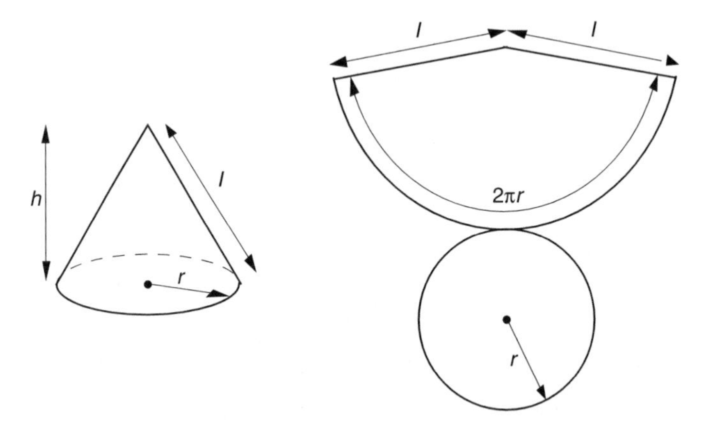

Figure 6.18

The curved surface of the cone opens out into a sector, with arc length equal to $2\pi r$ (the circumference of the base) and radius l (the slant height).

The area of the curved surface will be

$$\pi l^2 \times \frac{2\pi r}{2\pi l} = \pi l^2 \times \frac{r}{l} = \pi rl$$

Hence,

$$\text{the surface area} = \underset{\text{base}}{\underset{\uparrow}{\pi r^2}} + \underset{\text{curved surface}}{\underset{\uparrow}{\pi r l}}$$

Sometimes you know the height of the cone h, and not the slant height l. To find l, use the formula:

$$l = \sqrt{h^2 + r^2}$$

[*Note*: This is using Pythagoras' theorem, see Section 7.5.]

(v) Sphere

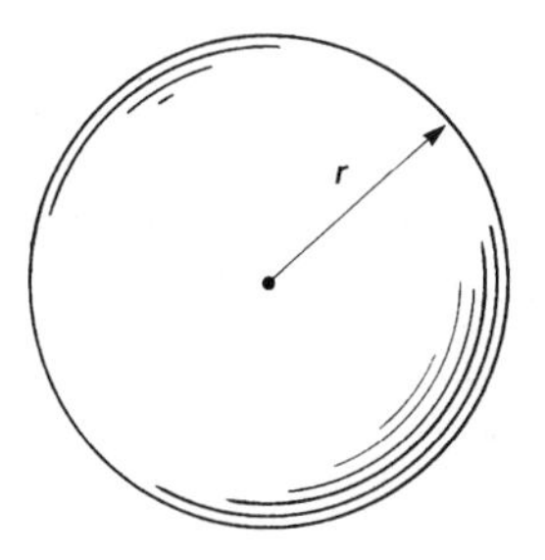

Figure 6.19

Surface area $= 4\pi r^2$

Example 6.6

Find the surface area of a cuboid which has a square base of side 4 cm, and height 2.5 cm.

Solution

Take $a = 4$, $b = 4$ and $c = 2.5$ as in Figure 6.15.

$$\begin{aligned}\text{Using surface area} &= 2ab + 2ac + 2bc \\ &= 2 \times 4 \times 4 + 2 \times 4 \times 2.5 + 2 \times 4 \times 2.5 \\ &= 32 + 20 + 20 = 72\end{aligned}$$

The surface area $= 72$ cm^2.

Example 6.7

100 identical spheres of radius 4 m are to be covered with a special anti-oxidising paint which costs £25 to cover 20 m^2. Find to the nearest £10 the cost of covering the spheres.

Solution

Surface area of 1 sphere $= 4\pi r^2 = 4\pi \times 4^2$

$= 201.1\ m^2$

Surface area of 100 spheres $= 100 \times 201.1\ m^2$

$= 20110\ m^2$

The total cost $= \dfrac{20110}{20} \times £25$

$= £25140$ (to nearest £10).

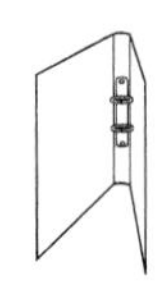

Exercise 6(b)

1 Find the surface area of a cuboid that measures 8 cm by 5 cm by 4.4 cm.

2 Find the total surface area of a cylinder of radius 4 cm, and height 3 cm.

3 Figure 6.20 shows the net for a square-based pyramid.

Find the total surface area of this pyramid.

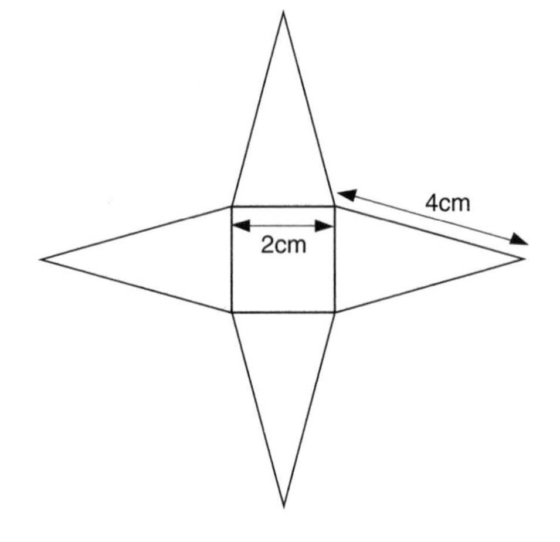

Figure 6.20

4 10 identical spheres of radius 7 cm are to be painted. What is the total area of surface to be painted?

5 A tin can, closed at both ends, has height 11 cm and base radius 4 cm. The metal is cut from a sheet which measures 12 cm by 50 cm. What percentage of metal remains, assuming no overlap is needed to solder the tin together?

6 The curved surface area of a cone is 100 cm^2. If the radius of the base is 4 cm, find the slant height of the cone.

6.4 Volume

Volume is a measure of the *amount* of space filled by an object.

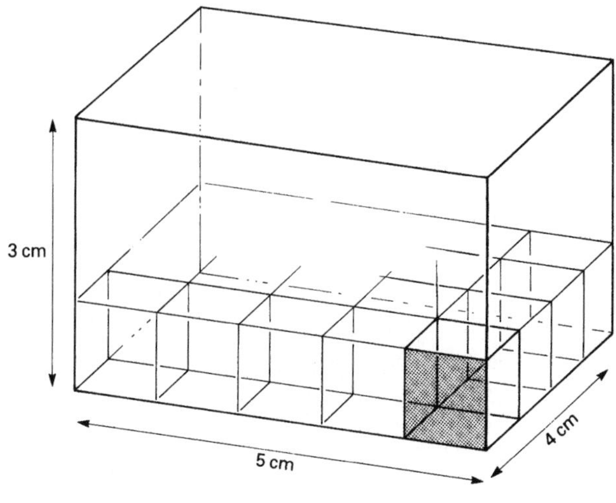

Figure 6.21

In Figure 6.21, a box measuring 3 cm by 5 cm by 4 cm is filled with one centimetre cubes.

There will be 3 layers, each with $5 \times 4 = 20$ cubes.

The total number of cubes $= 3 \times 20 = 60$.

Volume of box $= 60\ \text{cm}^3$.

(i) Cuboid

Referring to Figure 6.15, the volume $V = abc$.

(ii) Prism

A shape which has the same cross-section throughout is called a **prism**. A cylinder is in fact a **circular prism**, and a cuboid is a **rectangular prism**. Other types of prism are shown in Figure 6.22.

$$\begin{aligned} \text{The volume } V &= \text{area of cross-section} \times \text{length} \\ &= AL. \end{aligned}$$

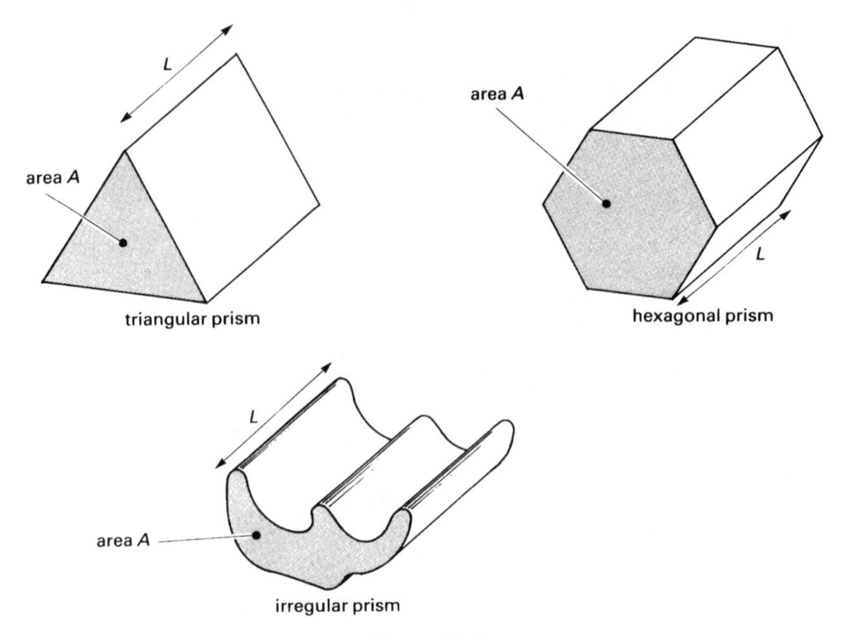

Figure 6.22

(iii) Cylinder

Since a cylinder is a prism, its volume V is given by

$$V = \pi r^2 \times h = \pi r^2 h \text{ (see Figure 6.16).}$$

(iv) Pyramid

In general, for any pyramid, it can be shown that the volume

$$V = \tfrac{1}{3}\text{base area} \times \text{height.}$$

For the cone in Figure 6.18, $V = \frac{1}{3}\pi r^2 h$.

(v) Sphere

Referring to Figure 6.19, the volume $V = \frac{4}{3}\pi r^3$.

Example 6.8

A metal block is in the shape of a cuboid measuring 4 cm by 4 cm by 3 cm. A hole of radius 1.5 cm is drilled through the block as shown in Figure 6.23. What percentage of the original metal remains?

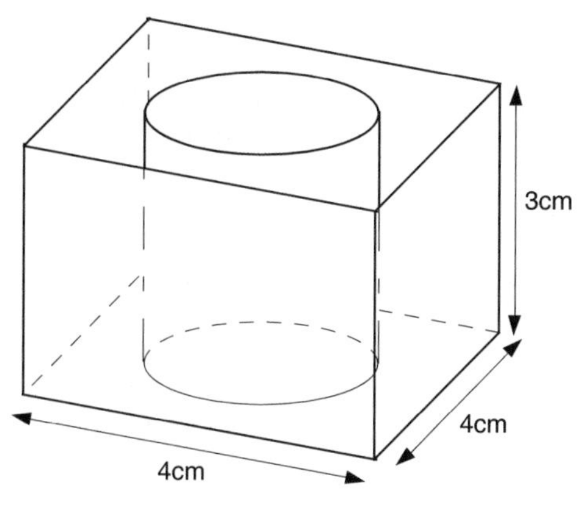

Figure 6.23

Solution

The volume of the block = 4 cm × 4 cm × 3 cm = 48 cm^3

The volume of the hole = $\pi \times 1.5^2$ cm^2 × 3 cm = 21.2 cm^3

The volume remaining = 48 cm^3 – 21.2 cm^3 = 26.8 cm^3

The percentage of the original $= \dfrac{26.8}{48} \times 100\% = 55.8\%$.

Example 6.9

A tin of paint holds 1.5 litres. The paint is spread evenly on a surface of area about 120 m^2. Find the thickness of the paint.

Solution

Although the paint layer is very thin, we can still think of the paint as a prism of area 120 m^2 and thickness d m.

The volume of the paint = $120 \times d = 120d$ m^3

Now 1 m^3 = 100 cm × 100 cm × 100 cm = 10^6 cm^3

So the volume of the paint = $120d \times 10^6$ cm^3

The volume in the tin = 1.5 litres (1 litre = 1000 cm^3)
= 1500 cm^3

So $1500 = 120 \times 10^6 \times d$

So $d = \dfrac{1500}{120 \times 10^6} = 1.25 \times 10^{-5}$

The thickness of paint = 1.25×10^{-5} cm.

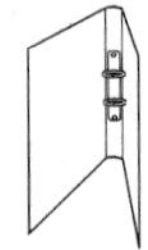

Exercise 6(c)

1 A rectangular tank is 50 cm long and 20 cm wide. Water is poured into the tank at the rate of 250 cm^3/second until the depth is 10 cm.

Calculate:

(i) the volume of water, in litres, poured into the tank;

(ii) the time taken, in seconds, for this water to be poured into the tank.

A sphere of radius 5 cm is lowered into the water and floats with exactly half of its surface above the level of the surface of the water. Taking π to be 3.14, calculate:

(iii) the volume in cm^3, of the sphere;

(iv) the rise of the water level in the tank.

2 Figure 6.24 shows the net of a solid prism. All lengths are given in centimetres.

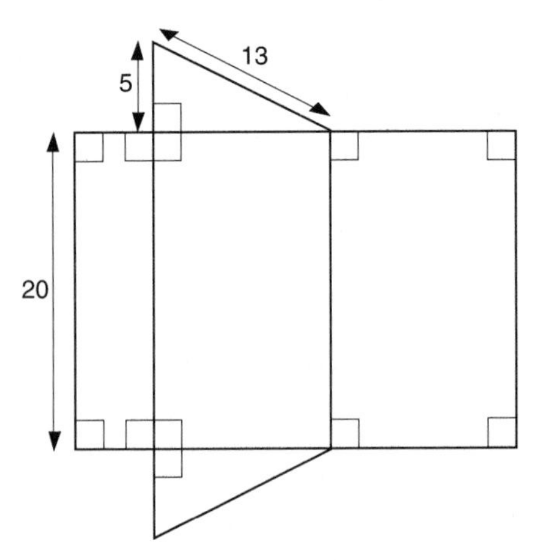

Figure 6.24

(i) Calculate the total surface area of the prism.

(ii) Calculate the total length of all the edges of the prism.

(iii) Calculate the volume of the prism, giving the units of your answer.

3 A swimming pool is of width 10 m and length 25 m. The depth of water in the pool increases uniformly from the shallow end, where the depth is 1.5 m to the deep end, where the depth is 2.5 m.

(i) Calculate the volume of water in the pool.

(ii) This water is emptied into a cylindrical tank of radius 3.5 m. Taking π as $3\frac{1}{7}$, calculate the depth of water in the tank.

4

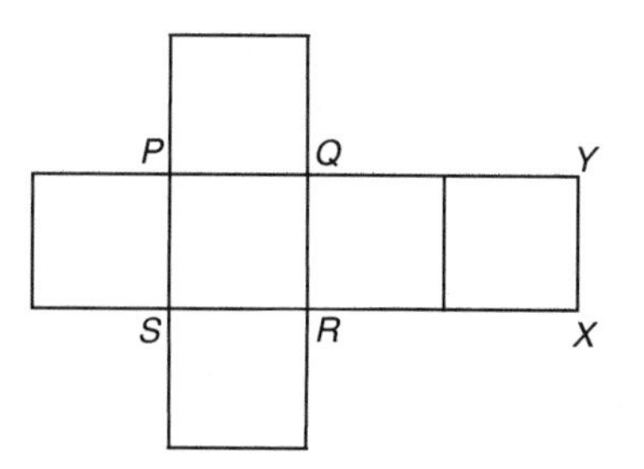

Figure 6.25

(i) Figure 6.25 represents the net of a cube of edge 6 cm. Find by accurate drawing:

  (a) the angle PXY when the net is folded to make the cube

  (b) the shortest distance on the surface of the cube from point X to point P.

(ii) Calculate the volume of the cube.

(iii) Calculate the total length of all the edges of the cube.

5

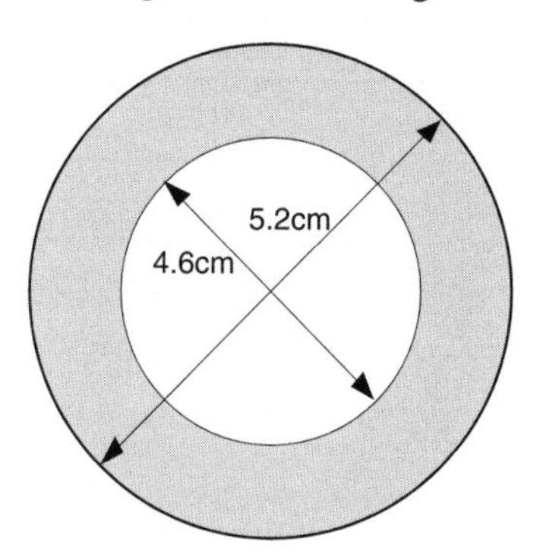

Figure 6.26

Figure 6.26 shows the uniform cross-section of a circular pipe. The external diameter is 5.2 cm and the internal diameter is 4.6 cm.

(i) Write down the radius of:

  (a) the larger circle

  (b) the smaller circle.

(ii) Calculate, in cm^2 to 2 decimal places, the area of the shaded region.

(iii) Calculate, in cm^3 to the nearest whole number, the volume of material required to construct a pipe of length one metre having this cross-section.

(iv) The mass of one cubic metre of the material used is 3000 kg. Calculate, in kg to the nearest whole number, the mass of an 80 m length of this pipe.

6.5 Similar shapes

(i) Area

Look at the two shapes in Figure 6.27. The lengths of the sides have been enlarged by a factor 3, they are in the ratio 1:3, but the area has been multiplied by 9, and hence the areas are in the ratio $1 : 9$ (i.e. $1^2 : 3^2$).

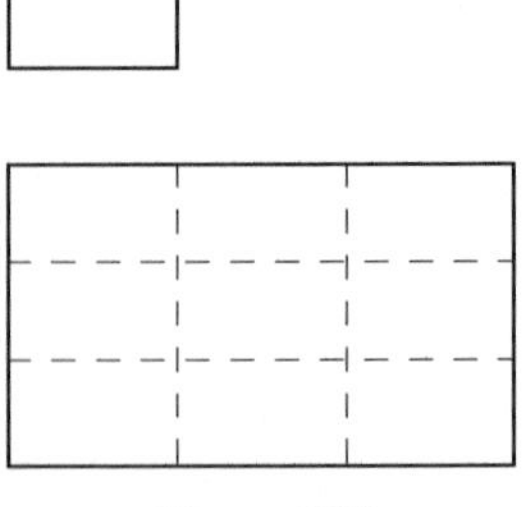

Figure 6.27

> **Rule**
>
> The ratio of the lengths is *squared* to find the ratio of the area
> (You square each number in the ratio)

Example 6.10

In Figure 6.28, $\triangle BCA$ is similar to $\triangle PQR$. $PR = 2$ cm, and $BA = 5$ cm. If the area of BCA is 20 cm^2, what is the area of PQR?

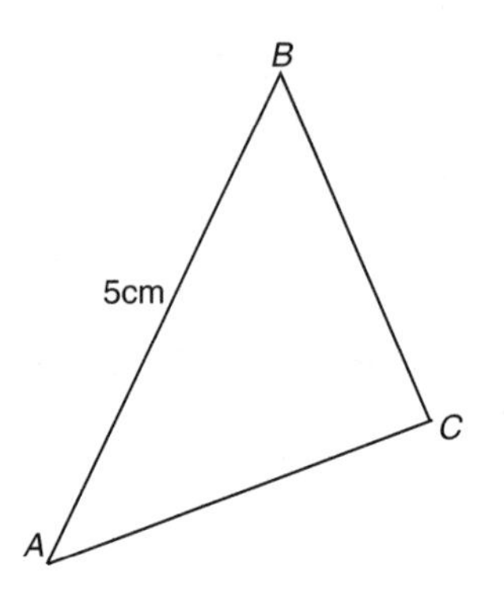

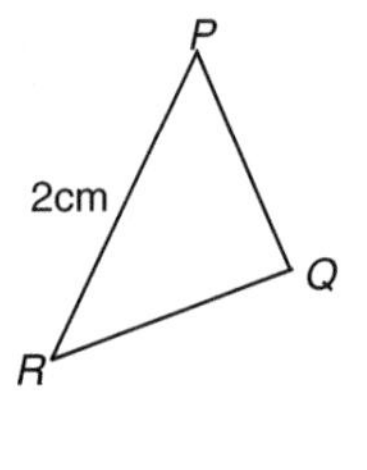

Figure 6.28

Solution

The ratio of the lengths is $5 : 2$

Hence the ratio of the areas is $25 : 4$

Dividing each number by 25 gives $1 : \frac{4}{25}$

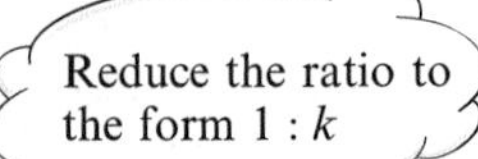

Multiply each number by 20, to give $20 : 20 \times \frac{4}{25} = 20 : 3.2$

Hence, the area of $\triangle PQR$ is 3.2 cm^2

Example 6.11

A map is drawn to a scale of 1 : 50 000.

(a) A lake has an area of 4 cm^2 on the map, what is the real area of the lake?

(b) A town covers an area of 8 km^2, what area would it cover on the map?

Solution

A scale of 1 : 50 000 means that 1 cm represents 50 000 cm = 0.5 km or 2 cm represents 1 km.

(a) So 4 cm^2 = 4 cm $\times$ 1 cm represents 2 km $\times$ 0.5 km = 1 km^2

(b) 8 km^2 = 8 km $\times$ 1 km is represented by 16 cm $\times$ 2 cm = 32 cm^2.

(ii) Volume

In Figure 6.29 can be seen two cubes whose lengths are in the ratio 1 : 3. It can be fairly easily seen that 27 of the smaller cubes can be fitted into the larger cube, hence the ratio of the volumes is $1 : 27 = 1^3 : 3^3$.

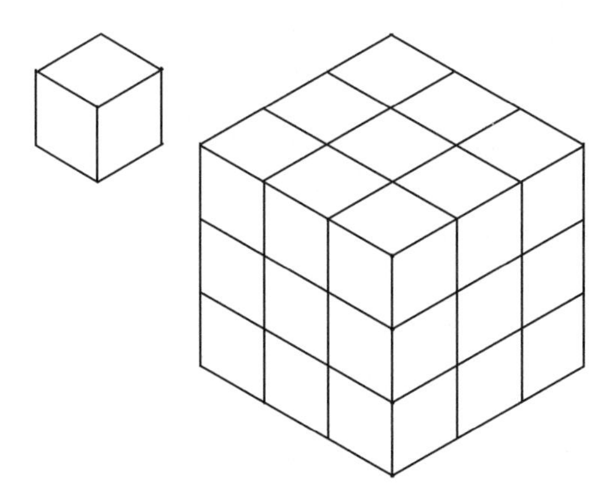

Figure 6.29

Rule

The ratio of the lengths is cubed to find the ratio of the volumes

(You cube each number in the ratio)

RULE

Example 6.12

Two similar solids have surface areas in the ratio 9:25. If the volume of the smaller solid is 12 cm^3, what is the volume of the larger solid?

Solution

The area ratio is $9 : 25 = 3^2 : 5^2$

Hence the length ratio is $3 : 5$

The volume ratio is $3^3 : 5^3 = 27 : 125$

$$\div 27 \text{ to give} \quad 1 : \frac{125}{27}$$

$$\times 12 \text{ to give} \quad 12 : \frac{500}{9} = 12 : 55$$

Hence the volume of the larger solid = 55 cm^3.

Exercise 6(d)

1 Calculate the ratios:

(i) of the surface areas

(ii) of the volumes

of two cubes with sides 6 cm and 10 cm.

2 A model car is built to a ratio of 1 : 144 and the fuel tank has a capacity of 12 mm^3. Calculate in litres the capacity of the fuel tank on the full size car.

3 A cylinder with a diameter of 6 cm has a height of 9 cm. Calculate:

(i) the radius and curved surface of a similar cylinder with a height of 12 cm

(ii) the volume of a similar cylinder which has a diameter of 9 cm.

4 Two similar cones have radii 60 mm and 15 mm. If the height of the smaller is 140 mm calculate:

(i) the height of the other cone

(ii) the ratio of their volumes.

5 A map is drawn to a scale of 1 : 20 000. Find:

(i) the area of a lake that measures 8 cm^2 on the map

(ii) the area on the map of a nature reserve that has an area of 16 km^2.

6 In Figure 6.30 the triangles ACB and AED are similar. Find the ratio of the area of triangle ACB to the area of the trapezium $BCED$.

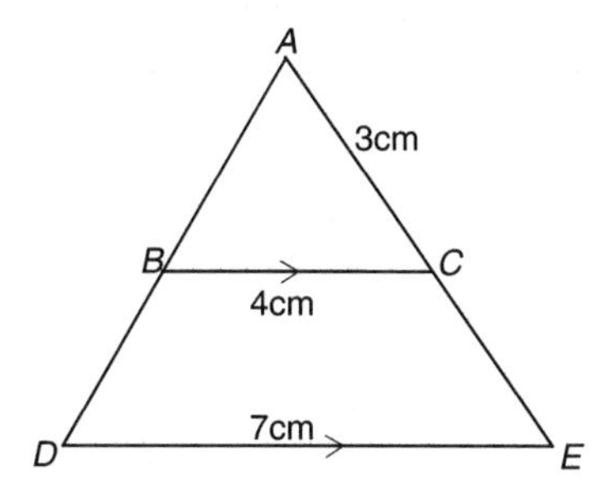

Figure 6.30

7 Two spheres have volumes in the ratio 8 : 27. What is the ratio of their surface areas?

8 Two similar cylinders have volumes in the ratio $a^3 : b^3$. The total surface area of the smaller cylinder is 20 cm^2. What is the total surface area of the larger? [Assume $a < b$.]

9 In triangle ABC, X is the midpoint of AB, and Y the midpoint of AC. Z is the point on XY such that $XY = 2ZY$. Find the ratio of the area of triangle AYZ to that of the trapezium $XYCB$.

10 A scale model of a hut is a cuboid of volume 12 cm^3. The height of the model is 2 cm, and the height of the real building is 25 m. What is the area of the floor of the building?

7 Trigonometry

Trigonometry is concerned with calculating angles and lengths in problems which can be reduced to looking at triangles. The study of these methods can be conveniently broken down into looking at right angled triangles and non-right angled triangles.

7.1 Tangent ratio

In Figure 7.1, ABC is a right angled triangle. The sides of the triangle are *named* in relation to angle A. The longest side AB is called the **hypotenuse**, the side next to the angle, i.e. AC, is called the **adjacent** side, and the third side BC is **opposite** angle A.

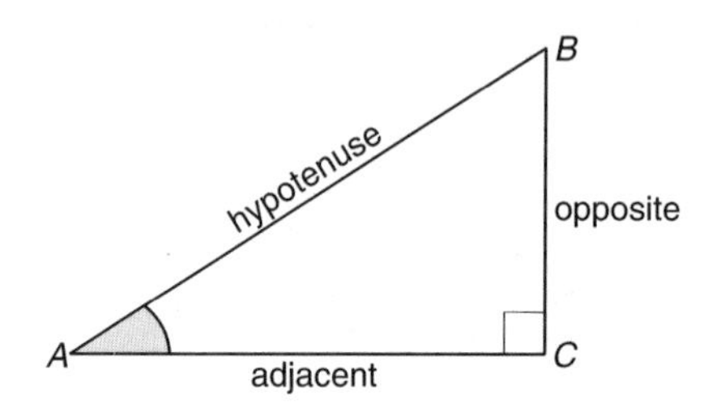

Figure 7.1

If you find the ratio $\frac{BC}{AC}$, you will find it always has the same value however large you draw the triangle, provided angle A remains the same (see Investigation XV). This ratio is called the **tangent** of angle A.

$$\text{i.e. tangent of } A = \frac{BC}{AC}$$

$$= \frac{\text{(opposite)}}{\text{(adjacent)}}$$

This is abbreviated to:

Rule

$$\tan A = \frac{\text{opposite}}{\text{adjacent}}$$

If the tangent is calculated as a decimal, the calculator can then be used to find the value of the angle A.

Example 7.1

Find the angles marked in the diagrams in Figure 7.2.

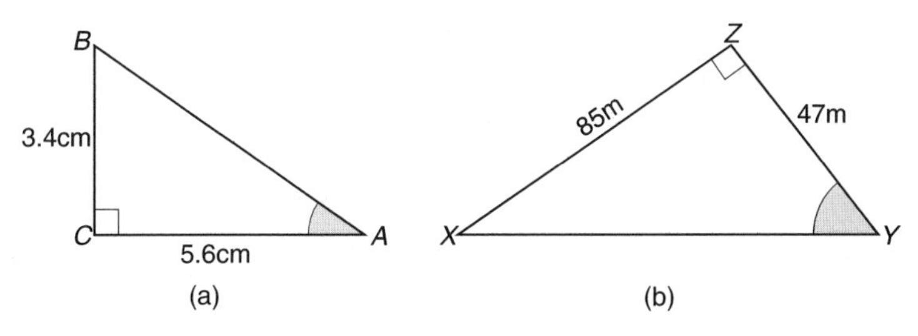

Figure 7.2

Solution

(a) Since AB is the hypotenuse, $AC = 5.6$ cm is the adjacent side, and $BC = 3.4$ cm is the opposite side

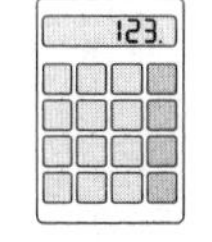

$$\text{Therefore} \quad \tan A = \frac{\text{opposite}}{\text{adjacent}} = \frac{3.4}{5.6} = 3.4 \div 5.6$$
$$= 0.6071.$$

To find the angle, using the calculator, the [INV] [tan] buttons should be pressed, so

			Display
[0.6071]	[INV]	[tan]	[31.2619375]

To 1 decimal place, angle $A = 31.3°$

[INV] [tan] is often referred to as the **inverse tangent**.

Common alternatives to [INV] [tan] are [2nd f] [tan] or just [$\tan^{-1}$]

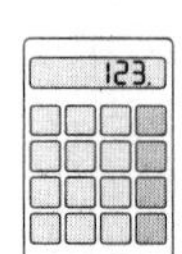

(b) You need to be careful here, as the triangle has been rotated. XY is the hypotenuse, $ZY = 47$ m is the adjacent side and $XZ = 85$ m is the opposite side

$$\text{So} \quad \tan Y = \frac{\text{opposite}}{\text{adjacent}} = \frac{85}{47} = 85 \div 47$$
$$= 1.8085 \text{ (4 d.p.)}$$

Display

1.8085	$\tan^{-1}$	61.05984458

To 1 decimal place, $Y = 61.1°$.

It is possible to carry out this type of calculation in one complete sequence, the answer is in fact slightly more accurate

Display

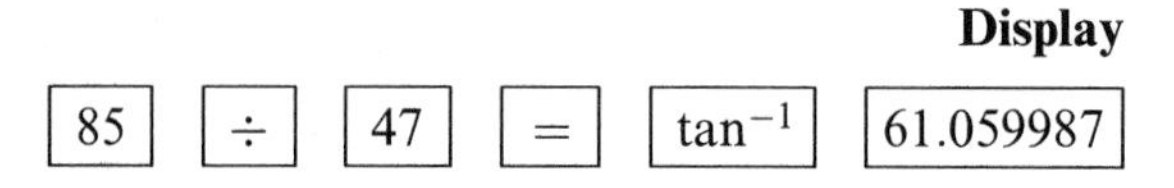

If the angle is known, then the sides can be found as in the following example.

Example 7.2

Find the value of x in the diagrams in Figure 7.3 (correct to 3 sig. fig.):

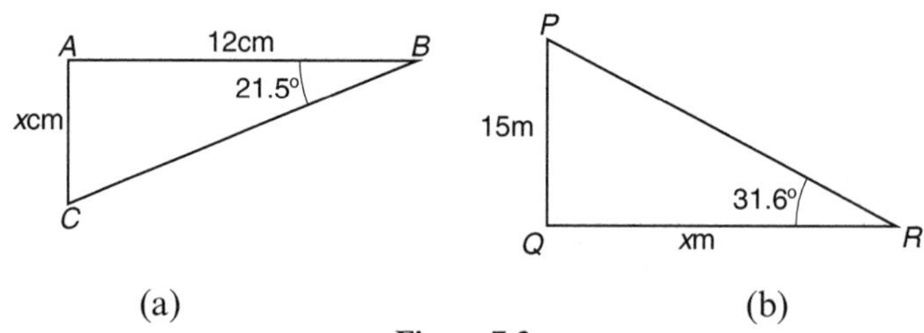

Figure 7.3

Solution

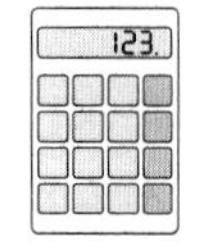

(a) Here BC is the hypotenuse, $AB = 12$ cm is the adjacent side and $AC = x$ cm is the opposite side

$$\text{So} \quad \tan 21.5° = \frac{\text{opposite}}{\text{adjacent}} = \frac{x}{12}$$

To find $\tan 21.5°$

Display

21.5	tan	0.393910475

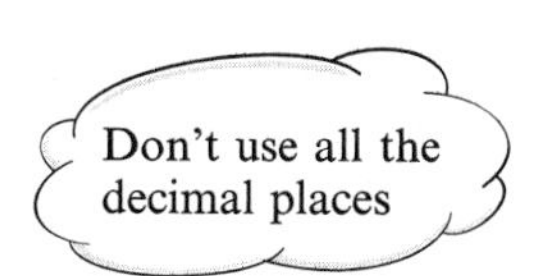

So $12 \times 0.3939 = x$

$x = 4.73$ cm (3 sig. fig.).

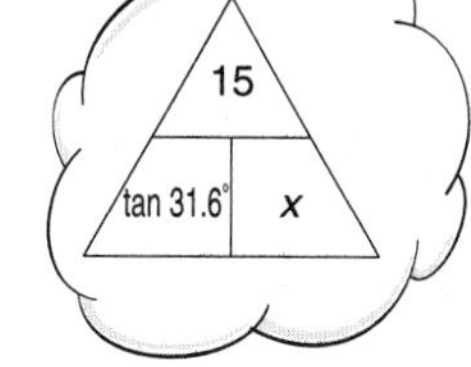

Or: in one sequence of operations:

Display

21.5	tan	×	12	=	4.7269257

(b) PR is the hypotenuse, $QR = x$ m is the adjacent side, and $PQ = 15$ m is the opposite side

$$\tan 31.6° = \frac{\text{opposite}}{\text{adjacent}} = \frac{15}{x}$$

Display

31.6	tan	0.615204104

and so $0.61520 = \dfrac{15}{x}$ (using 5 d.p.).

To solve x from this type of equation often causes difficulty, see Example 3.8(f).

$$x = \frac{15}{0.61520} = 15 \div 0.6152$$
$$= 24.4 \quad \text{(3 sig. fig.)}.$$

Exercise 7(a)

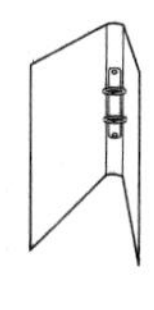

1 Find the marked angles in the diagrams in Figure 7.4:

(i)

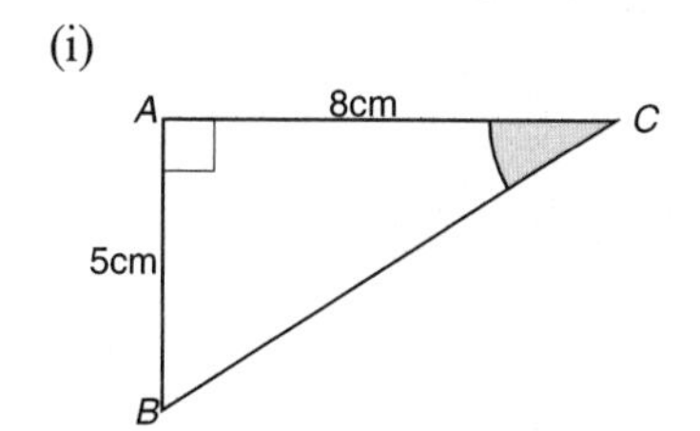

(ii)

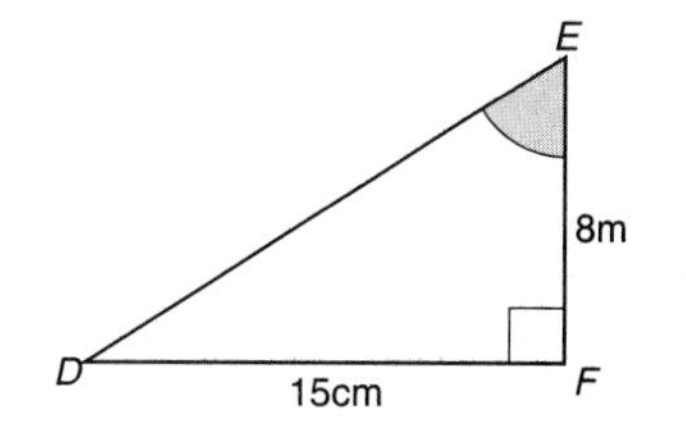

Figure 7.4

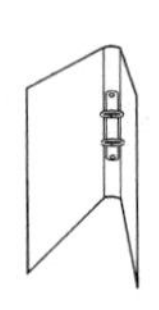

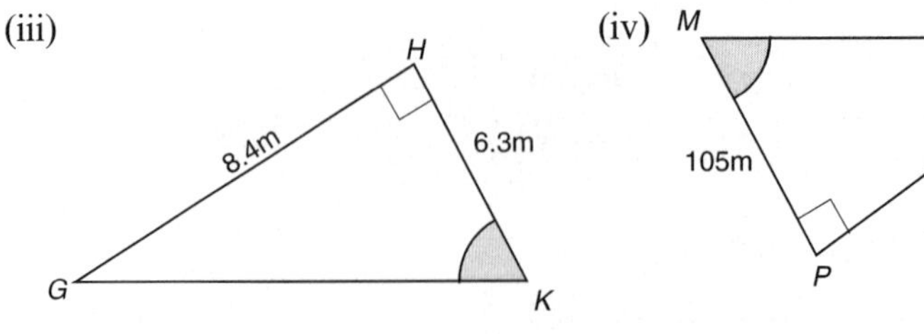

Figure 7.4

2 Find x in the diagrams in Figure 7.5:

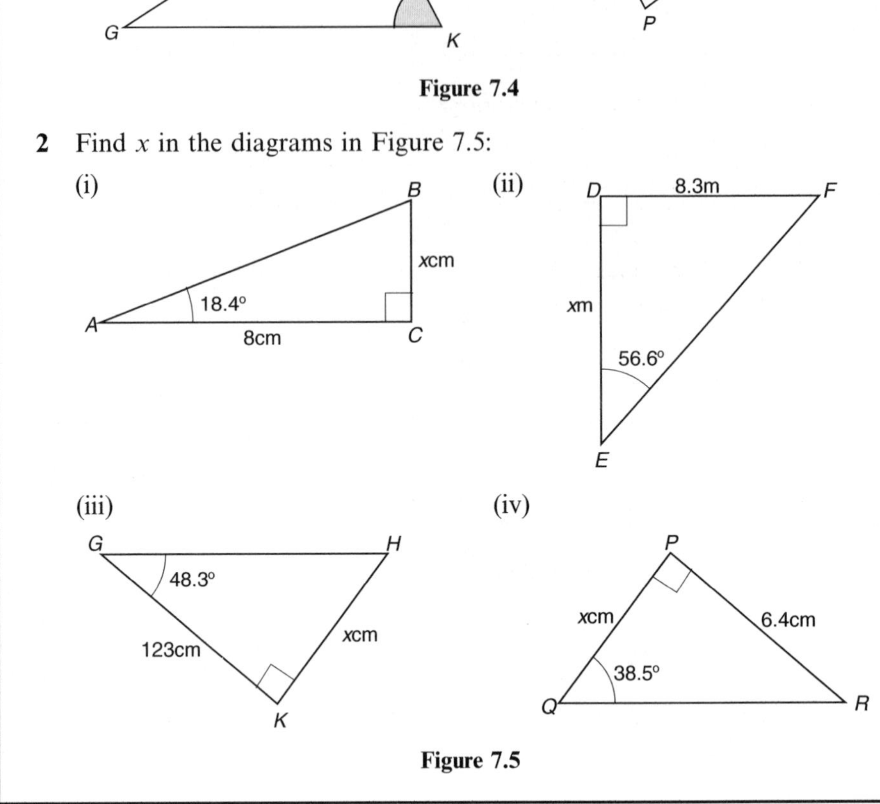

Figure 7.5

7.2 Sine ratio

Referring back to Figure 7.1, if you find the ratio $\frac{BC}{AB}$, you will find it also has a constant value provided the angle A remains unaltered (see Investigation XV). The ratio is called the **sine** of angle A, abbreviated $\sin A$.

We have

Rule

$$\sin A = \frac{BC}{AB} = \frac{\text{opposite}}{\text{hypotenuse}}$$

Example 7.3

Find the angles marked in the diagrams in Figure 7.6, correct to 1 decimal place:

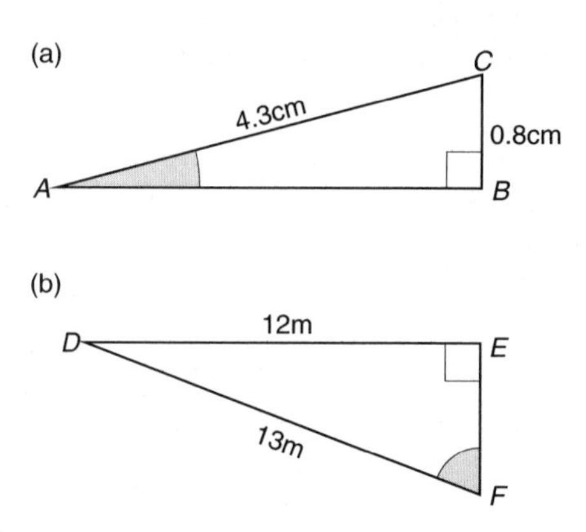

Figure 7.6

Solution

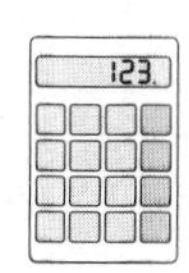

(a) $AC = 4.3$ cm is the hypotenuse, AB is the adjacent side, and $BC = 0.8$ cm is the opposite side

$$\sin A = \frac{\text{opposite}}{\text{hypotenuse}} = \frac{0.8}{4.3} = 0.8 \div 4.3$$
$$= 0.1860 \quad (4 \text{ d.p.}).$$

The [INV] [sin] button must be used (or [$\sin^{-1}$])

			Display
0.1860	INV	sin	10.71944015

So: $A = 10.7°$ (1 d.p.).

Or in one sequence of operations

					Display
0.8	÷	4.3	=	$\sin^{-1}$	10.72215241

(b) $DF = 13$ m is the hypotenuse, EF is the adjacent side, and $DE = 12$ m is the opposite side

$$\sin F = \frac{\text{opposite}}{\text{hypotenuse}} = \frac{12}{13} = 12 \div 13$$
$$= 0.9231 \quad (4 \text{ d.p.})$$

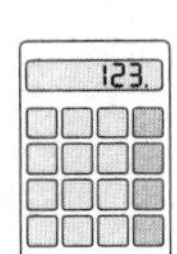

Display

0.9231	$\sin^{-1}$	67.38357305

Therefore $A = 67.4°$ (1 d.p.).

As with the tangent ratio, the sine can also be used to calculate sides if an angle is known. Look at the following example.

Example 7.4

Find the value of y in the diagrams in Figure 7.7.

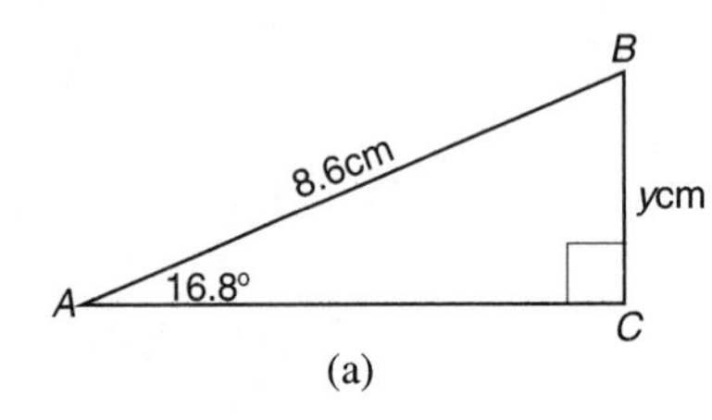

(a)

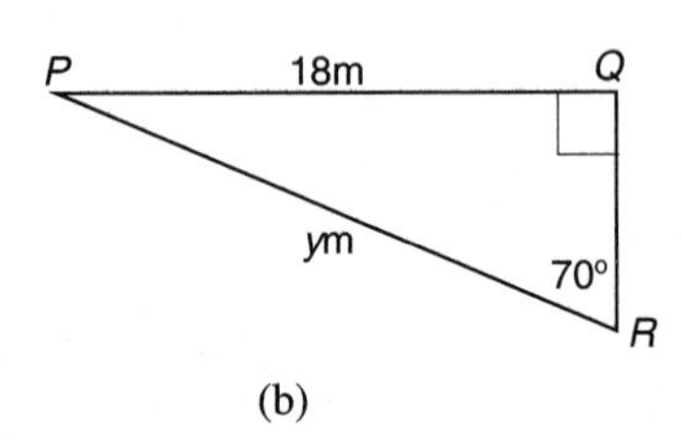

(b)

Figure 7.7

Solution

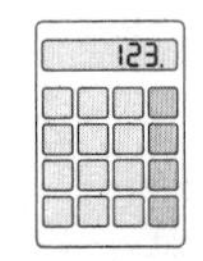

(a) $AB = 8.6$ cm is the hypotenuse, AC is the adjacent side, and $BC = y$ cm is the opposite side.

$$\sin 16.8 = \frac{\text{opposite}}{\text{adjacent}} = \frac{y}{8.6}$$

Display

16.8	sin	0.289031797

and so $0.28903 = \dfrac{y}{8.6}$ (using 5 d.p.)

So $y = 8.6 \times 0.28903 = 2.49$ (3 sig. fig.).

As one sequence of operations

Display

16.8	sin	×	8.6	=	2.485673454

(b) $PQ = 18$ m is the opposite side, QR is the adjacent side, and $PR = y$ m is the hypotenuse

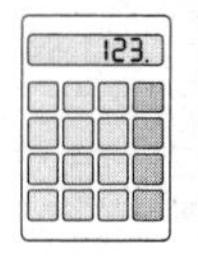

$$\text{Hence} \quad \sin 70^\circ = \frac{\text{opposite}}{\text{adjacent}} = \frac{18}{y}$$

Display

70	sin	0.93969262

$$\text{i.e.} \quad 0.9400 = \frac{18}{y}$$

$$\text{and} \quad y = \frac{18}{0.94} = 19.1 \text{ (3 sig. fig.).}$$

Exercise 7(b)

1 Find the angles marked in the diagrams in Figure 7.8:

(i)

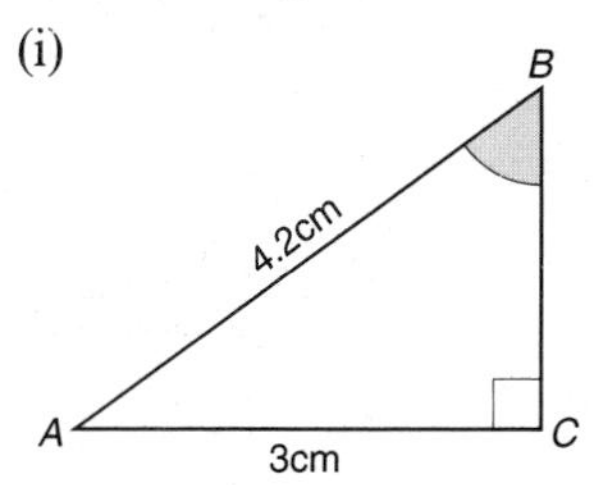

(ii)

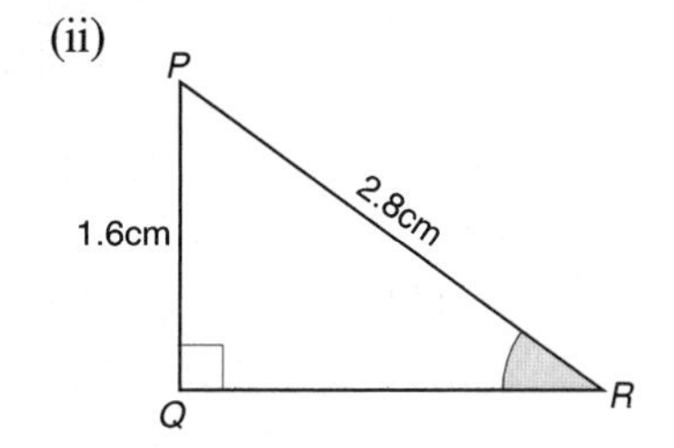

(iii)

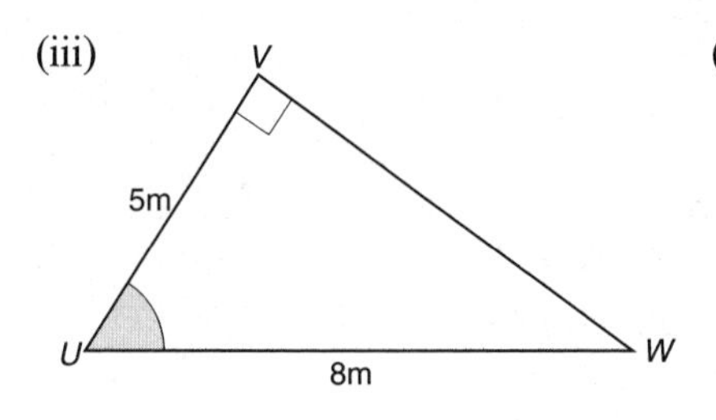

(iv)

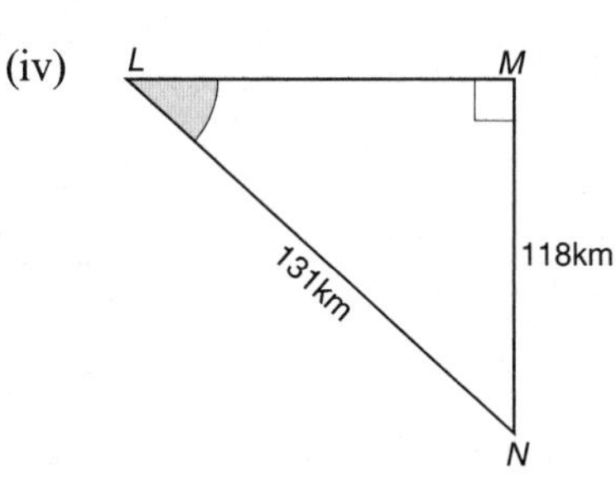

Figure 7.8

2 Find the sides marked x in the diagrams in Figure 7.9:

(i)

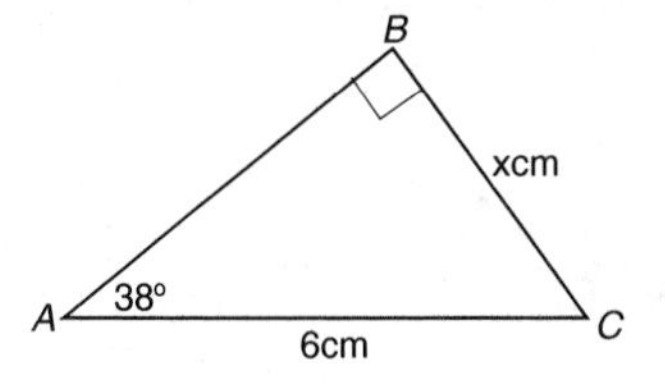

(ii)

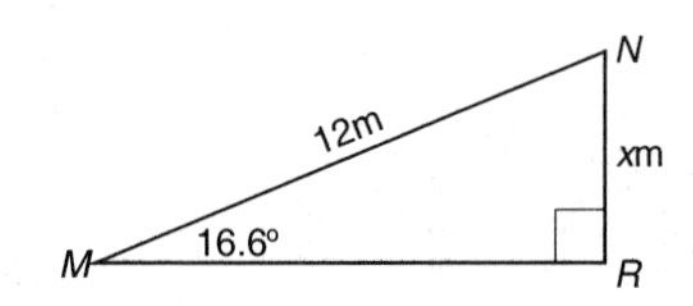

Figure 7.9

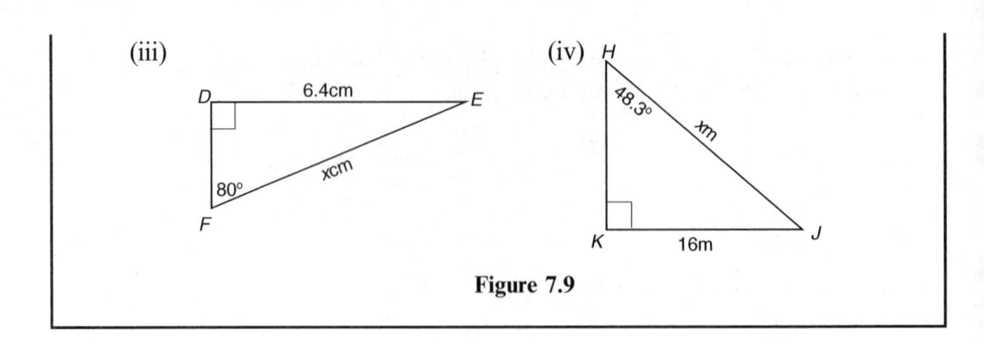

Figure 7.9

7.3 Cosine ratio

Again, referring back to Figure 7.1, if you find the ratio $\frac{AC}{AB}$, you will find it has a constant value provided that the angle A remains constant. This ratio is called the **cosine** of angle A, abbreviated $\cos A$.

Rule

$$\cos A = \frac{AC}{AB} = \frac{\text{adjacent}}{\text{hypotenuse}}$$

Example 7.5

Find the angles marked in the diagrams in Figure 7.10:

(a)

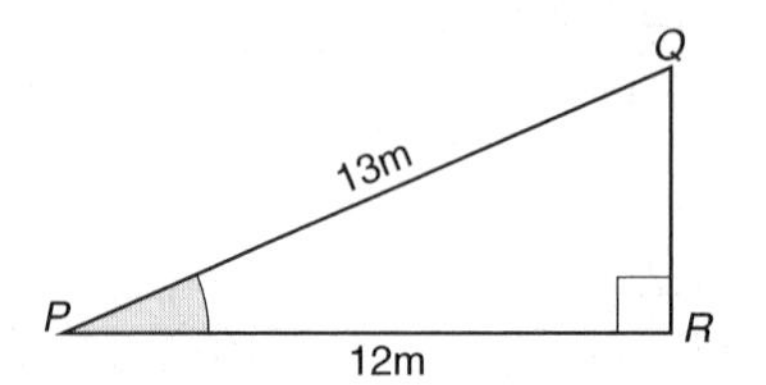

(b)

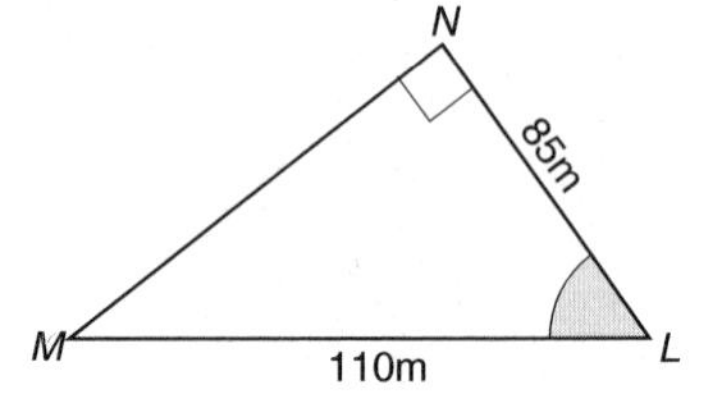

Figure 7.10

Solution

(a) $PQ = 13$ m is the hypotenuse, and $PR = 12$ m is the adjacent side

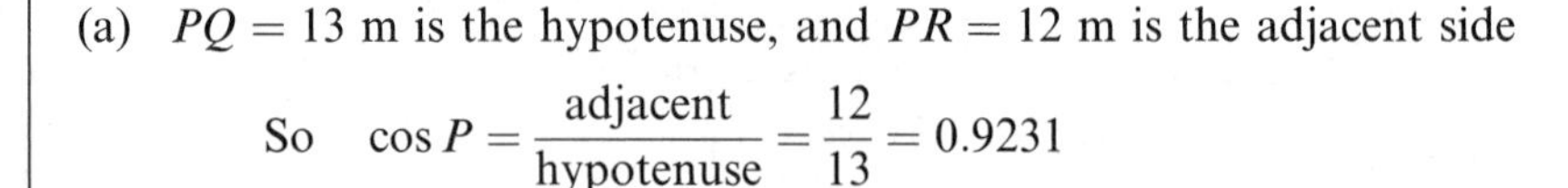

So $\quad \cos P = \frac{\text{adjacent}}{\text{hypotenuse}} = \frac{12}{13} = 0.9231$

Therefore

Display

0.9231	$\cos^{-1}$	22.616327

Hence $\quad P = 22.6°$ (3 sig. fig.).

(b) Here $ML = 110$ m is the hypotenuse, and $LN = 85$ m is the adjacent side

$$\text{So} \quad \cos L = \frac{\text{adjacent}}{\text{hypotenuse}} = \frac{85}{110} = 0.7727$$

Display

0.7727	$\cos^{-1}$	39.403031

So $L = 39.4°$ (3 sig. fig.).

As with tangent and sine, cosine can also be used to calculate sides of triangles.

Example 7.6

(a) (b)

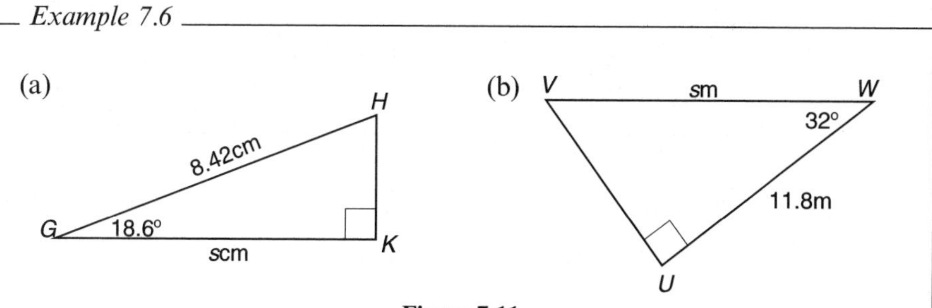

Figure 7.11

Solution

(a) $GH = 8.42$ cm is the hypotenuse, and $GK = s$ cm is the adjacent side

$$\cos 18.6 = \frac{\text{adjacent}}{\text{hypotenuse}} = \frac{s}{8.42}$$

Display

18.6	cos	0.947768

$$\text{Hence} \quad \frac{s}{8.42} = 0.9478$$

$s = 0.9478 \times 8.42 = 7.98$ (3 sig. fig.).

As one sequence

Display

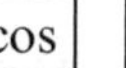

18.6	cos	×	8.42	=	7.9802100

(b) $VW = s$ m is the hypotenuse and $UW = 11.8$ m is the adjacent side

$$\cos 32^\circ = \frac{\text{adjacent}}{\text{hypotenuse}} = \frac{11.8}{s}$$

Display

32	cos	0.8480481

So $0.848 = \dfrac{11.8}{s}$

$s = 11.8 \div 0.848 = 13.9$ (3 sig. fig.).

Exercise 7(c)

1 Find the angles marked in the diagrams in Figure 7.12:

(i)

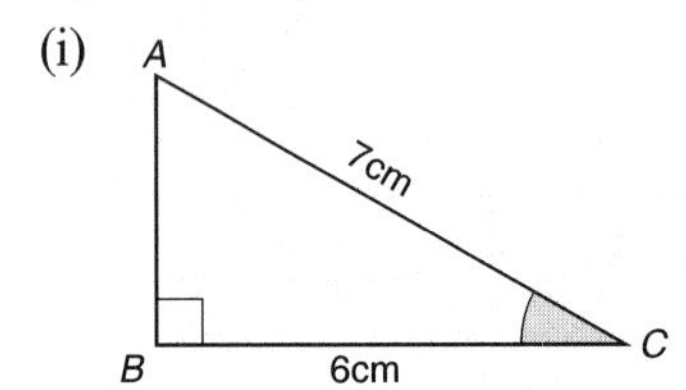

(ii)

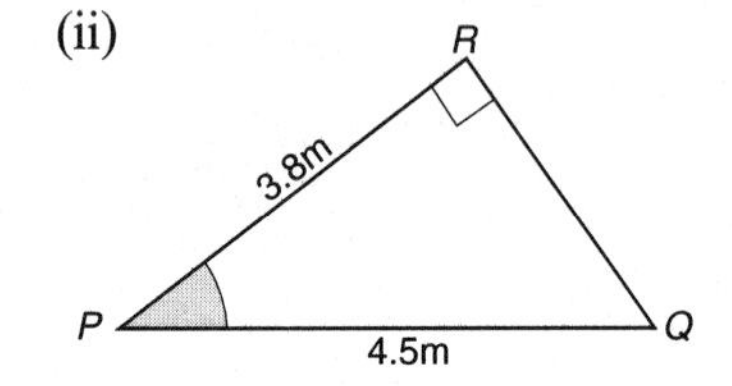

(iii)

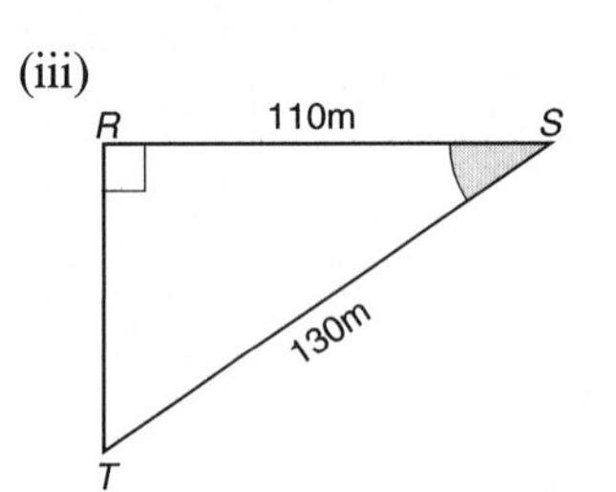

(iv)

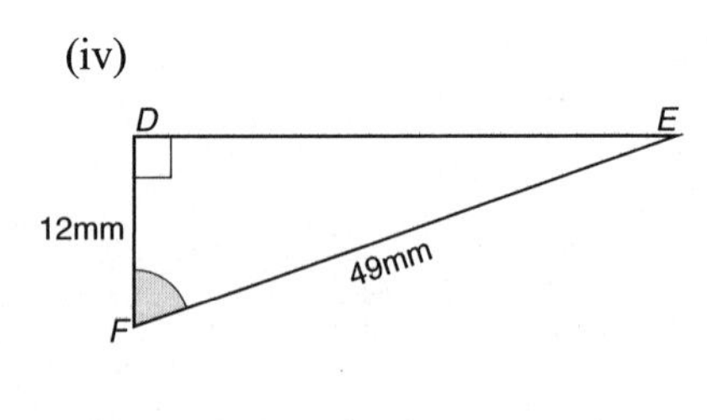

Figure 7.12

2 Find all the sides marked y in the diagrams in Figure 7.13:

(i)

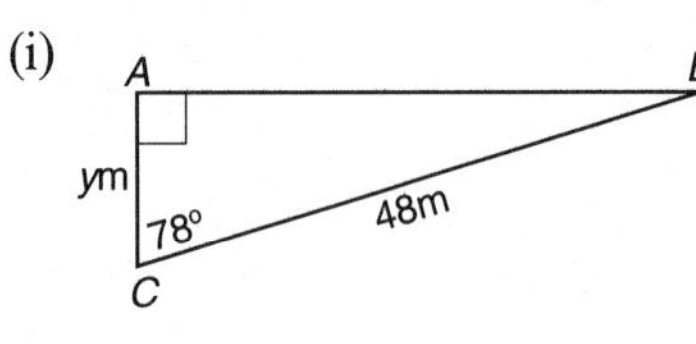

(ii)

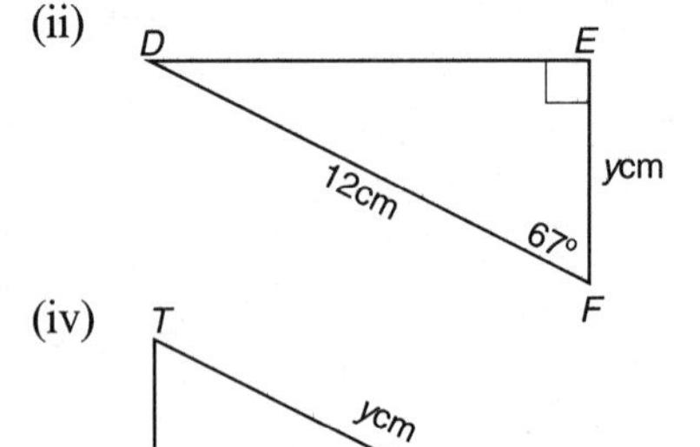

(iii)

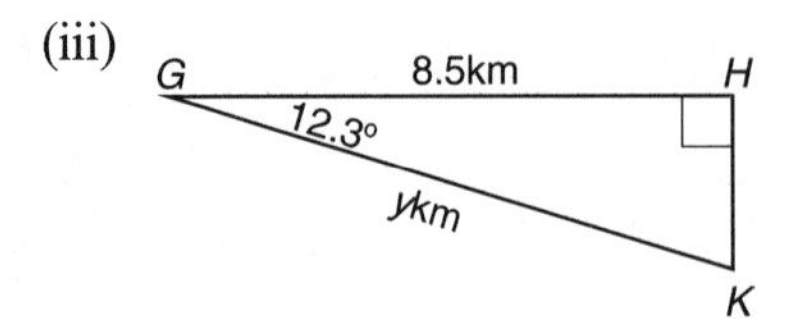

(iv)

Figure 7.13

Investigation XV: Trigonometrical ratios

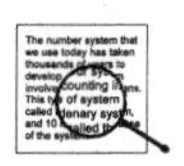

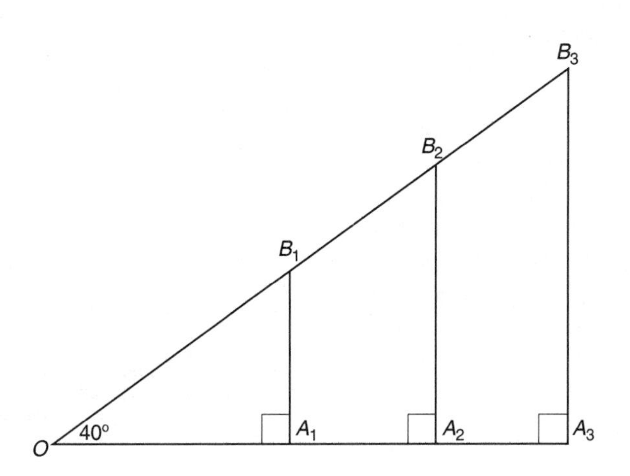

(i) Draw a line OA_3, about 10 or 12 cm long.

(ii) Draw the line OB_3, so that $\angle B_3OA_3 = 40°$, and $\angle B_3A_3O = 90°$.

(iii) Mark A_1 and A_2 anywhere on OA_3, and draw A_1B_1 and A_2B_2 at right angles to OA_3.

(iv) Calculate $\dfrac{A_1B_1}{OB_1}$ $\dfrac{A_2B_2}{OB_2}$ $\dfrac{A_3B_3}{OB_3}$.

Your answers should be approximately equal, showing that the sine of 40° does not depend on the size of the triangle. Take the average of the three values and compare with the value given by the calculator for sin 40°.

(v) Use your calculator to draw up a table of values of $\sin x$.

x	0	10	20	30	40	50	60	70	80	90
$y = \sin x$	0			0.50						

Plot these values on graph paper to obtain a sine curve. Two points are already plotted here.

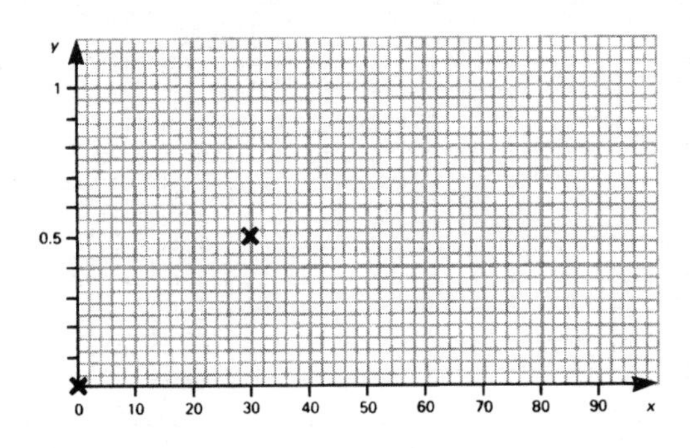

Extension: Carry out a similar investigation for cosine and tangent.

7.4 Angle of elevation (or depression), bearings

(i) Elevation and depression

In Figure 7.14, *BC* is a vertical wall, and *D* is the bottom of a window. The ground *GB* is horizontal, and *ED* is parallel to *GB*. We say that $\angle DGB$ is the **angle of elevation** of *D* from *G* (the angle you look *up* from the level) and $\angle EDG$ is the **angle of depression** of *G* from *D* (the angle you look *down* from the level).

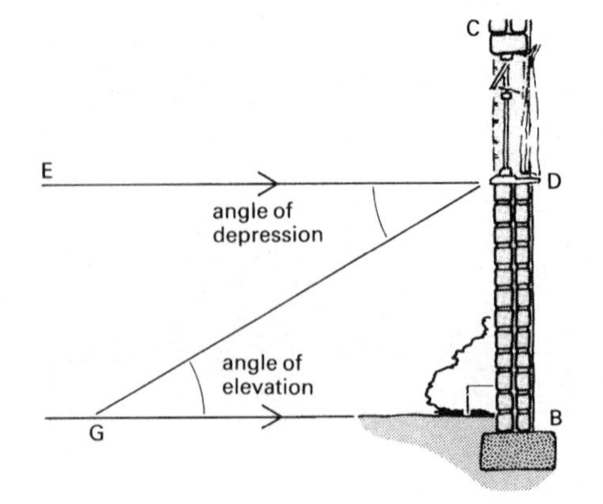

Figure 7.14

Example 7.7

At a point on the ground 20 m from the wall of a house, the elevation of the bottom of the window was 12°. From the same point, the elevation of the top of the window was 14°. Calculate the height of the window.

Solution

Figure 7.15 represents the problem. The best way to find *TB* is to find *TG* and then subtract *BG*.

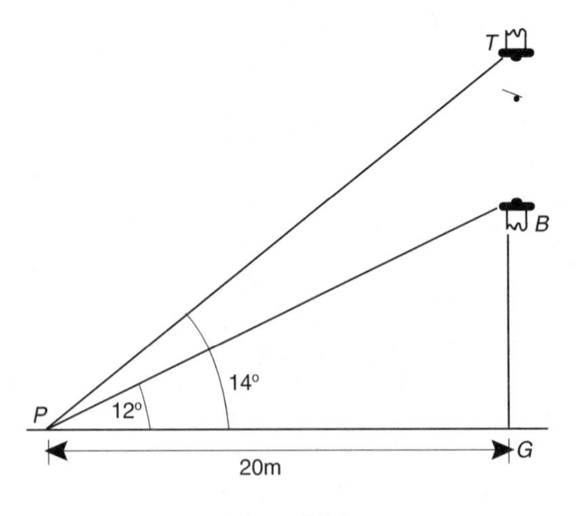

Figure 7.15

Now $\tan 12^\circ = \frac{BG}{20}$

So $BG = 20 \times \tan 12^\circ$

$= 20 \times 0.2126$

$= 4.252$

Also $\tan 14^\circ = \frac{TG}{20}$

Therefore $TG = 20 \times \tan 14^\circ$

$= 4.986$

So $TB = 4.986 \text{ m} - 4.252 \text{ m}$ ($TG - BG$, see Figure 7.15)

$= 0.734 \text{ m}$

The height of the window is 73.4 cm.

The next example illustrates an important use of the tangent ratio, but it is quite a difficult problem to solve.

Example 7.8

In Figure 7.16 A and B are two points 80 metres apart on horizontal ground due West of a tower which stands on top of a small hill. The angle of elevation of the top of the hill from A is 20.6° and from B is 28.4°. Find the height of the base of the tower above the level of AB.

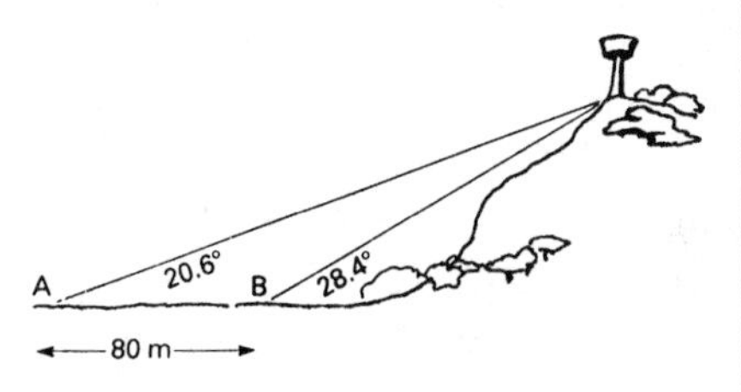

Figure 7.16

Solution

In Figure 7.17 N is the point vertically under the tower T and level with AB.

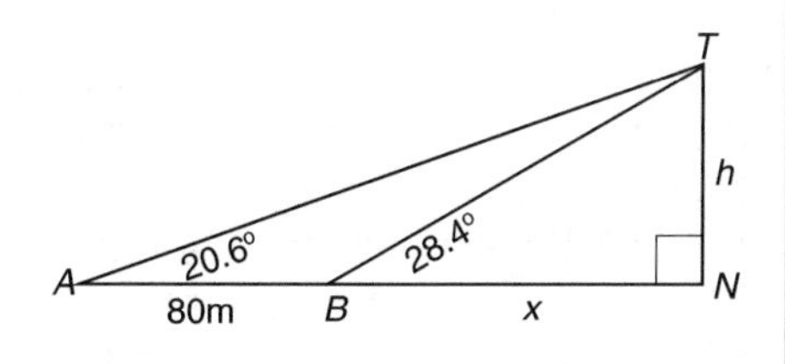

Figure 7.17

In triangle TAN, $\frac{h}{(80 + x)} = \tan 20.6$,

where $BN = x$ and $TN = h$

So $\dfrac{h}{80 + x} = 0.3759$

So $h = 0.3759 \times (80 + x)$

$= 30.072 + 0.3759x$ (removing the brackets)

In triangle TBN,

$$\frac{h}{x} = \tan 28.4$$

So $\dfrac{h}{x} = 0.5407$

So $h = 0.5407x$

The two expressions for h must be equal, which gives us the equation

$$30.072 + 0.3759x = 0.5407x$$

Hence

$$30.072 = 0.5407x - 0.3759x$$

$$= 0.1648x$$

$$x = \frac{30.072}{0.1648} = 182.4757$$

Now $h = 0.5407x$

$= 0.5407 \times 182.4757 = 98.7$

The height of the hill is 98.7 m.

(ii) Bearings

Another type of angle that is used quite frequently, is a **bearing** used in measuring directions. Figure 7.18 shows the compass points together with the bearing measured in a clockwise direction from North between 000° and 360° (always give three figures).

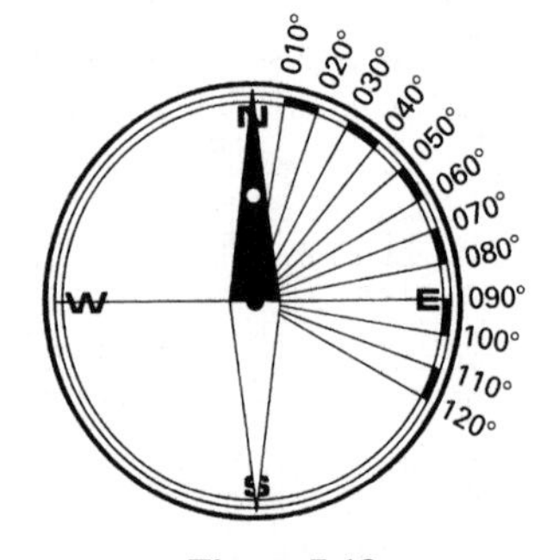

Figure 7.18

Bearings are always measured clockwise *from* a point O and you should imagine that the compass is placed at O. Referring to Figure 7.19, the bearing of A from O would be 150°, but the bearing of O from A would be the reflex angle marked at A, which is 330°. You should notice that the difference in these two answers is 180°. This will always be the case if you reverse the positions of the object being observed, and compass position.

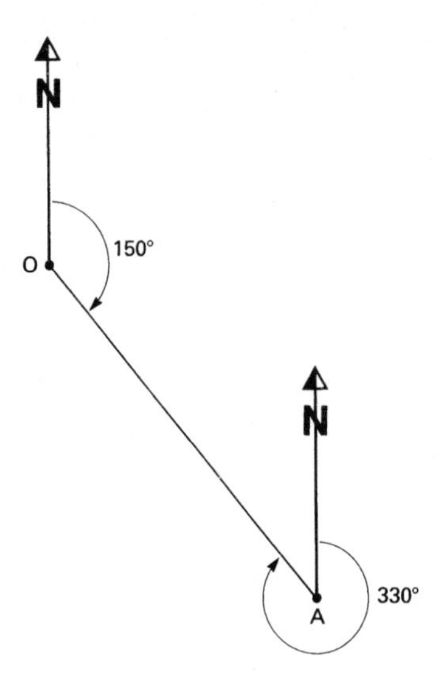

Figure 7.19

Example 7.9

A light aircraft flew from Shoreham airport a distance of 40 km on a bearing of 050°. It then changed directions and flew a further 40 km on a bearing of 110°, until it was over a large town. Find:

(a) the bearing of the town from Shoreham;

(b) the bearing of Shoreham from the town;

(c) the distance due East of the town from Shoreham.

Solution

Let X be the point where the aeroplane changes course (Figure 7.20).

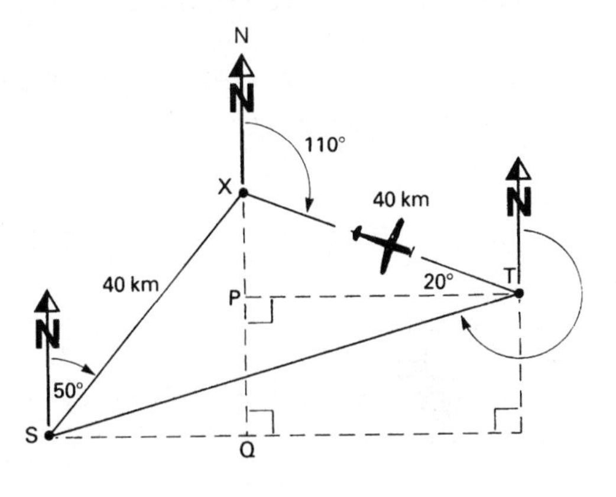

Figure 7.20

$\angle NXS = 180° - 50° = 130°$

$\angle SXT = 360° - 110° - 130° = 120°$

Hence since $\triangle SXT$ is an isosceles triangle

$\angle XST = \frac{1}{2}(180° - 120°) = 30°.$

(a) The bearing of T *from* $S = 050° + 030° = 080°$

(b) The bearing of S *from* $T = 080° + 180° = 260°$

(c) The total distance due East is $SQ + PT$

Referring to Figure 7.20, angle $XSQ = 90° - 50° = 40°$

Hence in triangle XSQ, $SQ = 40 \times \cos 40° = 30.6$ km

In triangle XPT, $PT = 40 \times \cos 20° = 37.6$ km The distance is 30.6 km + 37.6 km = 68.2 km due East.

Exercise 7(d)

1 From a point on the level ground 80 m from the wall of a block of flats, the angle of elevation of the roof was 62°. Find the height of the block of flats.

2 If the angle of depression of a windsurfer from the top of an almost vertical cliff is 28.5° and the cliff is 75 m high, roughly how far out to sea is the windsurfer?

3 Alton is 18 km North and 16 km West of Brigtown. Find:

(i) the bearing of Alton from Brigtown

(ii) the bearing of Brigtown from Alton.

4 Jodi is standing at the top of a cliff which is 65 m high. She observes a boat travelling directly away from her. When it leaves the foot of the cliff she starts timing. After 60 seconds, the angle of depression of the boat is about 30°. Estimate the speed of the boat in km/h.

5 A sailing ship left harbour A and travelled for 20 km on a bearing of 125°. It then changed course and sailed on a bearing of 035° for a distance of 10 km until it reached harbour B. Calculate:

(i) the direct distance AB

(ii) the bearing of B from A

(iii) the bearing of A from B.

7.5 Pythagoras' theorem

It has been known for a long time that if a piece of rope, knotted at equal intervals, is laid around three pegs A, B and C as shown in Figure 7.21 so that $BA = 3$ units, $AC = 4$ units and $BC = 5$ units, then angle $BAC = 90°$. This device would have been used in early surveying techniques of the Babylonians, or the Egyptians in building the pyramids. The Greek mathematicians were able to recognise the importance of this, and to produce the following theorem.

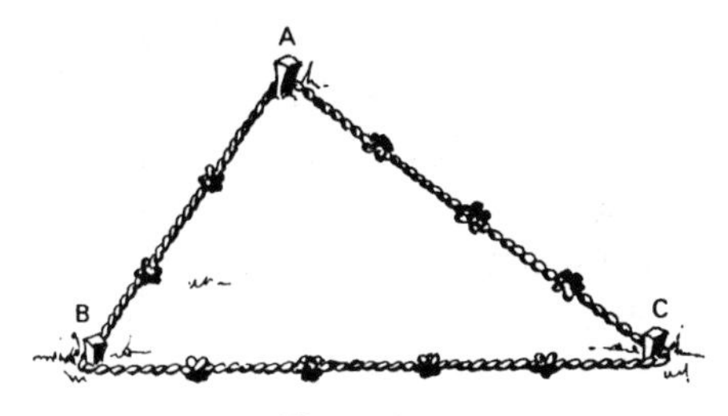

Figure 7.21

In Figure 7.22, ABC is a right angled triangle. Three squares I, II and III are constructed on the sides of the right angled triangle. It can be proved (see Investigation XVI) that

the area of III = area of I + area of II.

This can be written

$$AB \times AB = AC \times AC + CB \times CB$$

i.e. $$AB^2 = AC^2 + CB^2$$

This is known as **Pythagoras' theorem**.

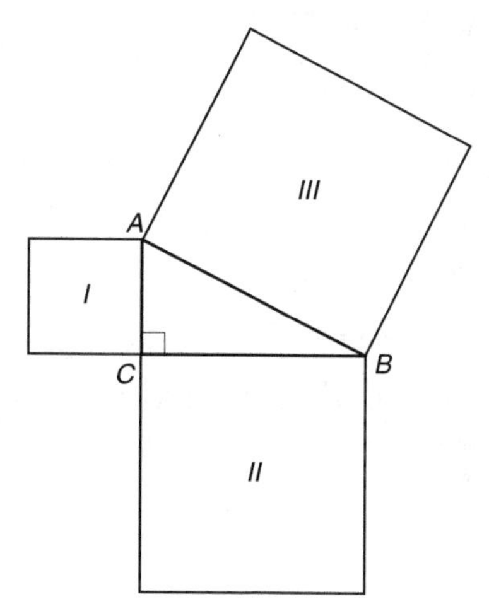

Figure 7.22

It is perhaps best to remember this theorem in words:

> **Rule**
>
> The area of the square on the largest side of a right angled triangle (the hypotenuse), is equal to the sum of the area of the squares on the other two sides.

Investigation XVI: Pythagoras' theorem

Take a piece of thin card, and draw the figure shown alongside where *I*, *II* and *III* are squares. Now see if you can cut square *II* in four pieces, which then fit with square *I* exactly inside square *III*. This demonstrates the truth of Pythagoras' theorem.

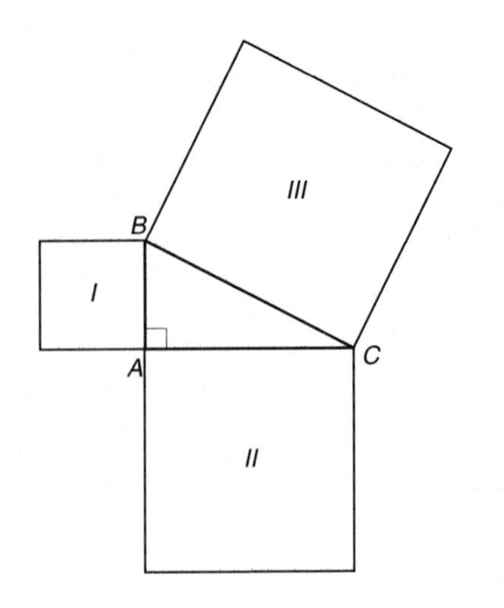

Pythagorean triples

The simplest right angled triangle with sides which are integers (whole numbers) is the 3, 4, 5 triangle. The numbers 3, 4, 5 are said to form a Pythagorean triple.

A similar set would be 5, 12, 13.

In each of these cases, the largest two numbers differ by 1. How many of these sets of three numbers can you find?

The following examples show how this theorem is used. Notice that no angles are involved.

Example 7.10

Find the length of x in the following cases (Figure 7.23):

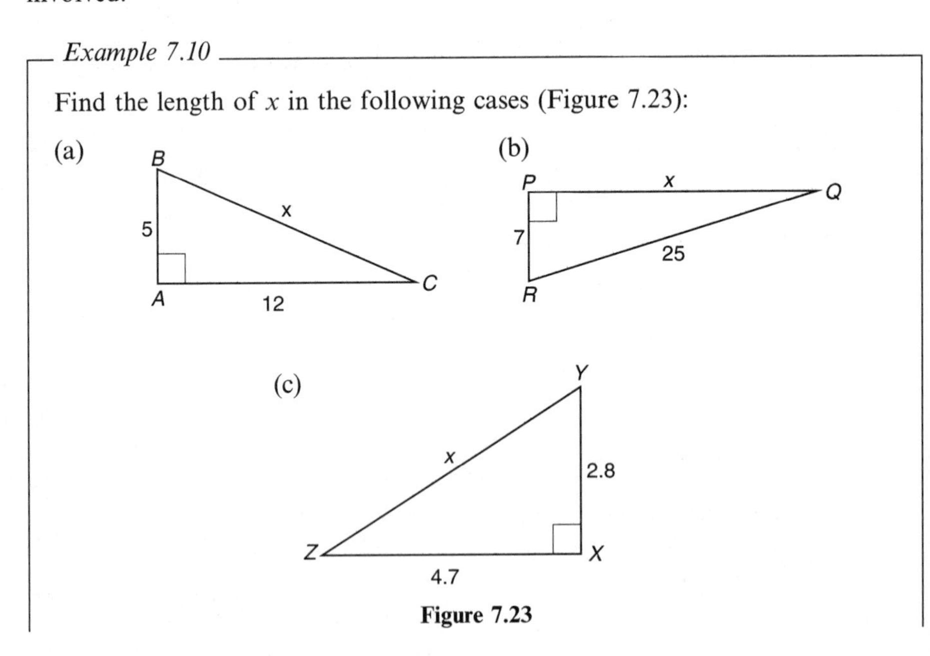

Figure 7.23

Solution

(a) x is the largest side

$$\text{So} \quad x^2 = 5^2 + 12^2 \quad (\text{remember } 12^2 \text{ means } 12 \times 12)$$

$$= 25 + 144 = 169$$

We need to find the *square root* of 169, hence

$$x = \sqrt{169}$$

Using the calculator:

Display

169	$\sqrt{}$	13.

Hence $x = 13$.

(b) Here 25 is the largest side, and so

$$25^2 = x^2 + 7^2$$

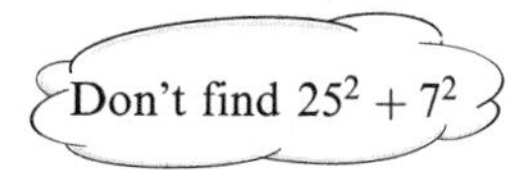

REMEMBER

You may need to use the x^2 button on the calculator in order to find the squares.

$$\text{Hence} \quad 625 = x^2 + 7^2$$

$$x^2 = 625 - 49 = 576$$

$$x = \sqrt{576} = 24.$$

(c) x is the largest side

$$\text{So} \quad x^2 = 2.8^2 + 4.7^2$$

$$= 7.84 + 22.09 = 29.93$$

$$x = 29.93 = 5.47 \text{ (3 sig. fig.)}.$$

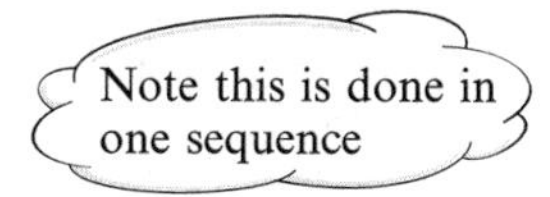

REMEMBER

Using the calculator:

Display

2.8	x^2	+	4.7	x^2	=	$\sqrt{}$	5.470831

Exercise 7(e)

Find all the sides marked with a letter in the diagrams in Figure 7.24:

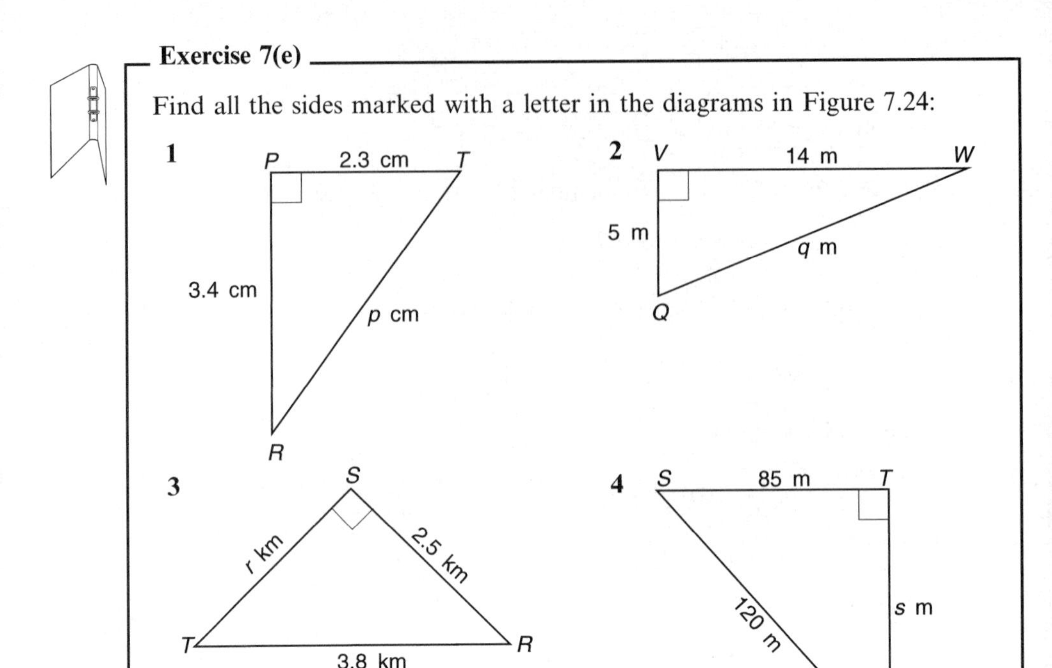

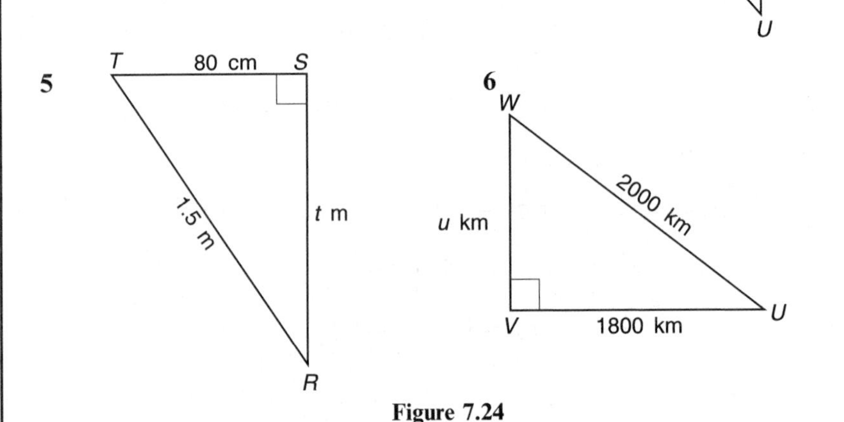

Figure 7.24

Most problems in trigonometry combine the trigonometrical ratios and Pythagoras' theorem. The following examples should indicate the type of problem that can be tackled.

Example 7.11

Figure 7.25 shows a goods yard crane. *PRQ* is an isosceles triangle, and *RSQ* is a right angled triangle. If $RS = 6$ m and $PQ = 4.6$ m, find (a) *RQ*; (b) the height of *R* above the ground.

Solution

(a) In triangle RSQ, RQ is the opposite side, and $RS = 6$ m is the adjacent side.

$$\text{So} \quad \tan 30^\circ = \frac{\text{opposite}}{\text{adjacent}} = \frac{RQ}{6}$$

$$\text{So} \quad RQ = 6 \times \tan 30^\circ = 3.46 \text{ m.}$$

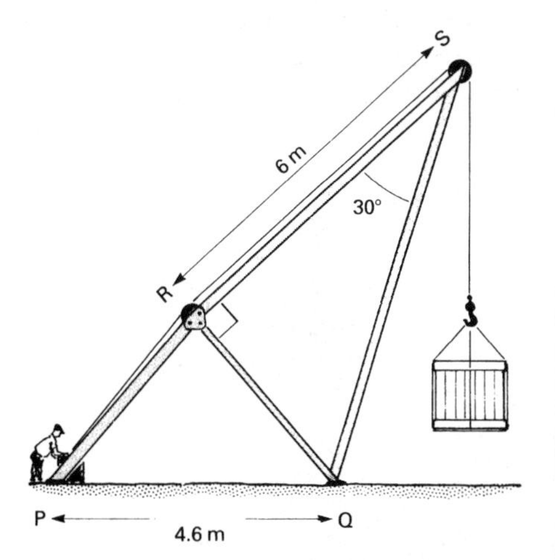

Figure 7.25

(b) An isosceles triangle can be split into two right angled triangles, as shown in Figure 7.26.

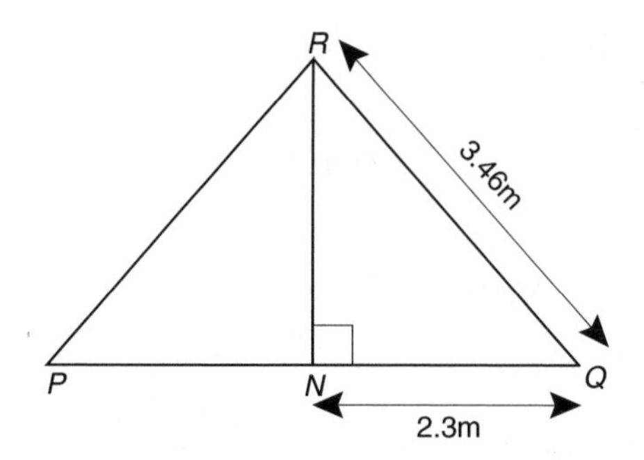

Figure 7.26

In triangle RQN, RQ is the largest side.

$$\begin{aligned}
\text{So} \qquad RQ^2 &= RN^2 + NQ^2 \\
\text{i.e.} \qquad 3.46^2 &= RN^2 + 2.3^2 \\
11.9716 &= RN^2 + 5.29 \\
\text{Hence} \qquad RN^2 &= 11.9716 - 5.29 \\
&= 6.6816 \\
\text{Hence} \qquad RN &= \sqrt{6.6816} = 2.58 \text{ (3 sig. fig.)}
\end{aligned}$$

The height of R above the ground is 2.58 m.

Example 7.12

Figure 7.27 shows the course for a yacht race. The boats start at S, travel 5 km on a bearing 020° to buoy B_1 where they change course and travel 6 km on a bearing of 070° to buoy B_2. At this point they then turn through 110° and sail 3 km due South to the finish at F.

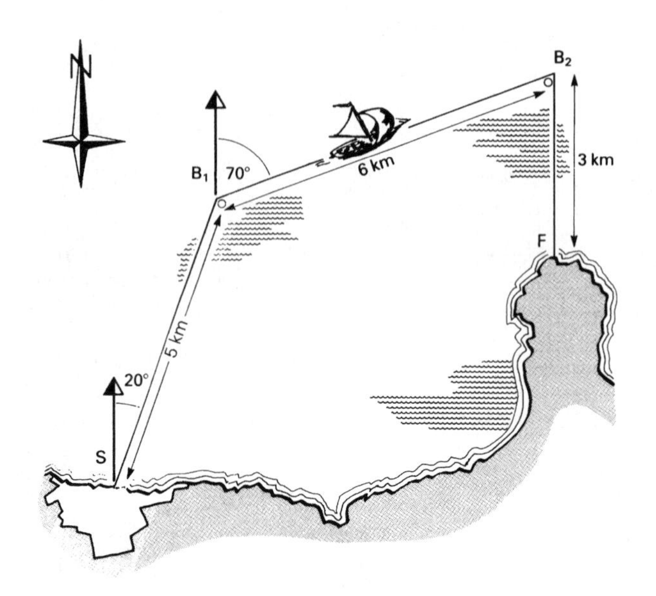

Figure 7.27

(a) How far due North and East is the finish F from the start S?

(b) What is the direct distance from S to F?

Solution

At first sight, it is not clear where the right angled triangles are. They have to be made (see Figure 7.28).

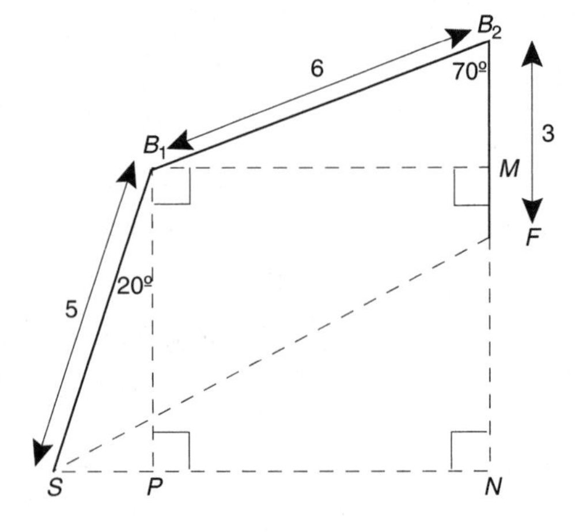

Figure 7.28

(a) In $\triangle B_1PS$

$$\frac{B_1P}{5} = \cos 20^\circ = 0.9397$$

So $\quad B_1P = 5 \times 0.9397 = 4.6985$

In $\triangle B_1B_2M$

$$\frac{B_2M}{6} = \cos 70^\circ = 0.342$$

So $\quad B_2M = 6 \times 0.342 = 2.052$

Hence $\quad B_2N = B_2M + B_1P = 2.052 + 4.6985 = 6.7505$

Hence $\quad FN = B_2N - 3 = 3.7505$

So $F = 3.75$ km due North of S.

In a similar fashion

$SN = SP + B_1M = SP + PN$
$SN = 5\sin 20^\circ + 6\sin 70^\circ = 7.35$

Hence F is 7.35 km due East of S.

(b) We can now use Pythagoras' theorem in triangle SFN

$$SF^2 = SN^2 + FN^2$$
$$= 7.35^2 + 3.75^2 = 68.085$$

So $\quad SF = \sqrt{68.085} = 8.25$ (3 sig. fig.)

The direct distance from start to finish is 8.25 km.

Exercise 7(f)

1 The diagram in Figure 7.29 shows part of a girder bridge. Find:

(i) $\angle RPQ$

(ii) RQ

(iii) SQ

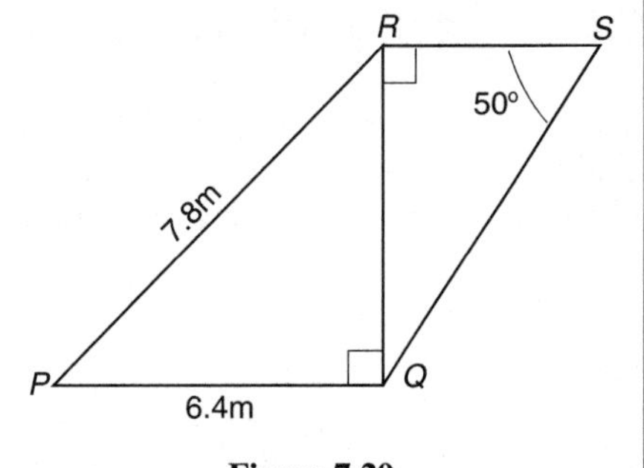

Figure 7.29

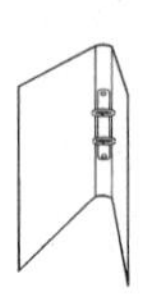

2 In Figure 7.30, $ABCD$ is a square of side 24 cm. $\angle EAC = 90°$. Calculate EC.

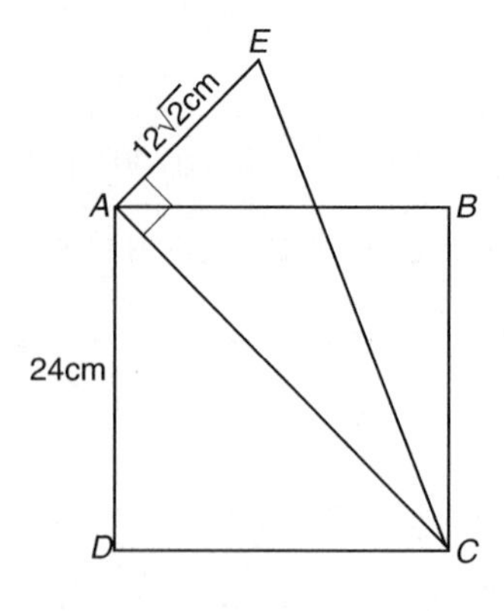

Figure 7.30

3 In the kite $ABCD$, where $AB = AD$, the diagonals BD and AC meet at E; P is the midpoint of AE and Q is the midpoint of CE. Given that $BD = 12$ cm, $AC = 13$ cm and $AE = 4$ cm:

(i) calculate AB and CB in cm

(ii) prove $\angle ABC = 90°$

(iii) calculate $\angle EBC$

(iv) calculate $\angle PBQ$.

4 The diagram in Figure 7.31 shows the cross-section of a cylindrical oil-storage tank of radius 2.6 m. Find the depth of the oil above the bottom of the tank when $\angle AOB = 109°$.

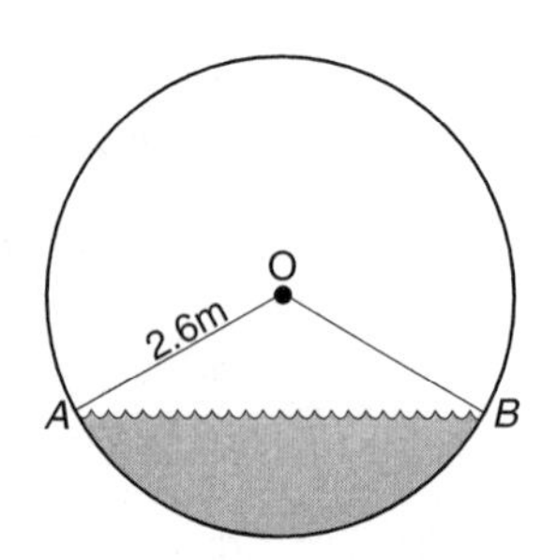

Figure 7.31

7.6 Three-dimensional problems

There are two main problems to solve in three-dimensional situations. The first is finding the angle between a line and a plane. The second is the angle between two planes. Constructing a simple model will often clarify the situation.

Example 7.13

In Figure 7.32, $ABCDEFGH$ is a cuboid, I is the midpoint of FG. Find the angle between the line AI, and the plane $BFGC$.

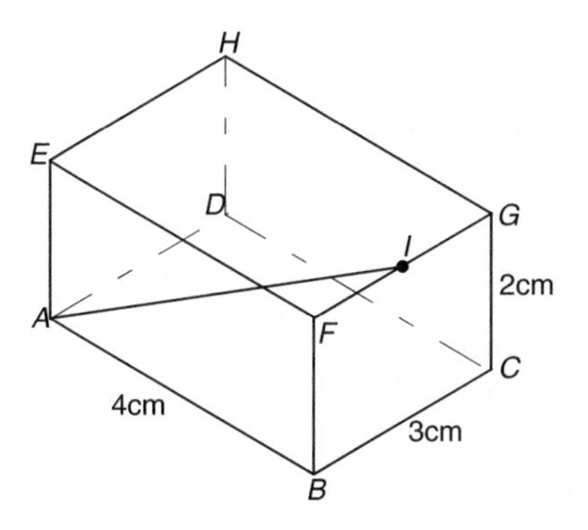

Figure 7.32

Solution

It is helpful to turn the solid over, so that the plane is the horizontal surface. The triangle ABI is right angled and stands upright.

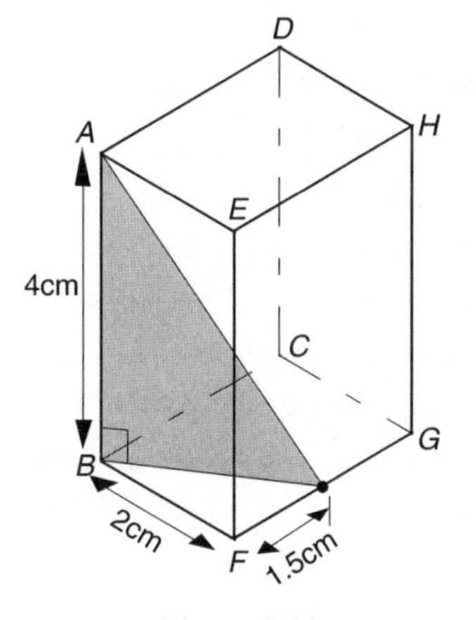

Figure 7.33

The required angle is $\angle AIB$.

We first need to find BI.

$$\text{In } \triangle BFI, \quad BI^2 = 2^2 + 1.5^2 = 6.25$$
$$\text{So} \quad BI = \sqrt{6.25} = 2.5$$

In $\triangle BAI$,

$$\tan \angle AIB = \frac{4}{2.5} = 1.6$$

$$\text{Hence} \quad \angle AIB = 58°.$$

Example 7.14

$PQRST$ is an upright square-based pyramid. The height of the pyramid is 3 m, and the base is of side 2 m. Find the angle between the side PST and the base.

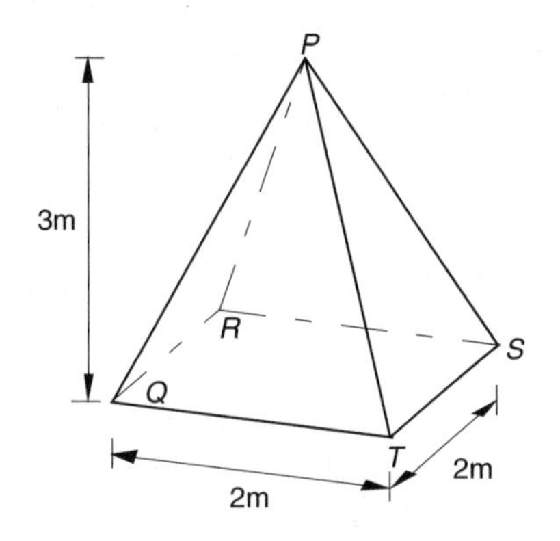

Figure 7.34

Solution

You must find the **common line** of the two planes. This is TS. A triangle must then be found which is at right angles to this line. If N is the midpoint of TS, and O is the midpoint of the base, then PON is the required triangle, and $\angle PNO$ is the required angle.

$$\tan \angle PNO = \frac{PO}{ON} = \frac{3}{1} = 3$$

$$\angle PNO = 71.6°.$$

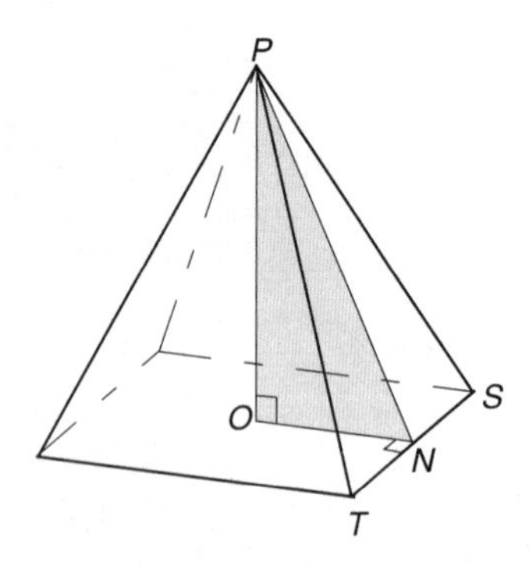

Figure 7.35

Exercise 7(g)

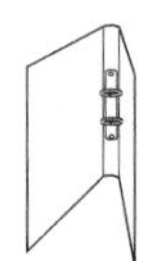

1 A, B and C are three points on level ground. B is due East of A, C is due North of A, and the distance AC is 160 m. The angles of elevation from A and B to the top of a vertical mast whose base is at C are respectively 30° and 28°. Calculate:

(i) the distance BC

(ii) the bearing of C from B.

2 Figure 7.36 shows a storage shed on a building site.

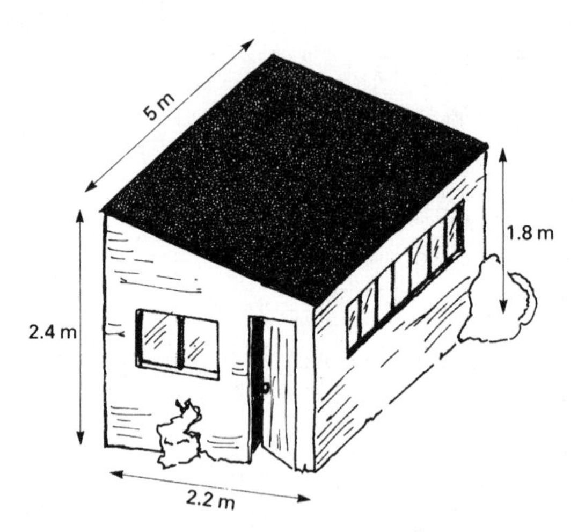

Figure 7.36

(i) Calculate the angle of elevation of the roof to the horizontal.

(ii) Calculate the length of the longest rod that can be stored in the shed.

3 Figure 7.37 shows a pyramid $TABC$ whose base ABC is an equilateral triangle of side 8 cm. L is the midpoint of BC. The sloping edges TA, TB and TC are each 7 cm long and the perpendicular from T to the base meets the line AL at P, where $AP = \frac{2}{3}AL$.

Calculate:

(i) the lengths of AL and AP
(ii) the angle between TA and the base
(iii) the angle between the face TBC and the base
(iv) the height TP.

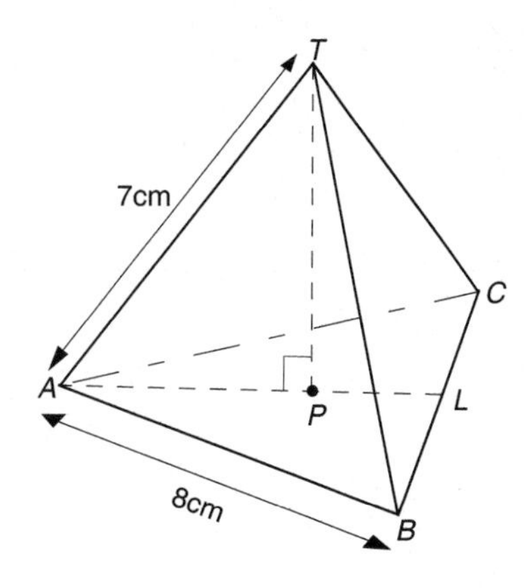

Figure 7.37

4 Referring to Figure 7.38, $ABCDEF$ is a regular hexagon of side 10 cm, and the triangles GAF etc. are all isosceles, each having its equal sides 12 cm long.

Calculate $F\hat{G}A$.

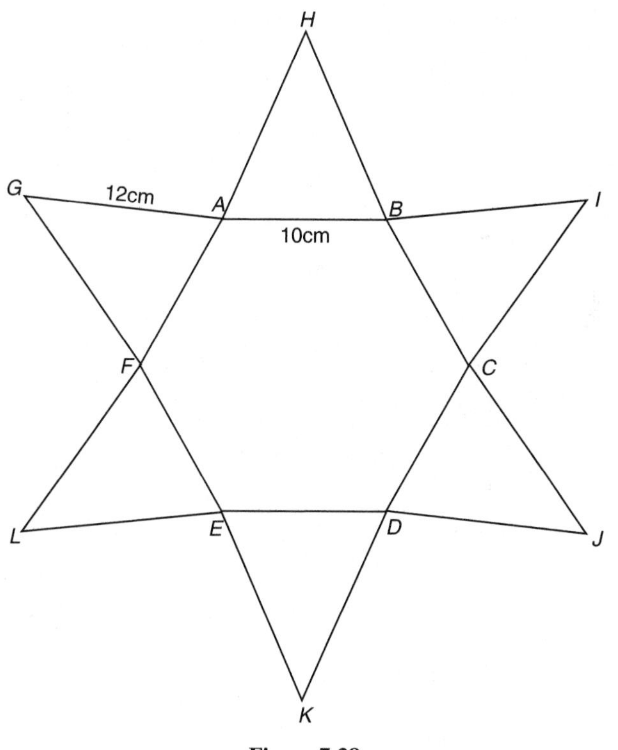

Figure 7.38

This shape is then used as a net for the construction of a pyramid. Let V denote the vertex of the pyramid. Calculate:

(i) the height of the pyramid

(ii) the volume of the pyramid.

8 Geometry and construction

8.1 Geometrical construction

A great deal of mathematics requires a diagram which often needs to be accurately drawn. There are a few simple skills involving the use of a ruler and pair of compasses (for drawing arcs of circles) which are given in this chapter.

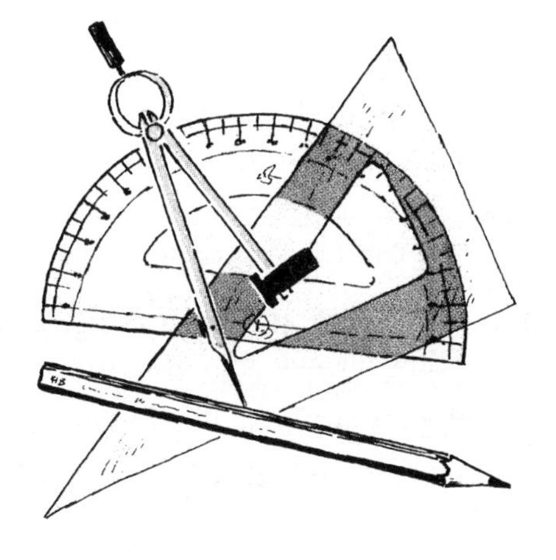

REMEMBER

(i) To bisect an angle

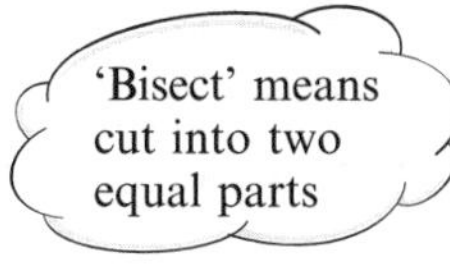

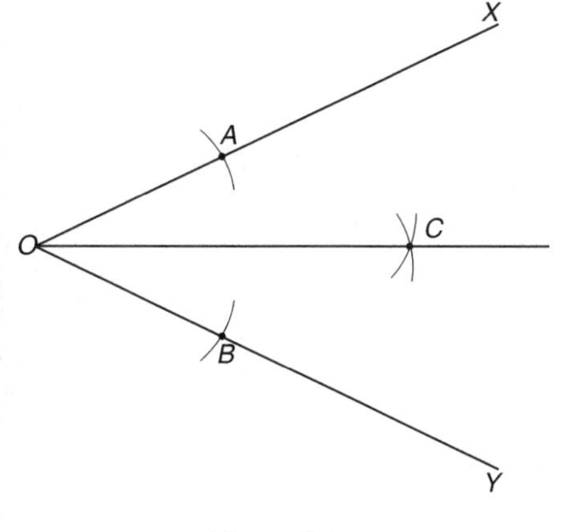

Figure 8.1

(a) Draw angle BOA. Place the point of the compasses at O, and with any convenient radius, draw equal arcs to cut OX at A and OY at B.

(b) Keeping the radius the same, using A and B as the centres, draw two arcs to cross at C.

(c) Joining OC gives the bisector of angle XOY.

All points along OC are equal distance (or **equidistant**) from OX and OY.

(ii) To bisect a line XY at right angles

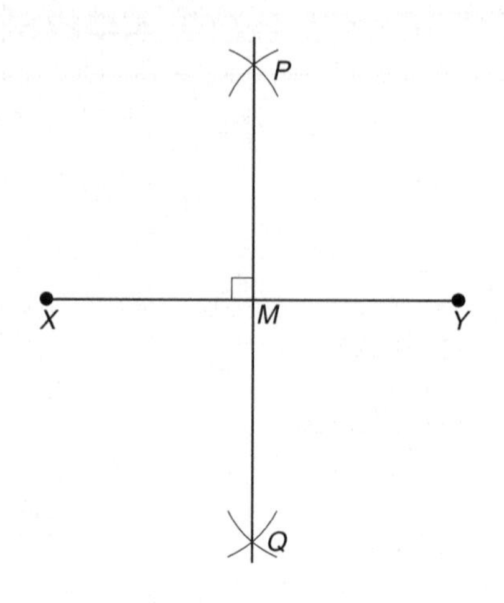

Figure 8.2

(a) With centre X and radius greater than half of XY, draw arcs above and below the line XY.

(b) Repeat this from Y, using the same radius, to give arcs crossing at P and Q.

(c) Join PQ, and M will be the middle of XY with PQ perpendicular to XY.

All points along PQ are equal distance (*equidistant*) from X and Y.

(iii) To draw a line from a point P perpendicular to a given line XY

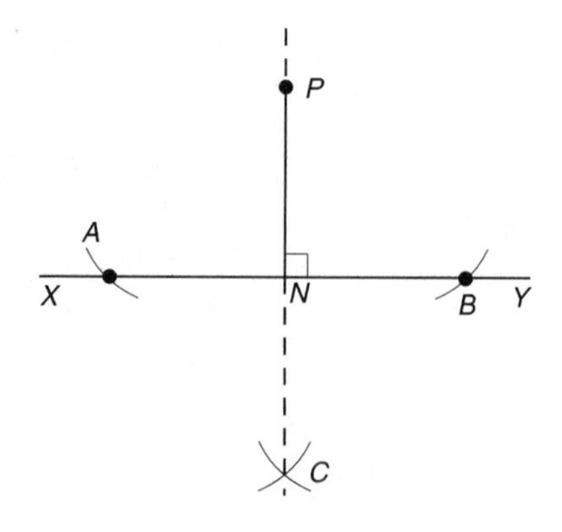

Figure 8.3

(a) With the compasses centred at P, draw two arcs to cut the line XY at A and B.

(b) Using the same radius, with centre A and then B, draw two arcs to meet below the line at C.

(c) Join PC to meet the line at N, and PN is the required perpendicular.

(iv) To copy a given angle

Referring to Figure 8.4, if it were necessary to copy exactly angle AOB, you would proceed as follows.

(a) Draw an arc of a reasonable radius, centre O to cut OA at X, and OB at Y.

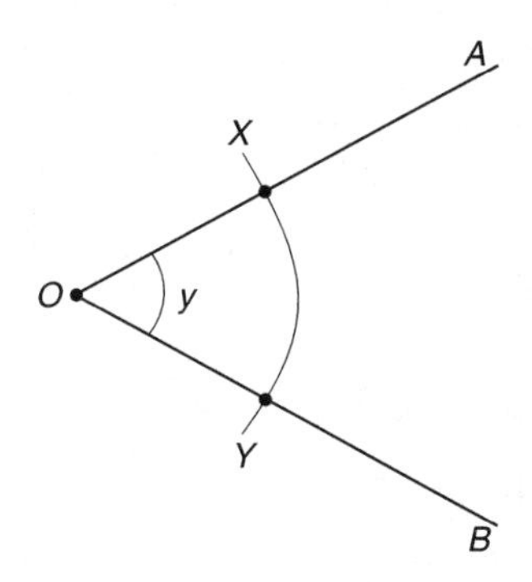

Figure 8.4

(b) If the angle is to be copied at P on line MN (see Figure 8.5) draw an arc of the same radius as in (a) with centre P, to cut MN at Q.

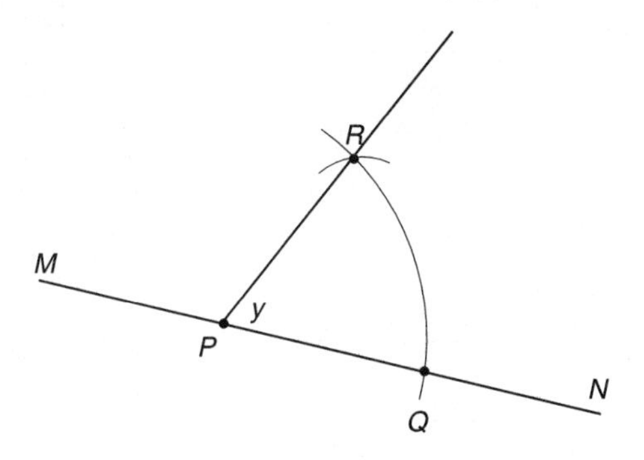

Figure 8.5

(c) Measure XY, and open the compasses to this radius. Draw an arc with this radius, centred at Q to meet the larger arc at R.

(d) Join PR, and angle RPN will be the same as angle AOB.

(v) Angles of 60° and 30°

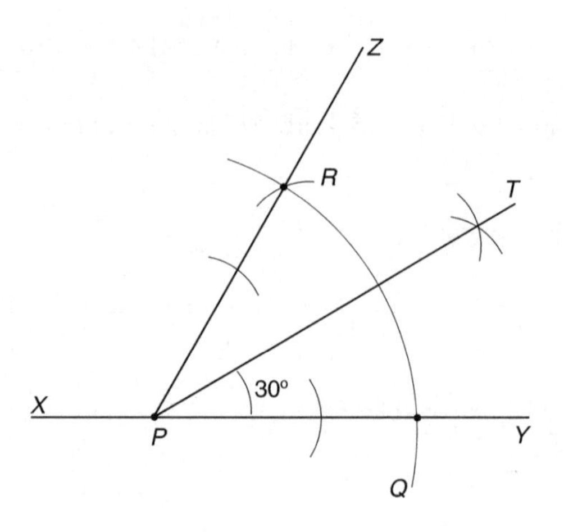

Figure 8.6

(a) With the compass point at P on XY in Figure 8.6, draw an arc to cut XY at Q.

(b) With centre Q and the same radius, draw an arc to cut this arc at R.

(c) Draw PZ through R, and you will find that angle $ZPY = 60°$.

(d) If the bisector of angle ZPY is found, i.e. PT in Figure 8.6, then clearly angle $TPY = \frac{1}{2}$ of $60° = 30°$. By continuing this process, we could find $15°$, $7\frac{1}{2}°$ and so on.

(vi) To construct an angle of 45°

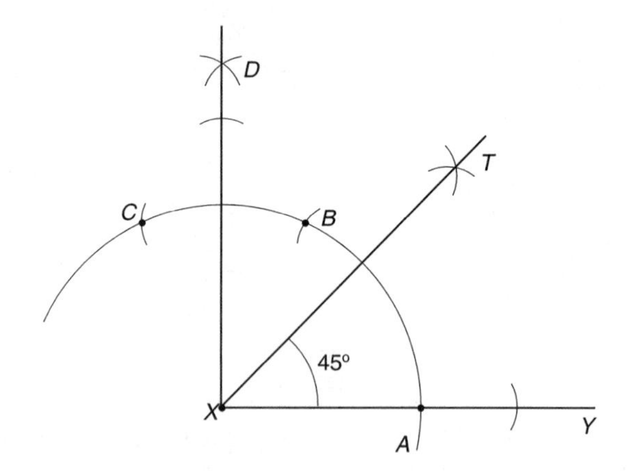

Figure 8.7

(a) A simple method would be to bisect an angle of 90°.

(b) If the angle is at the end of line say XY in Figure 8.7, then proceed as follows:

With centre X, draw an almost complete semi-circle to cut XY at A.

(c) With centre at A and the same radius, make an arc at B, and then with centre at B make an arc at C.

(d) With centre C and B and the same radius, make two arcs to meet at D. Join XD, and you will find that angle DXY is 90°.

(e) Bisect angle DXY with line XT, and hence angle $TXY = 45°$.

(vii) Parallel lines

Parallel lines can be drawn with either a pair of parallel rules, or two set squares as shown in Figure 8.8.

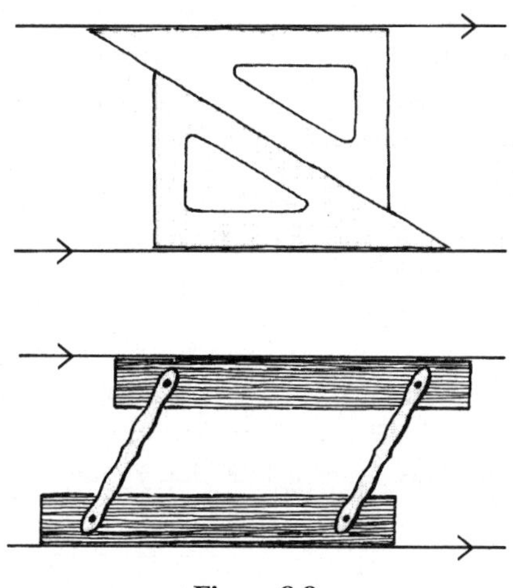

Figure 8.8

(viii) Circles associated with triangles

Example 8.1

Construct the triangle ABC, with sides $AB = 10$ cm, $AC = 8$ cm and $BC = 5$ cm. Construct also the circle which passes through the vertices of the triangle.

Solution

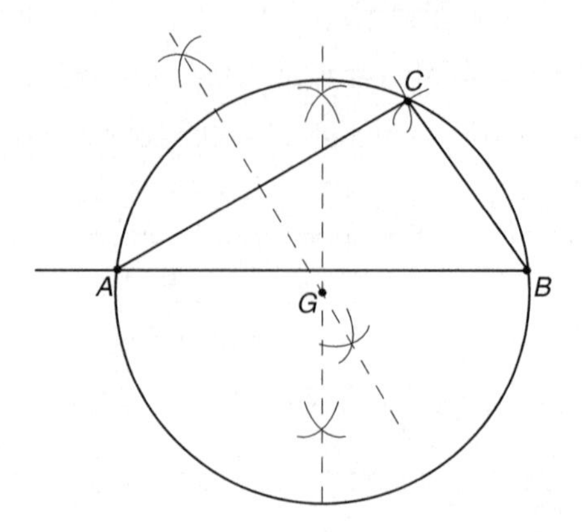

Figure 8.9

(a) Draw a line slightly larger than 10 cm, and mark the points A and B so that $AB = 10$ cm (Figure 8.9).

(b) With centre A draw an arc radius 8 cm, with centre B draw an arc radius 5 cm. These meet at C.

(c) The centre of the circle (G) lies on the perpendicular bisectors of the three sides (only *two* need be drawn) (see Figure 8.9).

This circle is called the **circumscribing** circle of the triangle. Notice that the centre of the circle is *outside* the triangle, in this case because the triangle is obtuse.

Example 8.2

Construct the triangle PQR, with $PQ = 12$ cm, $PR = 10$ cm, and angle $QPR = 60°$. Construct the circle inside the triangle which just touches all three sides.

Solution

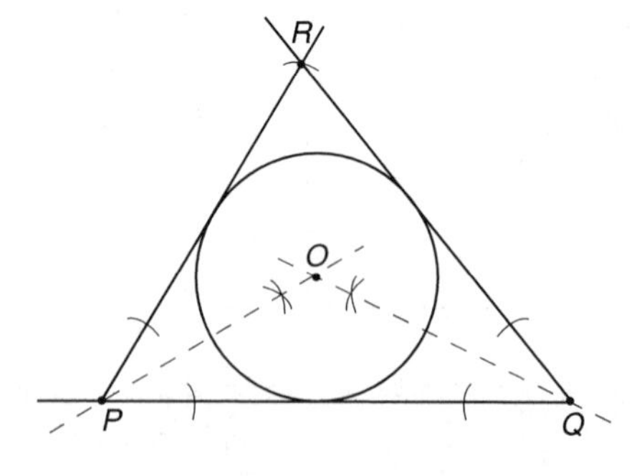

Figure 8.10

(a) Draw a base line and mark $PQ = 12$ cm.

(b) Construct an angle of 60° at P, and mark $PR = 10$ cm.

(c) Join RQ. The centre of the circle O is where the bisectors of the three angles meet (only two need be drawn). See Figure 8.10.

The compasses need to be opened out so that the circle just touches the sides. This circle is called the **inscribed** circle.

Exercise 8(a)

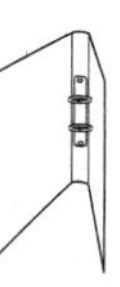

1 Construct the following triangles. You should use a protractor to measure the angles if they cannot be constructed. In each case, measure the sides and angles not given.

(i) $\triangle XYZ$: $XY = 8$ cm, $XZ = 4.5$ cm, $YZ = 6.5$ cm

(ii) $\triangle PQR$: $PQ = 12$ cm, $\angle RPQ = 60°$, $\angle RQP = 45°$

(iii) $\triangle ABC$: $AB = 6.3$ cm, $BC = 8.2$ cm, $\angle ABC = 125°$

(iv) $\triangle UVW$: $UV = 12.7$ cm, $\angle WUV = 30°$, $\angle UVW = 135°$.

2 Construct a right-angled triangle PQR, with $PQ = 5$ cm, $PR = 8$ cm, and $\angle PQR = 90°$. Bisect the hypotenuse and draw a circle with the midpoint O as centre and OR as radius. Mark any point X on the circle on the opposite side from Q, join PX and QX and measure $\angle PXQ$ and $\angle PRQ$.

3 Construct a rhombus having diagonals of length 150 mm and 80 mm. Measure the length of the side.

4 Construct a square $PQRS$ with sides 8.5 cm. Find X the midpoint of RS. Measure PX, QX and $\angle SPX$. Construct an equilateral triangle QRY, and measure the length of the median of triangle QRY which passes through Y. (A median joins a vertex to the midpoint of the opposite side.)

5 Construct, using ruler and compasses only, a quadrilateral $ABCD$ with base line $AB = 8$ cm, so that $\angle BAD = 60°$, $AD = 10.4$ cm, $BC = 9$ cm, and $DC = 2.8$ cm. What do you notice about the sides AB and DC?

[NEA]

6 Draw any acute-angled triangle BAC. Extend BC to D, where $CD = CA$. Extend CB to E where $BE = BA$. Construct the bisectors of angles ABE and ACD. Label the point of intersection of these two bisectors O. By drawing a circle through A with centre O, show that OB is the line of symmetry of the quadrilateral $OABE$.

8.2 Drawing three-dimensional shapes

In order to represent a three-dimensional shape on a flat piece of paper, it is necessary to distort the angles of that shape. However, as shown in Figure 8.11, any sides that are parallel in three dimensions should also be parallel on your drawing. The shape drawn is perhaps similar to a gravel bin at the side of a road or railway. Notice that the top is in fact sloping.

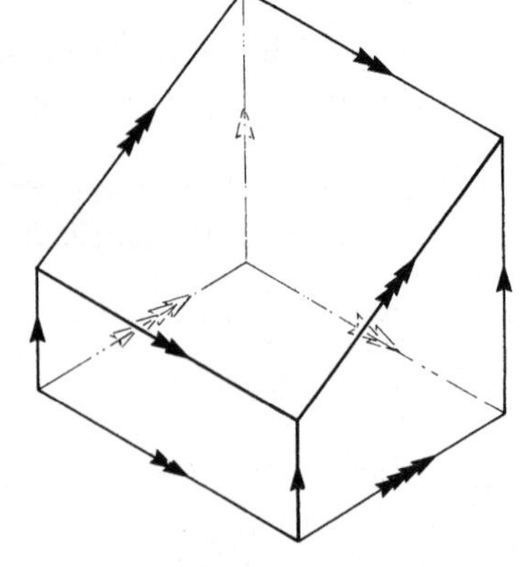

Figure 8.11

Sometimes (particularly in art) you might want to give the impression of distance. In this case, parallel lines going away from you are drawn so they converge to a point on the horizon (H), see Figure 8.12. This is known as drawing in **perspective**.

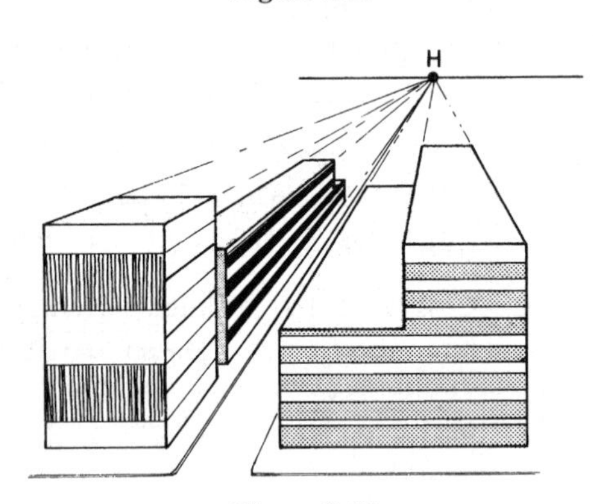

Figure 8.12

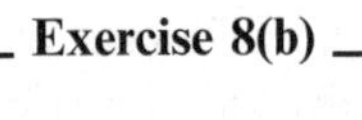

Exercise 8(b)

Make a neat copy of the shapes in Figure 8.13, and also try to draw them in perspective.

1

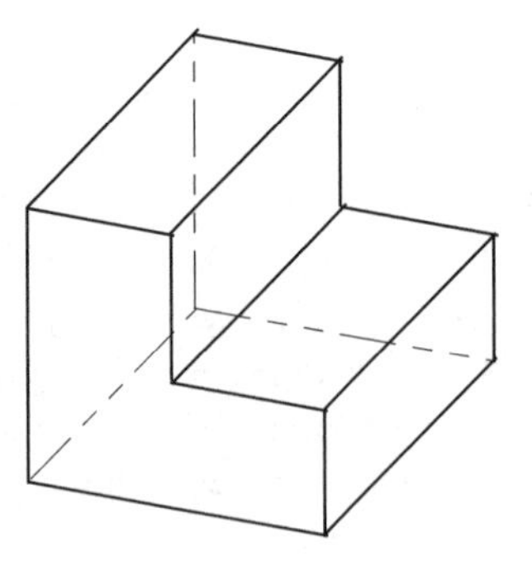

2

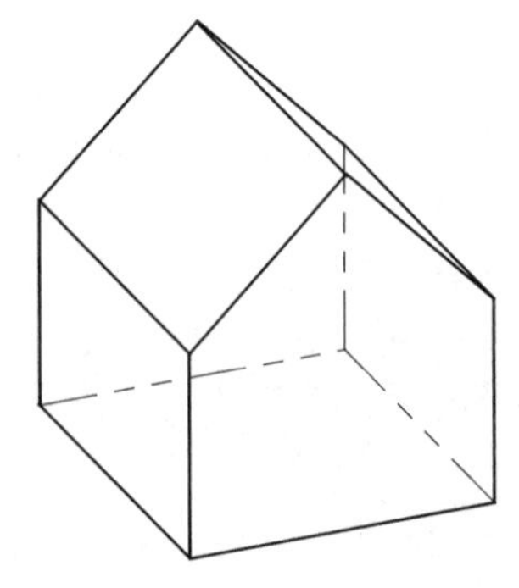

Figure 8.13

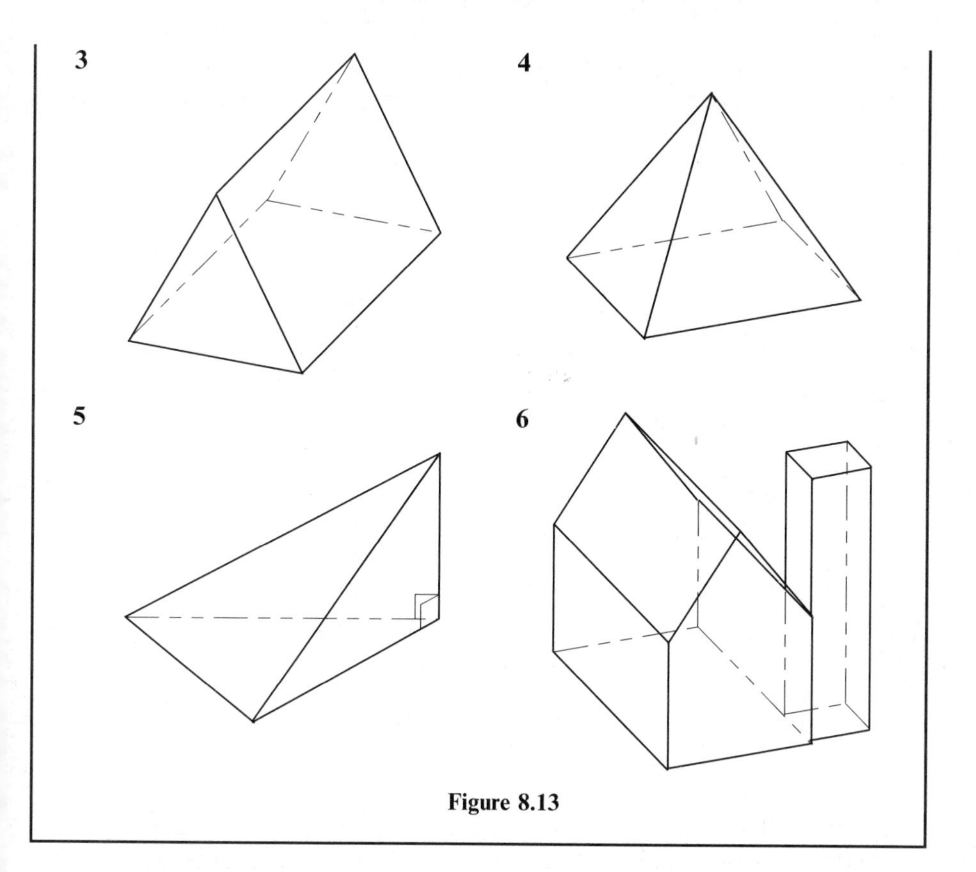

Figure 8.13

8.3 Plans and elevations

Three-dimensional models have to be represented in a different way to that of the last section, when we wish to make calculations. The basic idea behind the method is to visualise taking a photograph from various directions. The solid shown in Figure 8.14 is similar to the shape of a house. Hidden edges are dotted. Three different views of it are shown alongside.

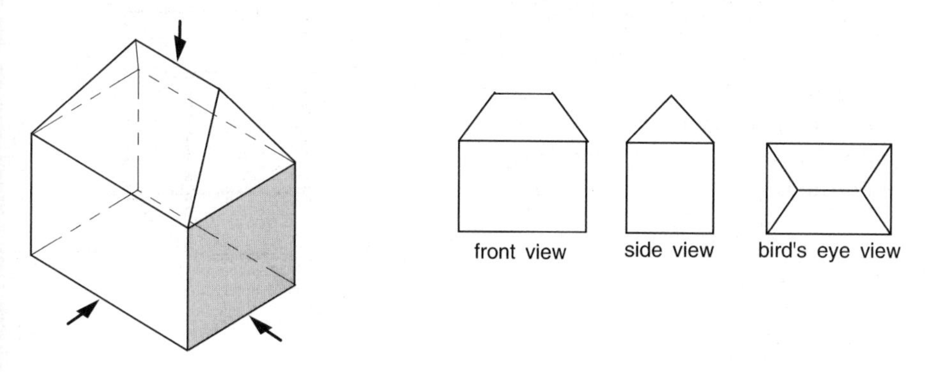

Figure 8.14

The front and side views are referred to as **elevations**, and the bird's eye view is called the **plan**. These can be combined together in a more technical fashion as shown in Figure 8.15.

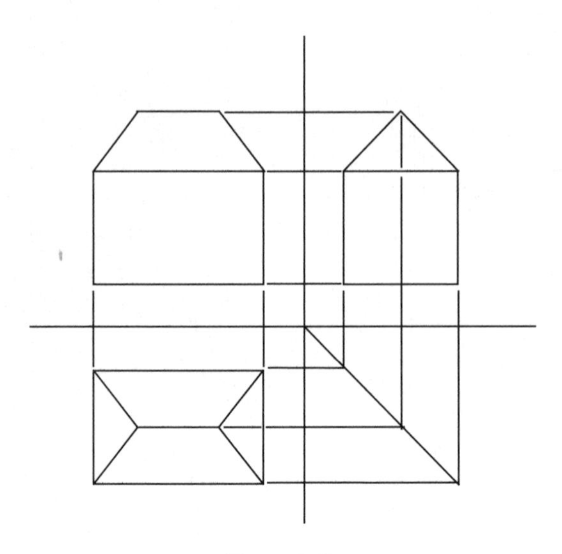

Figure 8.15

Diagrams of this nature are commonly used in engineering. They are very useful for finding distances measured through three-dimensional objects, as illustrated by the following example. In practice, most of these diagrams are now drawn using computer design packages.

Example 8.3

Figure 8.16 shows a framework made from 12 rods of length 8 cm, forming a cube $ABCDA'B'C'D'$. X is a point on $A'B'$ and $A'X = 3$ cm. Draw a plan and elevation from the direction of the arrow, and hence find the length of DX.

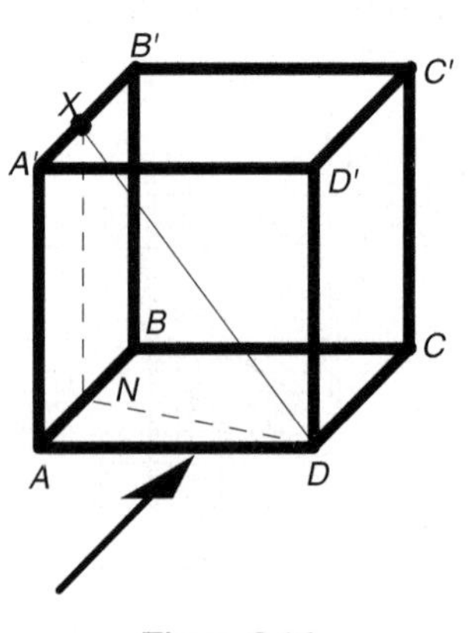

Figure 8.16

Solution

The plan and elevation are shown in Figure 8.17. In order to find the real distance DX, read the *horizontal* distance ND from the plan, and the *vertical* distance XN from the *elevation*.

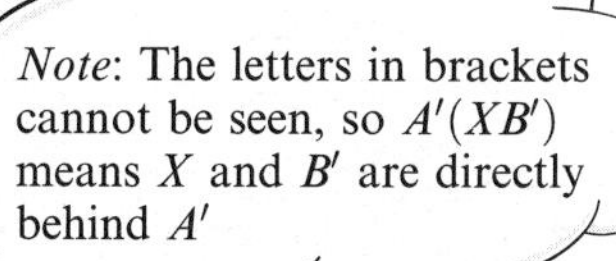

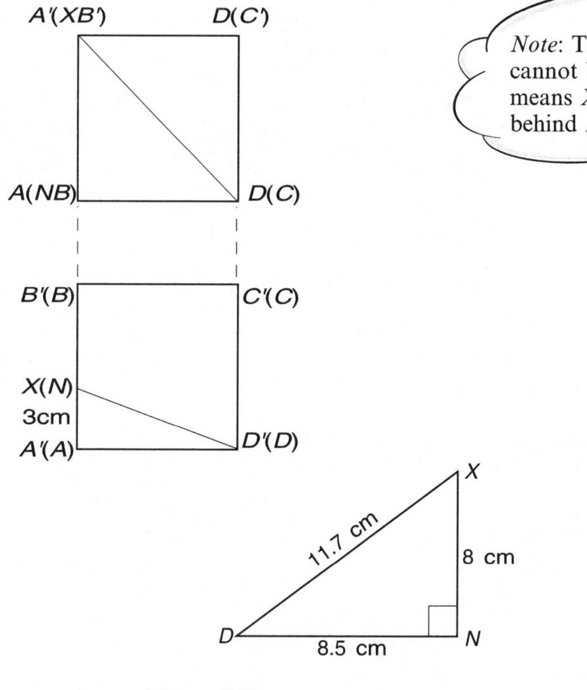

Figure 8.17

The triangle XDN can then be drawn, where N is the point vertically beneath X. The actual distance $XD = 11.7$ cm.

Exercise 8(c)

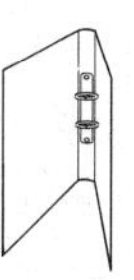

1 A pyramid with a square base stands on top of a cube. The sides of the pyramid are 50% of the length of the cube, and the height of the pyramid is the same height as the cube. Draw a suitable plan and front elevation of the combined solid.

2 A lamp shade is made by joining 3 circular hoops of wire of radius 20 cm, 25 cm and 30 cm, the narrowest being at the top, and all are held equally spaced in horizontal planes, a distance 10 cm apart, by three straight pieces of wire. Draw a plan and suitable elevation in order to find the length of the straight pieces of wire.

3 A rectangular block is cut in half so as to form two equal triangular prisms. The two prisms are then fastened together to form a new block, as shown in Figure 8.18.

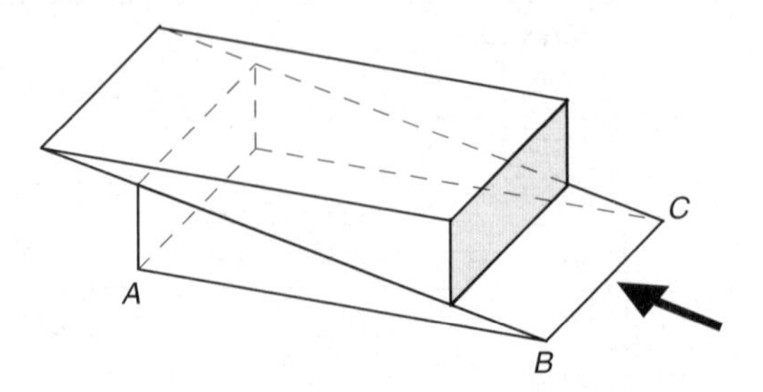

Figure 8.18

If the edges AB and BC are horizontal, sketch the elevation of the new block on a vertical plane parallel to BC as viewed

(i) in the direction of the arrow

(ii) in the opposite direction.

4 Figure 8.19 shows the front and side elevation of a solid object. Sketch a possible plan of this object.

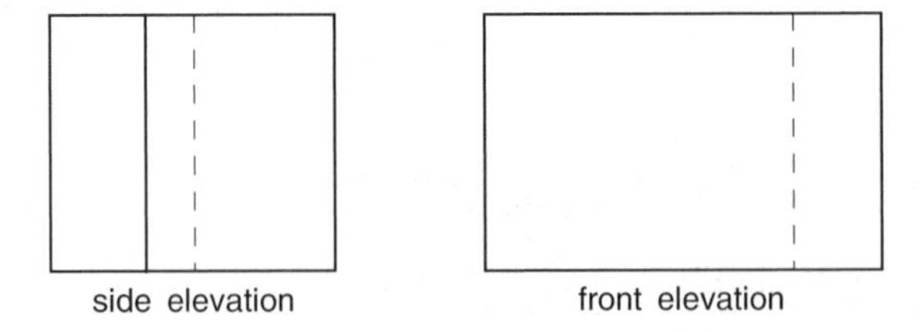

Figure 8.19

5 Three snooker balls are placed on a table, and a fourth rests on the top to touch all three. Draw a plan of the four balls.

8.4 Latitude and longitude

Another problem in three-dimensional work is the study of the earth's surface. It is considered as a sphere of radius 6370 km with a series of circles drawn on the surface.

(i) Any circle on the surface of the earth whose centre is at the centre of the earth is called a great circle. The **Equator** is a great circle, and also all circles of **longitude** (or **meridians**) which pass through the North or South Poles. All great circles have a radius of 6370 km.

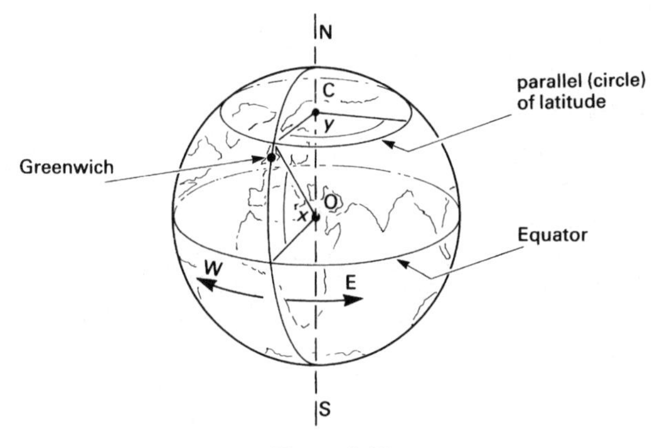

Figure 8.20

(ii) Circles on the surface which are parallel to the Equator are **parallels of latitude**. As you get nearer to the North or South Poles, the radius of a circle of latitude gets smaller.

(iii) The position of any point on the earth is given by the angle East or West of the circle which passes through Greenwich (the Greenwich meridian), and the angle North or South of the Equator. For example, (12°W, 25°N) has an **angle of longitude** 12°W, and **angle of latitude** 25°N.

(iv) A degree can be subdivided into 60 minutes (written 60′). On a great circle, each minute of arc is equal to a distance of 1 **nautical mile**.

(v) Figure 8.21 shows a cross-section through the earth. C is the centre of a circle of latitude (see also Figure 8.20). Since $\triangle ACO$ is a right angled triangle,

$$\frac{AC}{OA} = \cos x^\circ$$

So $AC = OA \cos x^\circ$

i.e. $r = R \cos x^\circ$.

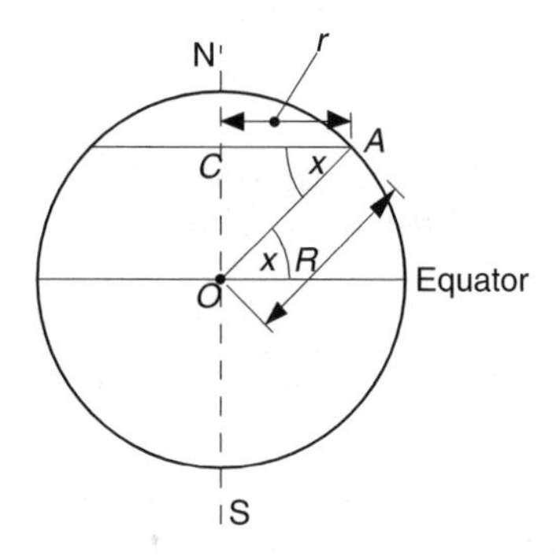

Figure 8.21

Rule

Hence the radius of the circle of latitude = radius of the earth × cosine (angle of latitude)

Example 8.4

(a) Find the distance in nautical miles between A(15°W, 63°20′N) and B(15°W, 44°18′N);

(b) Find the distance in nautical miles between E(16°W, 25°S) and D(18°20′E, 25°S).

Solution

(a) Since both points are 15°W, we can travel along a circle of longitude (see Figure 8.22).

Always produce 2 dimensional diagrams

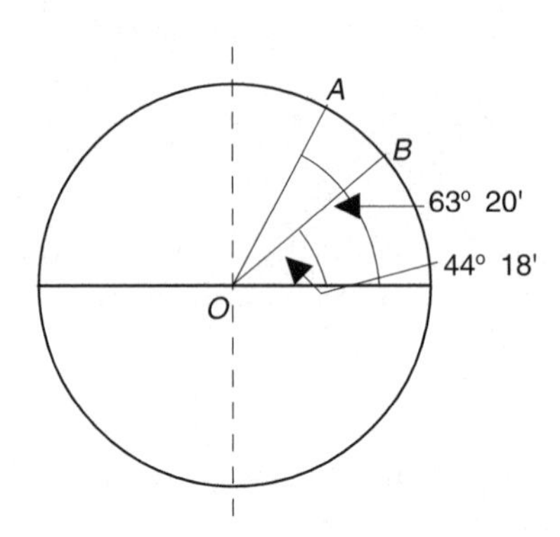

Figure 8.22

$$\text{The angle AOB} = 63°20' - 44°18'$$
$$= 19°2'$$

Changing this to minutes gives $19 \times 60 + 2 = 1142$

The distance = 1142 nautical miles.

(b) Here, both points are 25°S, so you can travel along a *circle of latitude*.

$$\text{The angle } ECD = 16° + 18°20'$$
$$= 34°20'$$

$$\text{Changing this to minutes} = 34 \times 60 + 20$$
$$= 2060$$

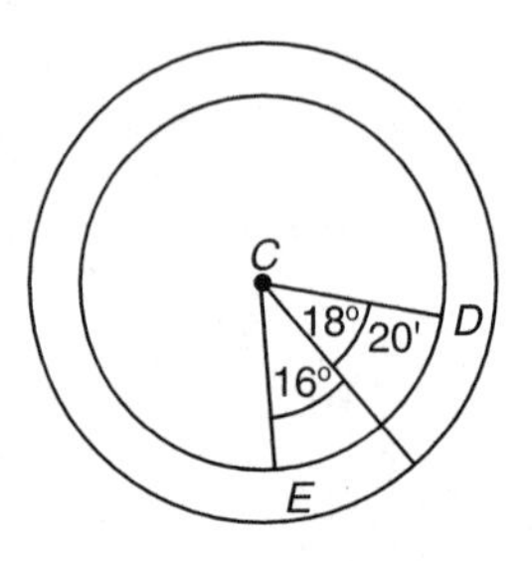

Figure 8.23

This circle is smaller than the Equator by a factor of cos (the angle of latitude) = cos 25°.

$$\text{Hence the distance} = 2060 \cos 25° \text{ nautical miles}$$
$$= 1867 \text{ nautical miles.}$$

Problems where the distances are not measured in nautical miles are slightly more difficult, and require the use of the length of an arc of a sector covered in Section 6.2(vi).

Example 8.5

(a) P is the point (12°E, 25°N) and Q is the point (12°E, 18°8′S). Find the distance between P and Q measured along a circle of longitude. Assume the radius of the earth is 6370 km.

(b) A is the point (18°E, 60°N), B is the point (12°W, 60°N). Find the distance in km measured along a circle of latitude between A and B.

Solution

(a) Since P and Q are on *opposite* sides of the Equator, the angle between P and Q is

$25^\circ + 18^\circ 8' = 43^\circ 8' = 43\frac{8}{60}{}^\circ$

$= 43.13^\circ$ (Figure 8.24).

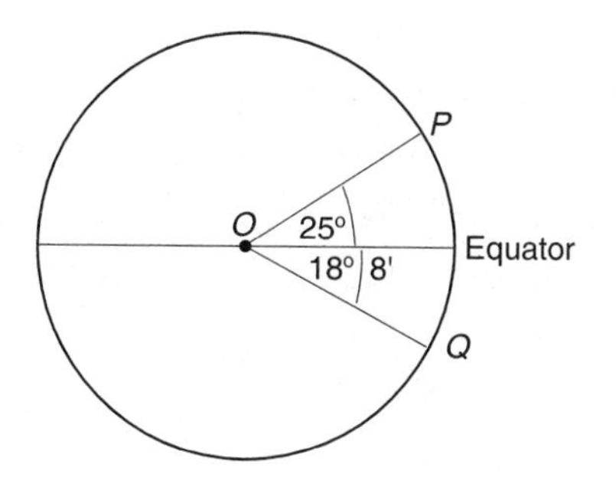

Figure 8.24

$$\text{The length of arc} = 2\pi \times 6370 \times \frac{43.13}{360}\ \text{km}$$
$$= 4795\ \text{km}.$$

(b) The angle between A and B is

$12^\circ + 18^\circ = 30^\circ$

Since we are on a circle of latitude 60°N, the radius of the circle of latitude

$= 6370 \cos 60^\circ$ km

$= 3185$ km

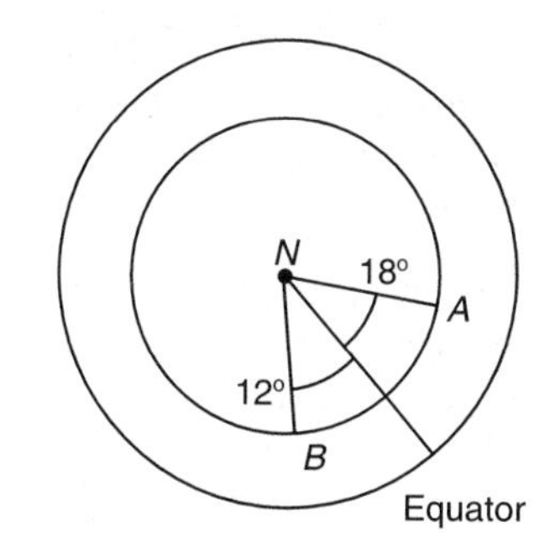

Figure 8.25

$$\text{The length of arc} = 2\pi \times 3185 \times \frac{30}{360}$$
$$= 1668\ \text{km}.$$

Exercise 8(d)

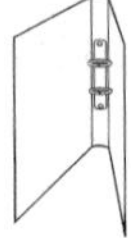

[*Note*: Take the radius of the earth = 6370 km]

1 Calculate the distance round the Equator in nautical miles.

2 Quito and Kampala lie on the Equator and the difference in their longitudes is 111°. Calculate the distance between them in nautical miles.

3 The distance between two places on the Equator is 500 km. Find the difference in their longitudes.

4 A ship sails due North at a speed of 30 knots. How long does it take for the ship to change its latitude by 3°? (1 knot = 1 nautical mile/hour).

5 Find the length of the Arctic Circle, latitude $66°32\frac{1}{2}'$N (i) in nautical miles; (ii) in km.

6 Find the distance along the circle of longitude between John O'Groats (3°W, 58°30′N) and Lyme Regis (3°W, 50°30′N) (i) in nautical miles; (ii) in km.

7 Calculate the distance between two places S and T which are on the same line of latitude of 30°N but whose longitudes differ by 20°.

8 The longest stretch of straight railway track in the world is from Nuringa (126°E, 31°S) to Ooldea (132°E, 31°S) in Australia. Calculate its length.

9 State the latitude and longitude of the place which is at the opposite end of a diameter of the earth from Greenwich (0°W, 51°N).

10 A ship sails 300 nautical miles due West and finds that her longitude has altered by 5°. What is her latitude?

11 Two places P and Q are both on the line of latitude 50°N. The longitudes of P and Q are 15°E and 70°W respectively. A plane leaves P and flies to Q at a speed of 300 knots. Calculate how long the journey takes.

12 A satellite circles the earth in such a way that it remains stationary above a point A on the Equator. B is a place due North of A in latitude 65°N. If B is the furthest point North at which the satellite can be seen with a telescope, what is the height of the satellite above the earth's surface?

13 Assuming the earth to be a sphere of radius 6370 km, calculate:

(i) the radius in km of the parallel of latitude passing through Paris (49°N, 2°E) and Vancouver (49°N, 123°W)

(ii) the length in km of the journey from Paris to Vancouver travelling due West

(iii) the shortest distance in nautical miles on the earth's surface from Paris to the North Pole

(iv) the latitude and longitude of a point, 1500 nautical miles from the North Pole, which is equidistant from Paris and Vancouver.

8.5 Locus

The word **locus** is the name given to a set of points that satisfy some conditions. In Figure 8.26, a piece of string with a paintbrush P on the end is tied to a fixed peg O in the ground. If we move the rope keeping it taut, the brush will trace out a set of points of the ground in the shape of a circle. Hence the locus of P which moves so that OP is constant and O is a fixed point, is a circle. If we could swing the rope freely in 3 dimensions keeping the rope taut, the brush would follow the surface of a sphere.

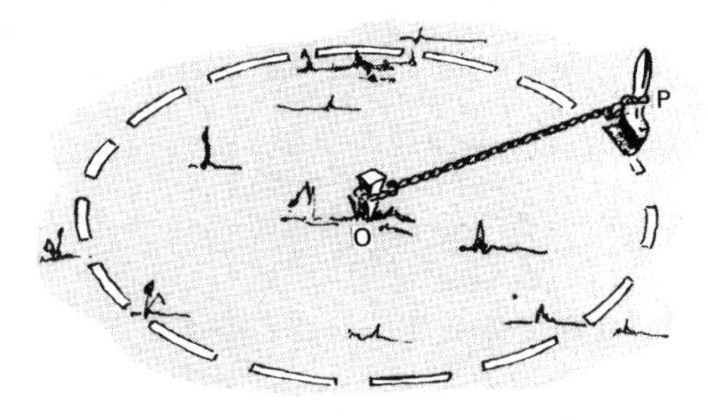

Figure 8.26

The locus of a point P which moves in 2 dimensions so that it is always the same distance from 2 fixed points A and B would be the **perpendicular bisector** of AB (see Figure 8.27). In 3 dimensions, this would be a plane.

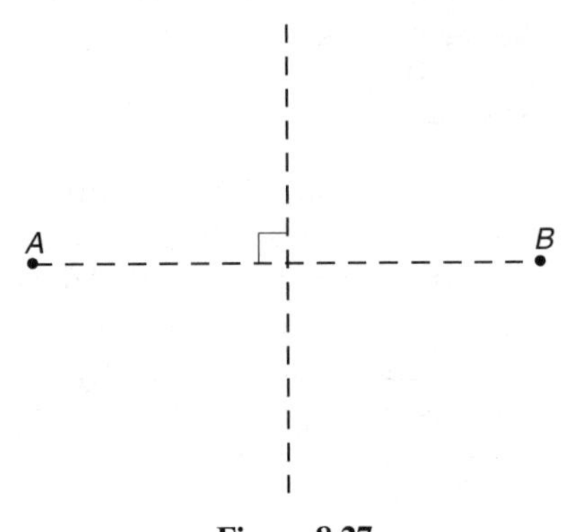

Figure 8.27

Sometimes, the exact form of the locus is not obvious, look at the following example.

Example 8.6

Draw a line AB 6 cm long. As accurately as you can, plot a series of points P on the paper so that $\angle APB = 60°$. Try and describe the locus of these points.

Solution

Since $\angle APB = 60°$, we need to draw several triangles so that $\angle PBA + \angle PAB$ is always 120°. Three points are shown in Figure 8.28, namely P_1, P_2 and P_3. Accurate plotting of other points would give two arcs of a circle. A proof of this will be given in Section 14.1(iii).

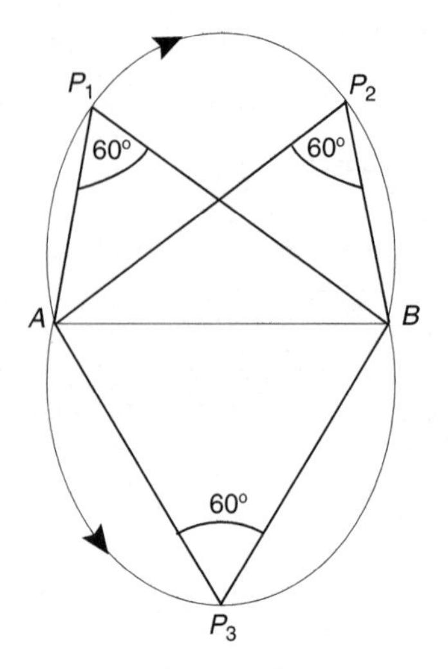

Figure 8.28

The word locus need not necessarily apply to points on a line, it may apply to points in a **region**.

Example 8.7

$ABCD$ is a square of side 5 cm. P is a point inside the square which satisfies two conditions: (i) it is nearer to A than B; (ii) it is nearer to B than D. Shade the region that P must be in.

Solution

Since P is nearer to A than B, it must be to the left of the line through the midpoints of AB and DC. It is also nearer to B than D, so it must be above the diagonal AC. The shaded region is the required region (see Figure 8.29).

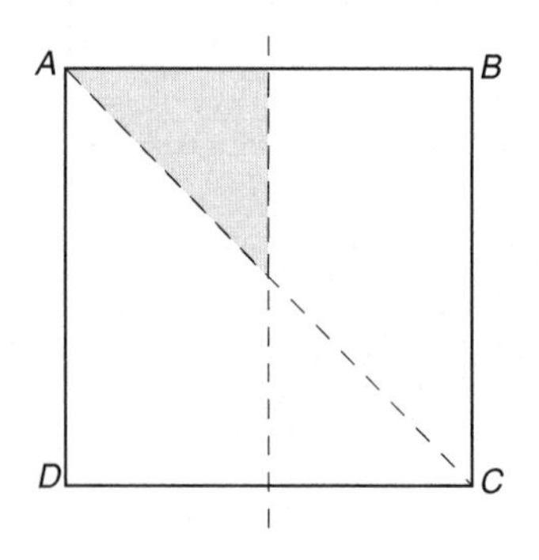

Figure 8.29

Exercise 8(e)

1 AB is a line of length 8 cm. Describe the locus of a point P which moves so that $\angle PAB = 30°$ (i) in two dimensions; (ii) in three dimensions.

2 Figure 8.30 shows a house in a garden. John (J) stands in the garden, and somewhere else in the garden is the cat Pedro, which cannot be seen by John. Make an accurate copy of the diagram and shade the region where Pedro is. Estimate the percentage of the garden that the cat could be in.

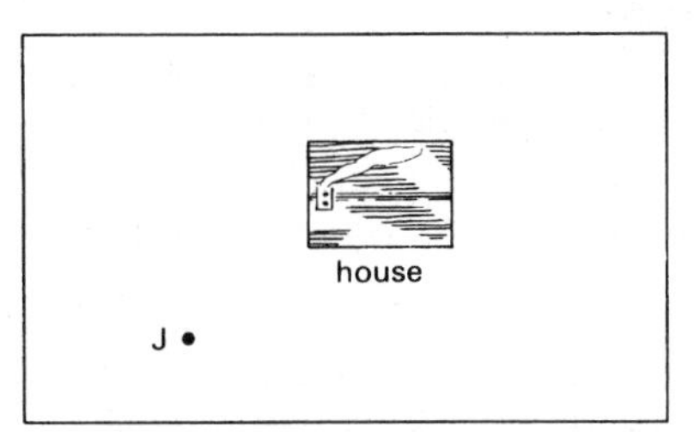

Figure 8.30

3 $ABCD$ is a rectangle with $AB = 5$ cm, and $BC = 4$ cm. Shade the region in which a point lies if it is never more than 1 cm from the centre of the rectangle, and it is nearer to A than C. Find the area of this region.

4 Ruler and compasses only may be used in this question. In a single diagram construct:

(i) a triangle PQR in which the base $QR = 7$ cm, $PQ = 11$ cm and $\angle PQR = 30°$

(ii) the locus of points equidistant from P and R

(iii) the circle which passes through P and R and has QR as the tangent at R.

Measure the radius of this circle.

5 A and B are two fixed points with $AB = 10$ cm. Sketch as accurately as you can the path followed by a point P which satisfies the condition that $PA + PB = 16$ cm.

6 How could you describe a solid cylinder using the ideas of locus developed in this section?

9 Statistics

9.1 Statistical data

Statistics is concerned with the collection and interpretation of numerical information. Numerical information is known as **data**. The collection of data is made by the use of taking a **sample** from the **population** being considered.

Information which can be collected by *counting* is called **discrete** data. For example, the number of children in a family, or the number of goals scored by a hockey team each week. **Continuous** data is collected by *measurement*, and includes such information as height or weight.

The general name given to the quantity being investigated is a **variable**. When information is first collected, it is usually obtained in no particular order, but written down as it is collected. In this form, it is known as **raw data** or **crude data**. When the data has been arranged in order or **grouped** into a small number of **classes** it is called **classified** data.

The word population in statistics does not necessarily refer to people, but to all possible values of the variable being studied. When taking a representative sample from a population, it is important that each member of the population should have an equal chance of being selected. Such a sample is called a **random sample**. Sometimes, it is important to select a sample in **layers**. For example, if you were investigating the average weekly wage of people in Great Britain, your sample would have to include the correct proportion of wage earners in all categories (high, medium and low wage earners). Such a sample is called a **stratified sample**.

Now look at the following Investigation.

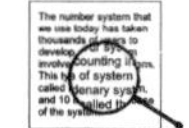

Investigation XVII: Surveys

If you are in a position to do so, it can be quite instructive to carry out a survey into a problem that interests you. There are many areas that can be looked at. Here are some suggestions.

(i) People's shopping habits
(ii) Car registration numbers
(iii) Likes in music or TV programmes
(iv) Wages earned in a particular type of employment
(v) Types of shops in a shopping centre
(vi) Speeds of cars

In order to carry out the survey, you need to look at a suitable *sample* (or cross-section). This sample must be as *random* as possible (everything has an equal probability of being in the sample), and it must be *representative*. This often needs the use of a *stratified* sample. For example, if you were looking at people's wages, you should include a certain proportion of low wage earners, middle wage earners and so on. (How do you think you would determine these proportions?)

The survey needs to be carried out by means of a *questionnaire*. This can either be filled in by you or by the person being questioned. Questions must be (i) clear; (ii) not leading questions; (iii) answerable in a few words. When you have collected your results, they can be written up and presented using the sort of diagrams discussed in this chapter. Try to draw a conclusion. Often you may find your results surprising. Good luck!

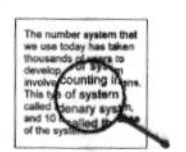

9.2 Statistical diagrams

(i) Bar graph (block graph or bar chart)

A bar graph consists of a series of parallel bars (horizontal or vertical) whose lengths are proportional to the number of items in the group.

Example 9.1

In studying a variety of pea, the number of peas in a sample of 40 pods was counted. The information was written on a piece of paper in the following way:

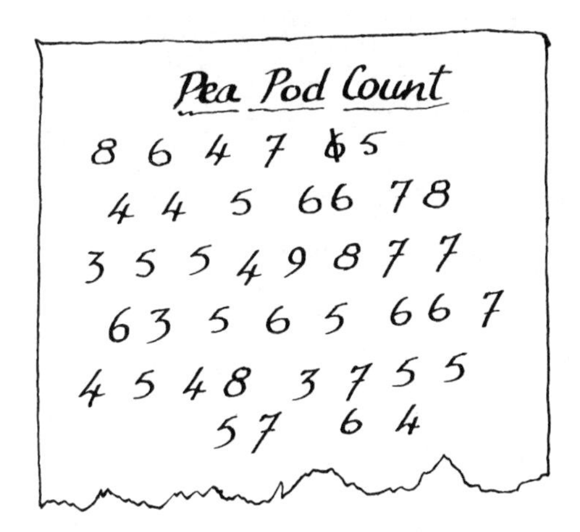

Figure 9.1

Classify this information into groups, and draw a bar graph to represent the information.

Solution

The information is best represented in the form of a table, called a **frequency** table (frequency means the number of times something happens). The table is obtained by a simple **tally**.

No. of peas	*Tally*	*Total frequency*
3	111	3
4	~~1111~~ 11	7
5	~~1111~~ ~~1111~~	10
6	~~1111~~ 111	8
7	~~1111~~ 11	7
8	~~1111~~	4
9	1	1

The bar graph can now be drawn (Figure 9.2) (the bars *do not have to touch each other*).

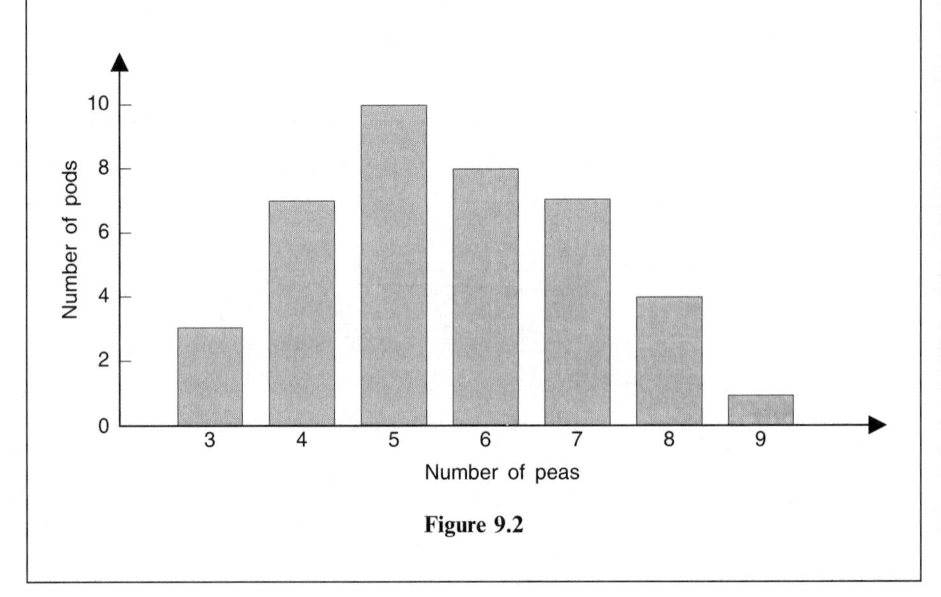

Figure 9.2

(ii) Pictograph (or pictogram)

A **pictograph** has much more visual effect. A small **motif** is used to represent a certain number of pieces of information.

Example 9.2

Represent the information in Example 9.1 by using a motif of [motif] to represent 2 pods.

Solution

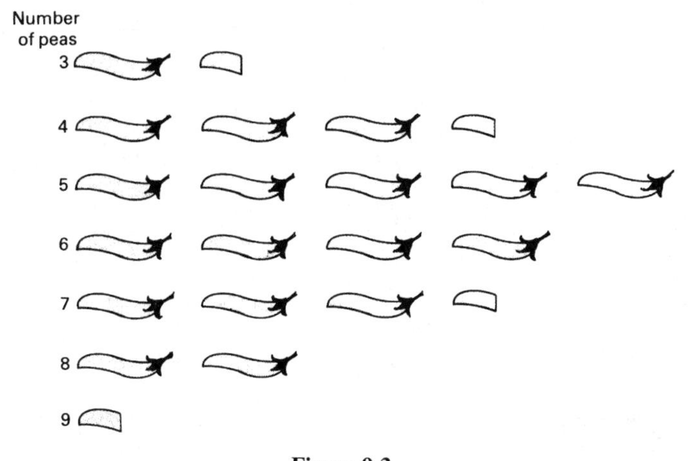

Figure 9.3

It can be seen that a major problem with this type of diagram is that when only part of the motif is needed, it is not clear how to divide it exactly. Very often, you have to guess what a part of the motif represents.

(iii) Pie chart

In this type of diagram, the total number of items of data is represented by the *area* of a circle. The size of each class is proportional to the angle of the sector representing it.

Example 9.3

The following table gives the results of friendly matches played by the local ladies' football team.

Result	*Win*	*Lose*	*Draw*	*Abandoned*
Frequency	7	10	5	2

Represent this information in a pie chart.

Solution

First find the total frequency $= 7 + 10 + 5 + 2 = 24$

Then find the proportion of 360° in each class.

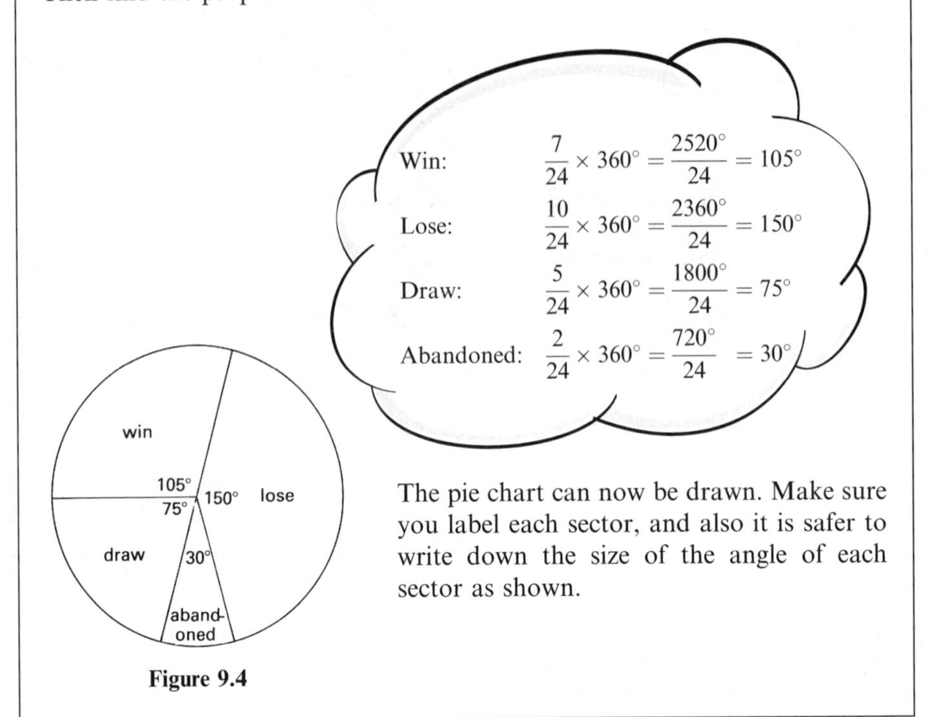

Figure 9.4

The pie chart can now be drawn. Make sure you label each sector, and also it is safer to write down the size of the angle of each sector as shown.

The following example shows you how to read a pie chart.

Example 9.4

A 5-litre tin of paint is made from the colours given in the pie chart in Figure 9.5. By careful measurement, estimate the volume of yellow needed.

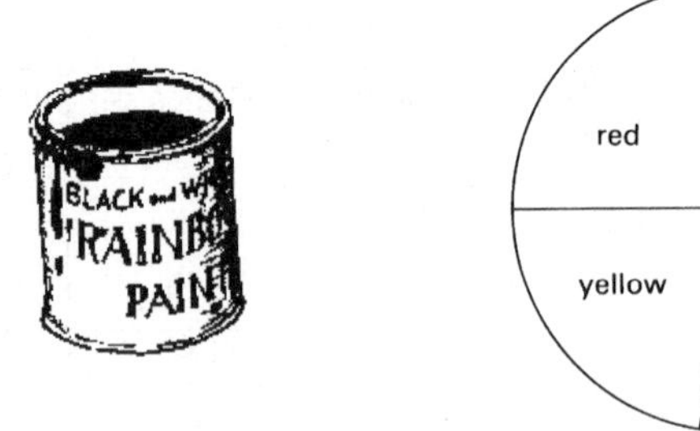

Figure 9.5

Solution

This question illustrates the problems with a pie chart. The diagram has probably not been drawn very accurately, and the angle for yellow appears to be about 80°.

Hence the volume of yellow

$$= \frac{80}{360} \times 5 \text{ litres}$$
$$= 1.1 \text{ litres.}$$

(iv) Histograms

A histogram (not to be confused with a bar chart) is the display of data in the form of a block graph where the *area* of each rectangle is proportional to the frequency. When the rectangles are the same width, their heights too are proportional to the frequency and the histogram is just a bar chart. An example of a histogram with unequal class intervals follows:

Example 9.5

The following table gives the distribution of the number of employees in 50 factories in the Midlands.

Number of employees	0–39	40–59	60–79	80–99	100–139
Number of factories	5	9	7	6	7

Construct a histogram to show the distribution.

Solution

(i) The widths of the five class intervals are

40 20 20 20 40

As the widths of the first and last intervals are twice the width of the other three, the heights of the rectangles are reduced in proportion, i.e. if the height of the second rectangle is 9 units, the height of the first rectangle is $\frac{5}{2} = 2\frac{1}{2}$ units and the height of the last rectangle is $\frac{7}{2} = 3\frac{1}{2}$ units.

Note: a histogram has a continuous scale along the horizontal axis. In this case, strictly speaking, the blocks should end at 39.5, 59.5, etc.

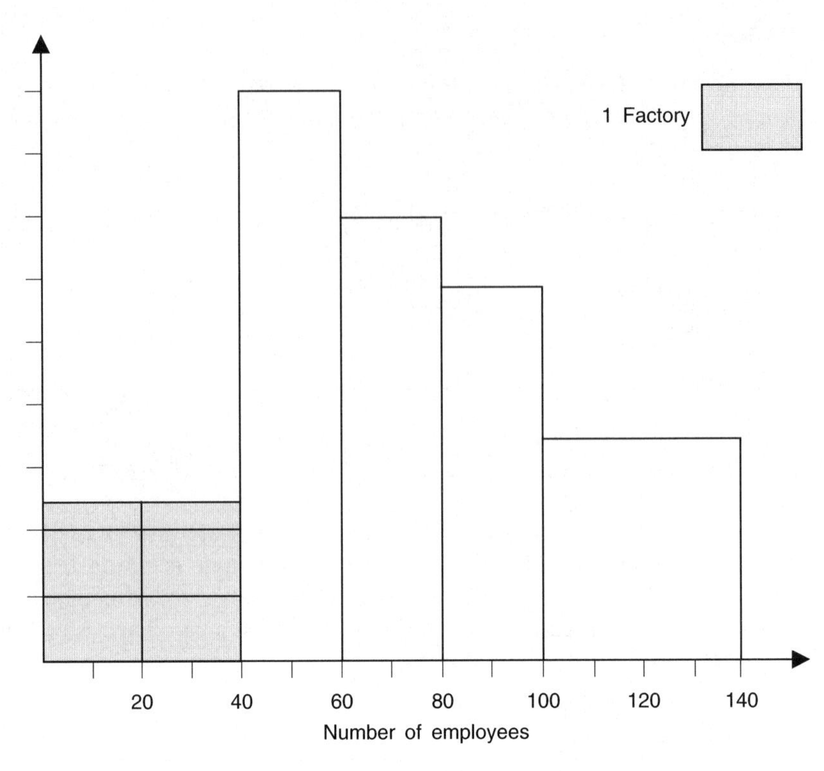

Figure 9.6a

You will notice that there is no scale on the vertical axis, but a key showing what area represents is given. You will see that 0–39 contains 4 units and two $\frac{1}{2}$ units, making 5 factories.

(ii) A more advanced way of labelling the vertical axis is to use the idea of *frequency density*.

Rule

Frequency density = frequency ÷ class width

In this case, the frequency densities will be

$$\frac{5}{40} = 0.125, \quad \frac{9}{20} = 0.45, \quad \frac{7}{20} = 0.35, \quad \frac{6}{20} = 0.3 \text{ and } \frac{7}{40} = 0.175.$$

This version of the histogram can be seen in Figure 9.6(b). To find the frequency for any of the blocks, you multiply the width of the block by its height.

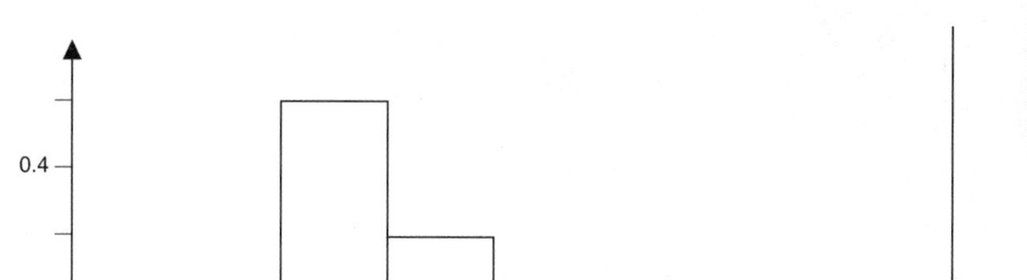

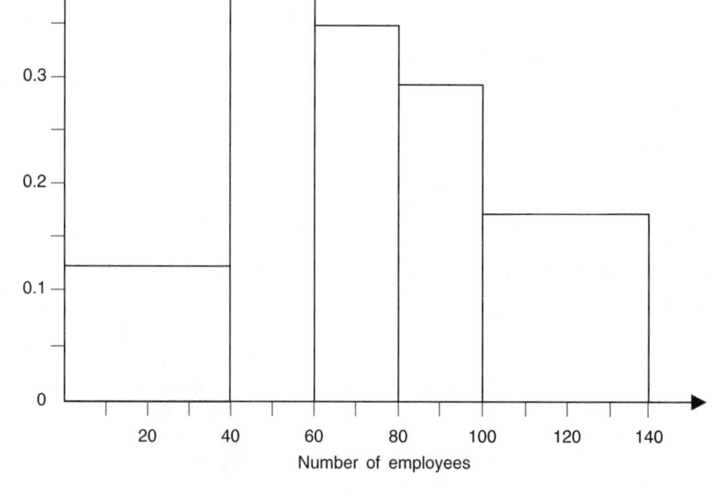

Figure 9.6b

Exercise 9(a)

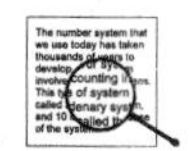

1 Of 120 pupils, 30 prefer French, 40 prefer German, 50 prefer Spanish. Illustrate this information by (i) a bar chart; (ii) a pie chart.

2 The number of newspapers sold daily in Saltash are:

Daily Mail	360	*Sun*	470
Guardian	80	*Daily Express*	400
The Times	70	*Daily Mirror*	420

Draw (i) a bar chart (ii) a pie chart stating clearly the angles for each sector to represent this information.

3 The strength of the armed forces in Moldavia is as follows:

Navy	*Army*	*RAF*	*Reserve*
1 700	14 000	4 800	3 500

Represent this information in (i) a bar chart; (ii) a pie chart. What percentage of the armed forces are on reserve?

4 1620 candidates took an examination, the results of which were graded 1, 2, 3 and Fail. If a pie chart was drawn, how many students failed if the angle representing 'Fail' was 78°?

5 The local supermarket in Washington sells five brands of washing powder. In one week the following number of boxes were sold:

Brand of washing powder	*A*	*B*	*C*	*D*	*E*
Number of boxes	130	78	265	126	221

Represent this information on a pie chart, marking in the sizes of the angles of the various sectors.

6 On one evening in a club, there are four interest groups: Sports, Music, Gymnastics and Drama. All members of the club are present and each member is in one and only one interest group.

The pie chart represents the numbers in each group.

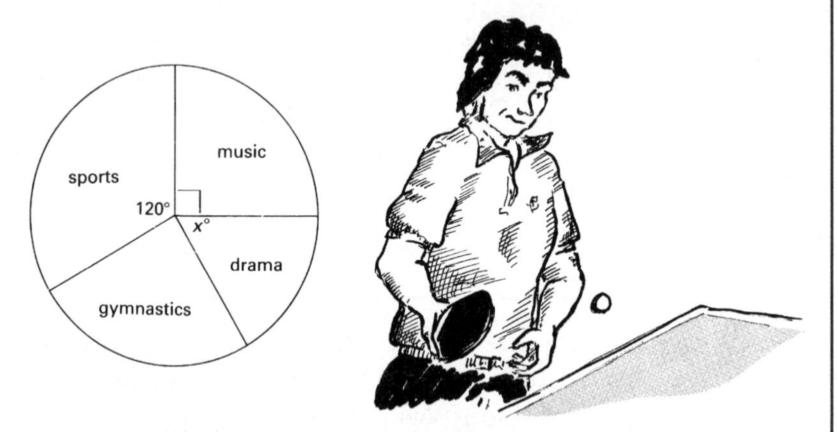

Figure 9.7

(i) There are 80 members in the Sports group. Calculate the total membership of the club.

(ii) What percentage of the total membership were in the Music group?

(iii) There were 48 members in the Drama group. Calculate the size of the angle marked $x°$.

(iv) Calculate the number of members in the Gymnastics group.

7 The results of goals scored by various teams of the football league one Saturday are given in the following table:

Goals scored	0	1	2	3	4	5
Number of teams	4	8	4	3	2	1

(i) How many matches were played by these teams on the given Saturday?

(ii) Find the total number of teams which scored at least 2 goals.

(iii) Find the total number of goals scored by all the 22 teams on the given Saturday.

(iv) Use a scale of 1 cm to represent 1 team and draw a bar chart to illustrate these results.

9.3 Statistical averages

Consider the following statements.

Gareth White, a cricketer: 'My batting average is 48.3'.

Mrs Peters, the owner of a shoe shop: 'The most popular size of women's shoe I sell is size 7'.

Ranjit, a fifth year student: 'I hope I finish in the top half of my mathematics group'.

All of these people are referring to different types of **average**. (An average is a value that represents a group of data.)

Gareth White is using the **arithmetic mean** (or just the **mean**).

To find the mean of the numbers 3, 9, 11, 28, 37, 20, proceed as follows:

$$\text{mean} = \frac{3+9+11+28+37+20}{6} = \frac{108}{6} = 18.$$

There are six numbers
$\therefore \div$ by 6

There are occasions when some of this working can be simplified. To find the mean of 505, 511, 525, 502, 517, we use what is called a **working mean** or **assumed mean** of 500.

Find the difference between the values and the working mean, and then we have:

$$\text{mean} = 500 + \frac{(5+11+25+2+17)}{5}$$

$$= 500 + \frac{60}{5} = 512.$$

Mrs Peters is concerned with the most popular value. This is known as the **mode**, or **modal value**.

Hence to find the mode of the numbers 2, 3, 3, 4, 2, 5, 4, 3, 4, 2, 3, since the number 3 occurs more than any other value, we have:

mode = 3.

Ranjit is using the idea of the middle value, or the **median**.

To find the median, the values have to be arranged in order of size. To find the median of the numbers 3, 7, 2, 9, 8, 11, 2, first arrange the numbers in order of size:

2 2 3 7 8 9 11

↓ middle (7)

So median = 7.

There is one small problem with finding the median. If you want to find the median of the numbers 5, 1, 25, 73, 2, 67, arranging in order of size, we have:

1 2 5 25 67 73

↓ middle (between 5 and 25)

There are two middle numbers, 5 and 25

The median is taken as $\frac{5+25}{2} = 15$.

Sometimes it is difficult to find the different forms of average from raw data, and it is helpful to tabulate the data in a frequency table.

Example 9.6

Find: (a) the mode (b) the median (c) the mean of the following set of figures:

3, 7, 9, 2, 4, 7, 5, 4, 5, 7,

8, 9, 7, 2, 3, 6, 5, 8, 3, 2,

4, 6, 8, 7, 3, 5, 5, 5, 6, 3.

Solution

First put the information into a frequency table:

(1) *No.*	(2) *Tally*	(3) *Frequency*	(4) *No.* × *Frequency*
2	111	3	6
3	~~1111~~	5	15
4	111	3	12
5	~~1111~~*11	6	30
6	111	3	18
7	~~1111~~	5	35
8	111	3	24
9	11	2	18
	Total	30	158

(a) Clearly the mode = 5.

(b) The starred position in the tally column is the median.
Therefore the median = 5 (not 5.5)

(c) To calculate the mean, we construct a further column to the table headed '*No.* × *Frequency*', and the mean is found by dividing the total of column 4 by the total of column 3:

$$\text{mean} = \frac{158}{30}$$
$$= 5.3 \text{ (1 d.p.).}$$

Very often, our frequency table does not consist of single values, but groups of values. The data is then put into a **grouped** frequency table.

Example 9.7

A group of people were asked to estimate a time interval of 60 seconds. Their guesses were as follows:

37	63	47	41	59	47	79	51	43	83
75	45	53	31	52	52	41	51	44	66
73	63	65	57	58	36	51	66	49	66

Put this data into groups 30–39, 40–49, etc. and estimate the average value.

Solution

Time	*Tally*	*Frequency*	*Middle value of class*	*Frequency × middle value*
30–39	111	3	34.5	103.5
40–49	~~1111~~ 111	8	44.5	356.0
50–59	~~1111~~ 1111	9	54.5	490.5
60–69	~~1111~~ 1	6	64.5	387.0
70–79	111	3	74.5	223.5
80–89	1	1	84.5	84.5
		Total 30		1645.0

The middle value of each group is the average of the lower and upper boundaries.

So $(30 + 39) \times 2 = 34.5$ is the middle value of the 30–39 group.

We are assuming each value in the group takes the middle value.

$$\text{mean} = \frac{\text{total of (frequency} \times \text{middle value)}}{\text{total frequency}}$$

$$= \frac{1645}{30} = 54.8.$$

Exercise 9(b)

1 Find the mean of the figures 8, 1, 0, 3, 15, 6.

Deduce the mean of the figures 58, 51, 50, 53, 65, 56.

2 Find the mean of the figures -3, -3, -11, -5, 7, -1.

Deduce the mean of the figures 217, 217, 209, 215, 227, 219.

3 Calculate the mean of each of the following:

(i) 79, 85, 81, 78, 82, 88, 80

(ii) 52.3, 52.1, 52.9, 52.7, 51.8, 51.2

(iii) 5248, 5253, 5249, 5243, 5256, 5240, 5251

(iv) -0.95, -0.88, -0.91, -0.84, -0.95, -0.88.

4 Find the median of each following set of figures:

(i) 41, 26, 27, 64, 72, 65, 85, 20, 41

(ii) 32, 72, 99, 44, 57, 71

(iii) 80, 74, 74, 91, 66, 7, 51, 26, 59, 83

(iv) −13, 3, −2, 12, −11, −3, 4, 3.

5 Calculate: (i) the mean (ii) the median (iii) the mode of the following sets of figures:

(a) 3, 4, 4, 4, 5, 8, 8, 9, 9

(b) 10, 19, 17, 14, 12, 19, 18, 14, 16, 17, 19, 11.

6 The mean age of a class of 20 boys is 12 years 6 months. Four new boys join the class, their average age being 12 years. By how much is the average age of the class lowered?

7 Write down a set of five integers:

(i) whose mean is 7 and whose median is 5

(ii) whose median is 7 and whose mean is 5.

8 The number of trains per hour through Stenworth station for a period of 24 hours were as follows:

3 5 2 1 2 5 7 4

2 6 3 5 6 2 5 1

4 7 3 2 3 2 5 6

Find (i) the mean (ii) the mode (iii) the median of the distribution.

9 The results of an examination taken by 100 candidates are given in the following table:

Mark obtained	0–19	20–39	40–59	60–79	80–99
Number of candidates	10	15	35	27	13

(i) Draw a histogram to illustrate these results

(ii) Compute an estimation for the mean.

10 The following is a set of nine numbers:

4, 7, 16, 13, 2, 7, 14, 7, 20

(i) State the mode of this set.

(ii) Find the mean of the set of nine numbers.

(iii) If another number x is included in the set, the mean of the ten numbers is $10\frac{1}{2}$. Find the value of x.

11 The following table gives the score of 50 throws at a dartboard:

38,	27,	32,	17,	45,	46,	31,	28,	78,	54
47,	51,	39,	5,	12,	19,	63,	58,	47,	47
56,	23,	31,	28,	25,	42,	3,	81,	56,	37
32,	29,	27,	3,	67,	54,	48,	21,	18,	22
17,	47,	35,	34,	34,	28,	57,	48,	36,	33

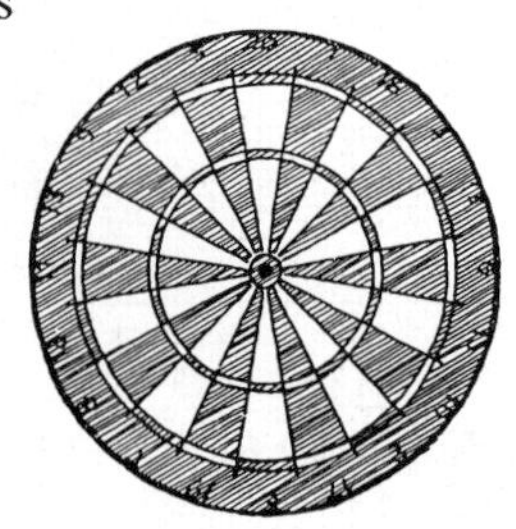

(i) Construct a histogram for the data using class intervals 0–9, 10–19, 20–29, etc. What is the modal class of this distribution?

(ii) Calculate an estimate for the mean score for the throw of a dart.

9.4 Cumulative frequency

A market research team did a door-to-door survey of 40 families to see how many children they had. The results were as follows:

0	2	3	1	2	4	0	1	3	3
3	3	4	2	3	1	3	2	1	2
2	3	3	2	5	2	0	4	3	1
5	3	2	3	1	3	2	2	1	3

We have already seen how to group these figures in a frequency table and how to construct a histogram. Now let us add an extra column to the frequency table headed 'running total' or **cumulative frequency**. The figure in a given row of the cumulative frequency column is computed from all the figures in the frequency column up to and including the corresponding row, e.g. $21 = 11 + 7 + 3$.

No. of children	*Tally*	*Frequency*	*Running total (cumulative) frequency*
0	111	3	3
1	1111 11	7	$7 + 3 = 10$
2	1111 1111 1	11	$11 + 10 = 21$
3	1111 1111 1111	14	$14 + 21 = 35$
4	111	3	$3 + 35 = 38$
5	11	2	$2 + 38 = 40$

If we plot the number of children against the running total, we have a cumulative frequency curve (see Figure 9.8).

Without considering the raw data we can estimate the median from the graph by reading off the half-way mark on the vertical scale. In the above graph we follow the dotted lines to see that the estimate for the median is 1.9, which would be given as 2 in this case in order to make sense.

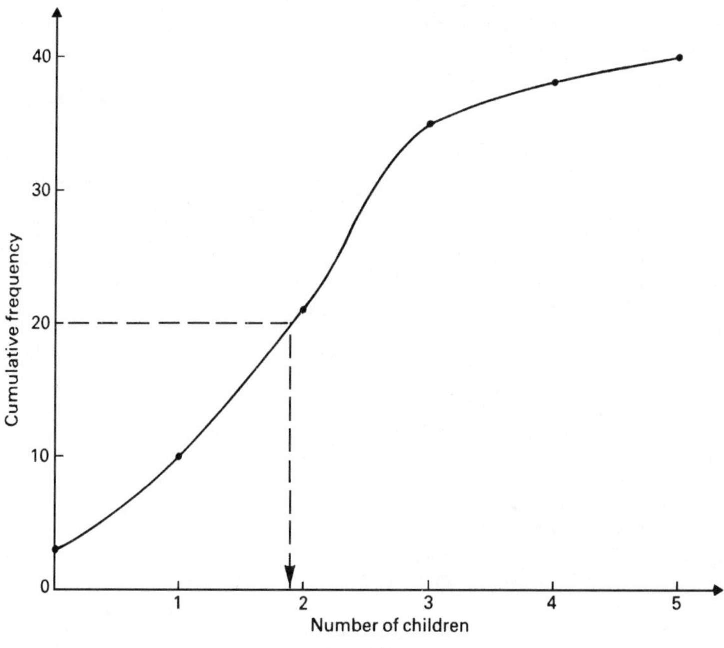

Figure 9.8

Example 9.8

100 commuters were interviewed to find the distance travelled to work daily to the nearest kilometre. From the results, the following table was constructed:

Distance in km	*Number of commuters*	*Cumulative frequency*
0–5	6	6
6–10	11	17
11–15	17	34
16–20	25	59
21–25	21	80
26–30	12	92
31–35	8	100

To be able to draw the cumulative frequency curve, we need to know the coordinates to plot. Taking the first class interval, 6 commuters travel a distance of between 0 and 5 km. As the distance is measured to the nearest km, the 6 commuters all travel less than 5.5 km.

Therefore the point to plot is (5.5 km, 6). Similarly, the next point is (10.5 km, 17) that is, 17 commuters travel less than 10.5 km.

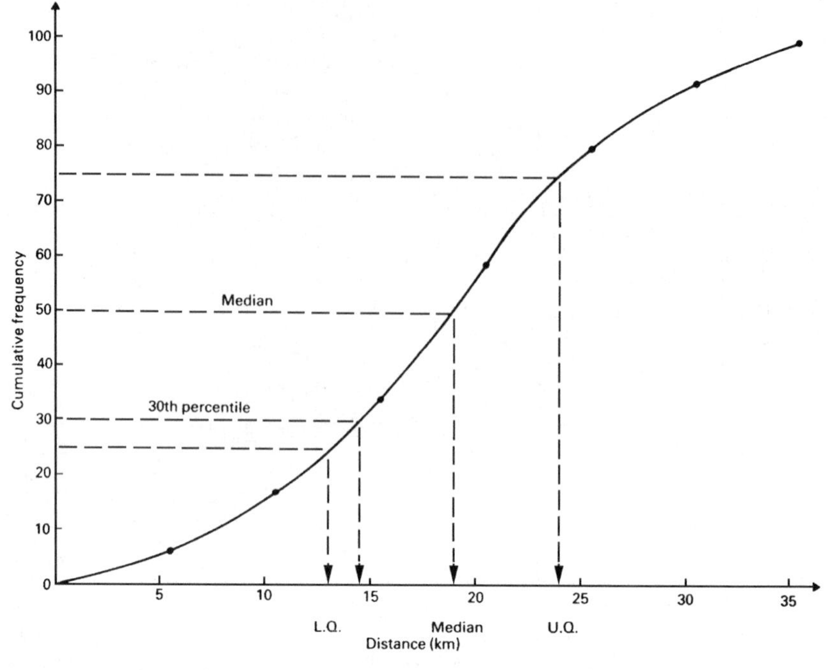

Figure 9.9

If we divide the number of commuters into *quarters*, then the *second* quarter position gives the median. An estimate for the median is 19 km. The distance travelled by the 25th commuter, 13 km, is known as the *lower quartile* (LQ). The distance travelled by the 75th commuter, 24 km, is known as the *upper quartile* (UQ). In general, we can find any *percentile* as follows. To find the 30th percentile in a sample of 100, you read from the cumulative frequency of 30, to give a value of 14.5. Hence 30% of the commuters travelled 14.5 km or less.

The spread of information could be measured using the **range** which is the difference between the highest and the lowest value. In this case:

$$\text{range} = 35 \text{ km} - 0 \text{ km} = 35 \text{ km}.$$

Sometimes the *interquartile range* is used to measure spread, and is defined as:

$$\begin{aligned}\text{interquartile range (IQR)} &= \text{upper quartile} - \text{lower quartile}\\ &= 24 \text{ km} - 13 \text{ km} = 11 \text{ km}.\end{aligned}$$

The semi-interquartile range (SIQR)

$$\begin{aligned}&= \text{half the interquartile range}\\ &= \tfrac{1}{2} \times 11 \text{ km} = 5.5 \text{ km}.\end{aligned}$$

The following investigation gives you a chance to use some of the techniques developed so far in this chapter.

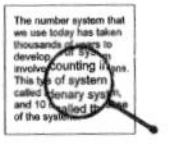

Investigation XVIII: Newspapers

THE TIMES

The Bishop of Durham, the Rt Rev David Jenkins, said that a thanksgiving service with any element of triumphalism would be obscene. In an interview with BBC Radio 4's *Sunday* programme, he said that any such service should be 'focused on repentance and looking forward' and should concentrate on the number of deaths, the destruction of the Iraqi infrastructure, and the environmental damage. He added: 'At the moment there is all this euphoria over a great victory and we should never have got into it.'

Church leaders are anxious to avoid a rift with the state similar to that perceived after the Archbishop of Canterbury's Falklands war sermon.

DAILY STAR

Quiet! For God's sake

THE Bishop of Durham was at it again yesterday.

This time we were spared his doubts about the religion he is paid to uphold. Instead, he sneered at the thought of Gulf War thanksgiving ceremonies.

It would be 'obscene' if they concentrated on victory, he tut-tutted. They should be full of 'repentance' for fighting the war at all.

What pompous poppycock from the spiritual leader of the Soft-on-Saddam Peacenicks!

Why shouldn't we celebrate such an amazing triumph? We can't be blamed if Iraq had more mouth than muscle. And what do we have to repent? Freeing a helpless little country from the clutches of a despot?

Yet again, the silly old buffer is out of touch and out of order.

Given his views, the Bishop probably cannot share in the nation's pride.

The diagrams show two extracts from well-known newspapers. The style of these newspapers is clearly different. The purpose of this investigation is to try and *quantify* the difference between newspapers. You should investigate some of the following ideas (or some of your own). The results should be presented in clear diagrams, enabling you to make quick comparisons.

(i) Pages allocated to different types of article
(ii) Number of pictures
(iii) Size of print
(iv) Average sentence length
(v) Number of pages.

Look at as many newspapers as you can. Write a short report as to whether your findings were what you expected or not.

Exercise 9(c)

1 Five coins were tossed 500 times, and the number of heads were recorded as follows:

No. of heads	0	1	2	3	4	5
Frequency	40	60	140	170	70	20

(i) Construct a cumulative freqency curve.

(ii) Estimate the median number of heads.

2 The number of people waiting at bus stops was as follows:

No. of people	0	1	2	3	4	5	6
No. of stops	3	8	12	15	13	7	2

Construct a cumulative frequency table and draw the cumulative frequency curve. From the curve find the median number of people per stop.

3 The following results were obtained in an examination:

Marks	0–20	21–40	41–60	61–80	81–100
No. of candidates	18	37	86	40	19

Construct a cumulative frequency curve and obtain, to the nearest whole number, the median mark and the semi-interquantile range.

4 An experiment was performed on 200 animals to estimate their surface area. The results obtained were as follows:

Area (cm^2)	0–2	3–5	6–8	9–11	12–14	15–17
Frequency	12	21	50	76	30	11

Rewrite this information as a cumulative frequency table and from it draw the cumulative frequency curve. Use this to estimate:

(i) the median area

(ii) the percentage of animals whose area was less than 10 cm^2

(iii) the semi-interquartile range of the areas.

5 The table below shows the cumulative mark distribution of 560 candidates in an examination. Construct a cumulative frequency curve, labelling your axes clearly. How many candidates fail if the pass mark is 45? What must the pass mark be if 60 per cent of the candidates are to pass? Determine the median of the distribution.

Mark	10	20	30	40	50	60	70	80	90	100
Cumulative frequency	18	43	78	130	240	372	462	523	552	560

6 The weights of 100 new-born babies are given, to the nearest 10 g, in the following table:

Weight (kg)	*Frequency*
1.00–1.49	3
1.50–1.99	8
2.00–2.49	12
2.50–2.99	22
3.00–3.49	27
3.50–3.99	18
4.00–4.49	9
4.50–4.99	1

Weight	*Cumulative frequency*
Not more than 0.995 kg	0
Not more than 1.495 kg	3

(i) Complete the cumulative frequency table above.

(ii) On graph paper, draw a cumulative frequency diagram to illustrate the data.

(iii) From your diagram, and making your methods clear, estimate

(a) the median weight

(b) the probability that a new-born baby chosen at random from this sample weighs between 2.4 kg and 3.2 kg.

[NEA]

9.5 Scatter diagrams

Another aspect of statistical work is to compare the *relationship* between one quantity and another. This can be done by a simple graph known as a scatter diagram. On this diagram is drawn a *line of best fit* (usually a straight line) and predictions can often be made from this line.

Example 9.9

An intelligence test was given to 8 different students of various ages (in months). The score obtained (S) and their age (A) are given in the following table:

Student	1	2	3	4	5	6	7	8
Age (A)	124	128	130	130	138	141	150	155
Score (S)	70	73	75	80	82	85	90	90

(a) Plot this data on a scatter diagram

(b) Find the average age and the average score

(c) Plot your answers to part (b) on the diagram, and hence draw the line of best fit

(d) Use your graph to estimate the score of a pupil aged 12 years.

Solution

(a)

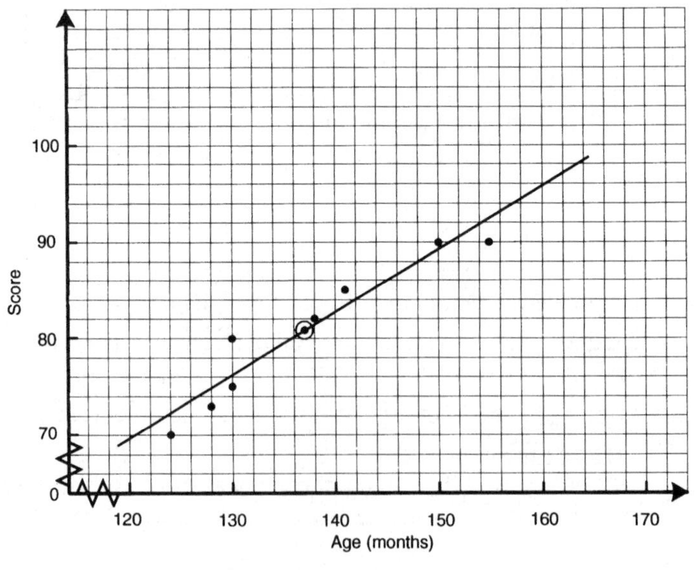

Figure 9.10

It doesn't really matter which way round the axes are drawn, although quantities involving time are usually plotted on the horizontal axis.

(b) Average age $= \frac{1}{8}(124 + 128 + 130 + 130 + 138 + 141 + 150 + 155)$
$= 137.$

Average score $= \frac{1}{8}(70 + 73 + 75 + 80 + 82 + 85 + 90 + 90)$
$= 80.6.$

(c) The average point has been plotted as a circled dot. The straight line is then drawn through this point so that the crosses are roughly evenly spread on either side of the line (practise makes perfect).

(d) 12 years = 144 months.

Reading from the line, 144 months gives a score of 85.

If the points are close to the line of best fit, then we say there is a high degree of correlation (cause and effect) between the two quantities. If both sets of data are increasing, the correlation is positive, if one quantity increases while the other decreases, the correlation is negative. Figure 9.11 shows examples of correlation.

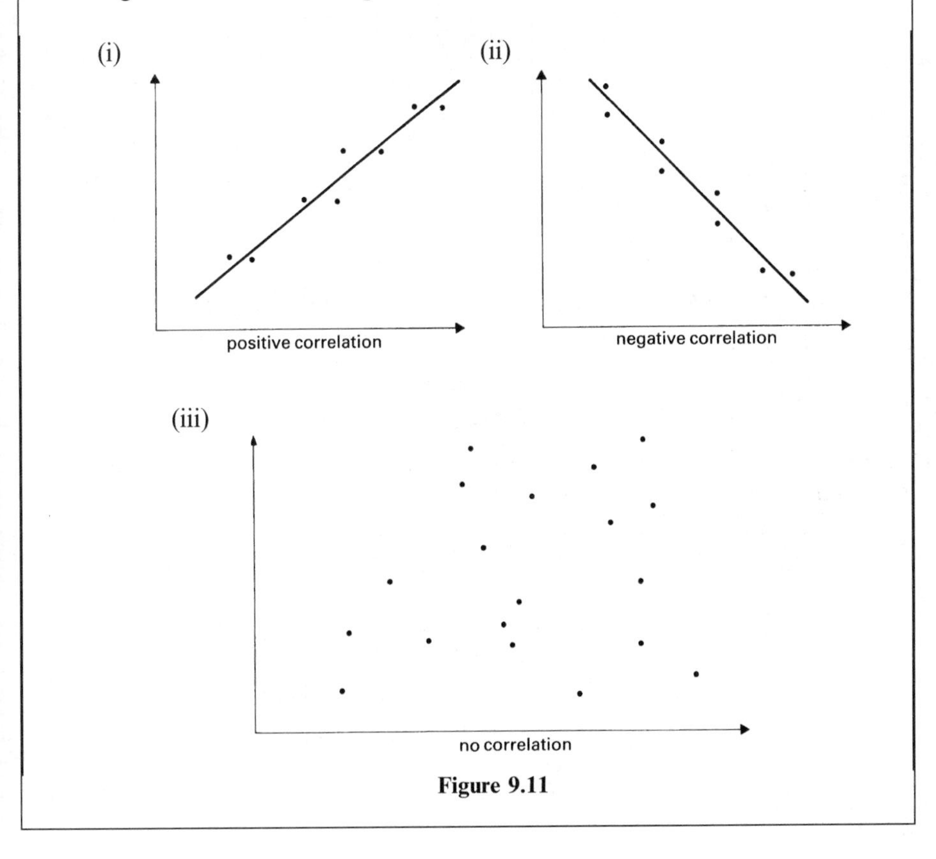

Figure 9.11

Exercise 9(d)

1 Salespeople are often assessed according to a grading scheme. An aptitude test was given to 10 sales staff and the results together with their annual sales are shown in the following table.

Salesperson	*A*	*B*	*C*	*D*	*E*	*F*	*G*	*H*	*I*	*J*
Aptitude score	8	9	12	11	8	11	14	16	12	9
Annual sales (£000)	20	25	30	29	22	28	32	40	25	26

(i) Draw a scatter diagram to illustrate this information

(ii) Draw the line of best fit

(iii) Estimate the annual sales of a new employee who scored 10 on the aptitude test.

2 The county hockey league consists of ten teams. At the end of the season, the number of points scored (two for a win and one for a draw) and goals conceded for nine of the teams were as follows.

Team	*A*	*B*	*C*	*D*	*E*	*F*	*G*	*H*	*I*	*J*
No. of points scored	34	28	27	26	20	15	8	5	4	4
No. of goals conceded	18	19	19	25	28	33	31	40	49	56

(i) Draw a scatter diagram to illustrate the information

(ii) Draw the line of best fit

(iii) How many points did team *K* score?

(iv) Estimate the number of goals conceded in the season.

3 The trading profits of a sweet-manufacturing firm from 1990 to 1994 are shown in the table below.

Year	1990	1991	1992	1993	1994
Profit £ (millions)	46	49	57	73	94

(i) Represent this information on a suitable diagram.

(ii) Hence estimate the profit to be made by the manufacturer in 1995, if this trend continues.

[NEA]

4 Paul measured the height of some seedlings in his greenhouse, together with the number of buds on the plants. His results are summarised in the following table:

Height (mm)	15	28	24	18	30	32	48	40	40	16
No. of buds	8	12	10	10	11	12	17	14	12	9

(i) Draw a suitable diagram to show this information.

(ii) Describe the type of correlation shown.

(iii) Draw a line of best fit through your results, and estimate the number of buds on a seedling height of (a) 10 mm (b) 35 mm. How valid do you think your results are?

5 A department store opens a snack bar and during the first four weeks records are kept of the number of cups of coffee sold each day. The store is open six days each week, Monday to Saturday. The results are as follows:

	M	*T*	*W*	*T*	*F*	*S*
Week 1	50	150	201	203	249	302
Week 2	148	240	301	321	375	421
Week 3	260	349	400	402	460	540
Week 4	380	440	460	495	520	580

(i) Plot the data for these four weeks on a scatter diagram, labelling the points carefully. Draw in lines to indicate the trends for Mondays and for Saturdays.

(ii) Use your trend lines to estimate the sales for Monday and for Saturday of the next week. What do you think will happen in future weeks?

6 The table below shows the income and expenditure of some charities in 1994 (million £).

Charity	*Income*	*Expenditure*
Oxfam	29.1	27.7
RNLI	22.7	17.6
Cancer Research	21.8	20.1
Save the Children	17.1	15.3
Help the Aged	13.2	10.6
Spastics Society	27.6	26.8
Christian Aid	11.2	11.7

(i) Plot a scatter graph to show the income and expenditure for these charities

(ii) Calculate the mean income

(iii) Calculate the mean expenditure

(iv) On your scatter graph draw in a line of best fit

(v) Write down a conclusion that you can draw from the graph.

[LEAG]

10 More algebra

10.1 Fractional and negative indices

(i) Rules

We have already looked at the meaning of a power when it is a positive or negative whole number, and we shall now look at the rules for using indices in algebra. We can summarise the work we have done so far as follows:

Rules

(1) $a^m \times a^n = a^{m+n}$

(2) $(a^m)^n = a^{mn}$

(3) $a^m \div a^n = a^{m-n}$

(4) $a^{-n} = \dfrac{1}{a^n}$

(5) $a^0 = 1$

Suppose now that the power is *not* a whole number, for example, what does $9^{1/2}$ mean? According to the rules above

$$9^{1/2} \times 9^{1/2} = 9^{1/2+1/2} = 9^1 = 9 \qquad \textbf{using rule (1)}$$

So $9^{1/2}$ must be 3, since $3 \times 3 = 9$. It follows that $9^{1/2}$ means $\sqrt{9}$.

Similarly $27^{1/3}$ means the cube root of 27, or $\sqrt[3]{27} = 3$.

In general, $a^{1/n} = \sqrt[n]{a}$ (the nth root of a).

We can now look at an expression such as $8^{2/3}$. This is the same as $\left(8^{1/3}\right)^2$ using rule (2).

$$\text{So} \quad 8^{2/3} = \left(8^{1/3}\right)^2 = (2)^2 = 4$$

It is useful to realise that another way of writing $8^{2/3}$ is $\left(8^2\right)^{1/3} = (64)^{1/3} = 4$.

We can summarise this as follows:

Rules

(6) $a^{p/q} = (a^{1/q})^p = (\sqrt[q]{a})^p$

(7) $a^{p/q} = (a^p)^{1/q} = \sqrt[q]{a^p}$

Example 10.1

Simplify the following:

(a) $1000^{2/3}$ (b) $4^{-1/2}$ (c) $25^{-3/2}$

(d) $(\frac{1}{4})^{-1/2}$ (e) $4x^3 \div x^4$ (f) $(2x^2)^3 \div (x^{-1})^2$

(g) $(ab^2)^2 \times 2ab$

Solution

(a) $1000^{2/3} = (1000^{1/3})^2 = 10^2 = 100$

(b) $4^{-1/2} = \dfrac{1}{4^{1/2}} = \dfrac{1}{2}$

(c) $25^{-3/2} = \dfrac{1}{25^{3/2}} = \dfrac{1}{(25^{1/2})^3} = \dfrac{1}{5^3} = \dfrac{1}{125}$

(d) $(\frac{1}{4})^{-1/2} = \dfrac{1}{(\frac{1}{4})^{1/2}} = \dfrac{1}{(\frac{1}{2})} = 2$ (1 divided by $\frac{1}{2}$)

(e) $4x^3 \div x^4 = 4(x^3 \div x^4) = 4(x^{3-4}) = 4x^{-1}$ or $\dfrac{4}{x}$

(f) $(2x^2)^3 \div (x^{-1})^2 = 2^3x^6 \div x^{-2} = 8(x^6 \div x^{-2}) = 8(x^{6--2}) = 8x^8$

(g) $(ab^2)^2 \times 2ab = ab^2 \times ab^2 \times 2ab = 2a^3b^5$.

(ii) The calculator

The buttons most likely to appear on your calculator for powers are $\boxed{x^y}$ or $\boxed{y^x}$ and $\boxed{x^{1/y}}$ or $\boxed{y^{1/x}}$, sometimes needing $\boxed{\text{INV}}$ to operate them. You can work out any problem about powers with these buttons.

Example 10.2

Use your calculator to evaluate:

(a) 16^5 (b) $1728^{1/3}$ (c) $64^{-2/3}$.

Solution

(a) $\boxed{16}$ $\boxed{x^y}$ $\boxed{5}$ $\boxed{=}$ **Display** $\boxed{1048576.}$

(b) $\boxed{1728}$ $\boxed{x^{1/y}}$ $\boxed{3}$ $\boxed{=}$ **Display** $\boxed{12.}$

(c) $\boxed{64}$ $\boxed{x^{1/y}}$ $\boxed{3}$ $\boxed{=}$ $\boxed{x^y}$ $\boxed{2}$ $\boxed{+/-}$ $\boxed{=}$ **Display** $\boxed{0.0625}$

We have had to write this as $(64^{1/3})^{-2}$.

If your calculator has a fraction button, you could proceed as follows:

$\boxed{64}$ $\boxed{x^y}$ $\boxed{2}$ $\boxed{a^{b/c}}$ $\boxed{3}$ $\boxed{+/-}$ $\boxed{=}$ **Display** $\boxed{0.0625}$

Exercise 10(a)

Work out the following:

1 $27^{1/3}$ **2** $49^{1/2}$ **3** $(\frac{1}{4})^0$

4 $8^{-1/3}$ **5** $125^{2/3}$ **6** $(x^4y^2)^{1/2}$

7 $(\frac{1}{4})^{-1/2}$ **8** $1000^{-1/3}$ **9** $9^{3/2}$

10 $(\frac{4}{9})^{-3/2}$ **11** $(4x^2)^{3/2}$ **12** $(2x)^2 \div 4x^3$

13 $64^{5/6}$ **14** $(\frac{1}{4})^{-3/2}$ **15** $(x^6)^{2/3}$

16 $169^{-1/2}$ **17** $(x^4)^{1/2} \div x^3$ **18** $27^{5/3}$

19 $\sqrt[3]{125^{-1}}$ **20** $(\frac{25}{9})^{-5/2}$

Use a calculator to evaluate the following (to 3 sig. fig.)

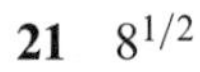

21 $8^{1/2}$ **22** $25^{1/3}$ **23** $1000^{-5/3}$

24 $\sqrt{0.875^3}$ **25** $\sqrt[3]{180}$ **26** $15^{-1/3}$

27 $(\frac{2}{3})^{-1/2}$ **28** $24^{-1/2}$ **29** $\sqrt[3]{0.01}$

30 $85^{1/5}$

10.2 Algebraic fractions

Since algebra is the use of letters to replace numbers, then working with algebraic fractions involves the *same* methods as working with ordinary fractions.

(i) Addition and subtraction

Remember *each* fraction must be expressed with the *same* denominator.

(a) $\frac{x}{a}+\frac{y}{b}=\frac{xb}{ab}+\frac{ay}{ab}=\frac{xb+ay}{ab}$ (The common denominator is ab)

(b) $\frac{1}{a}+\frac{1}{a^2}=\frac{a}{a^2}+\frac{1}{a^2}=\frac{a+1}{a^2}$ (The common denominator is a^2)

(c) $\frac{1}{x}-2=\frac{1}{x}-\frac{2x}{x}=\frac{1-2x}{x}$ (The common denominator is x).

(ii) Multiplication

(a) $\frac{a}{b}\times\frac{2}{x}=\frac{2\times a}{b\times x}=\frac{2a}{bx}$

(b) $\frac{2}{x}\times\frac{x}{t}=\frac{2\cancel{x}^{1}}{{}_{1}\cancel{x}t}=\frac{2}{t}$ (By cancelling)

You can also cancel before multiplying.

(c) $\frac{x^2}{y}\times\frac{2x}{3y}=\frac{2x^3}{3y^2}$ (This cannot be simplified).

(iii) Division

(a) $\frac{2}{x}\div\frac{4}{x}=\frac{2}{x}\times\frac{x}{4}$ (The second fraction is turned upside down (see Section 2.1(iv) *Division*))

$=\frac{2x}{4x}=\frac{1}{2}$

(b) $\frac{x}{a}\div\frac{2x}{b}=\frac{x}{a}\times\frac{b}{2x}=\frac{xb}{2ax}=\frac{b}{2a}$ (By cancelling)

(c) $4\div\frac{3}{x^2}=\frac{4}{1}\times\frac{x^2}{3}=\frac{4x^2}{3}$.

Exercise 10(b)

1 Find the LCM and the HCF of the following:

(i) a, b (ii) a, a^2 (iii) $4x, 2x^2$

(iv) t^2, at^3 (v) $2xy, xy^2$ (vi) abc, bcd.

2 Evaluate and simplify where possible:

(i) $\frac{1}{a}+\frac{1}{b}$ (ii) $\frac{2}{a}+\frac{3}{ab}$ (iii) $\frac{4}{t}-\frac{3}{t^2}$

(iv) $\frac{x}{2}+\frac{x}{3}+\frac{x}{4}$ (v) $\frac{4}{y}-\frac{2}{3y}+\frac{5}{y}$ (vi) $\frac{4x}{3y}-\frac{5y}{7x}$

(vii) $4x+\frac{1}{x}$ (viii) $\frac{(2a+5b)}{4}+\frac{(3a+2b)}{8}$

(ix) $\frac{x+2}{3}-\frac{5}{6}$ (x) $\frac{x}{y}\times\frac{2}{x}$ (xi) $\frac{a}{b}\times\frac{b^2}{a}$

(xii) $\frac{a}{b}\times\frac{b}{c}\times\frac{c}{d}$ (xiii) $\frac{x^2}{y}\div\frac{x}{y}$ (xiv) $\frac{2x}{y}\div\frac{1}{2}$

(xv) $\frac{2}{3}\times\frac{1}{x^2}$ (xvi) $5\div\frac{3}{x}$ (xvii) $\frac{2}{3x}\div 5$.

10.3 Quadratic expressions

In Figure 10.1 a rectangle $ABCD$ is divided into four rectangles by the lines EF and GH, which cross at X.

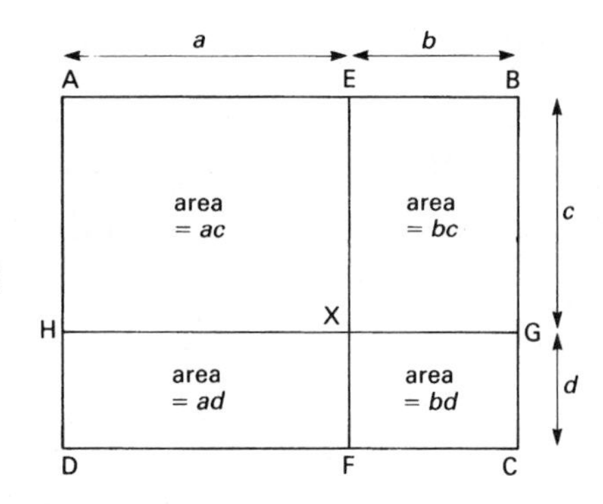

$$\text{The area of } ABCD = AB \times BC$$
$$= (a+b)(c+d)$$
$$\text{The area of } AEXH = ac$$
$$\text{The area of } EBGX = bc$$
$$\text{The area of } HXFD = ad$$
$$\text{The area of } XGCF = bd$$

Hence $(a+b)(c+d) = ac + bc + ad + bd$.

To remove the brackets, then, you multiply *each* term in one bracket by *each* term in the other bracket, and add the results. If the brackets contain negative signs a certain amount of care is needed. You should now study the following example.

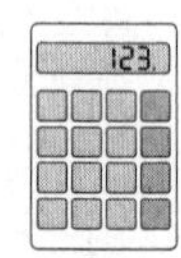

Example 10.3

Remove the brackets from the following expressions and simplify where possible:

(a) $(2x+y)(3x+y)$ (b) $(2t-3q)(t+2q)$

(c) $(x-2)(x-4)$.

Solution

(a) $(2x+y)(3x+y) = 6x^2 + 3xy + y^2 + 2xy$ ($3xy + 2xy = 5xy$)

$= 6x^2 + 5xy + y^2$

(b) $(2t-3q)(t+2q) = 2t^2 - 3qt - 6q^2 + 4tq$

$= 2t^2 + qt - 6q^2$

(c) $(x-2)(x-4) = x^2 - 2x + 8 - 4x$ ((-2×-4))

$= x^2 - 6x + 8$

An expression such as $2x^2 + 3x + 5$ is called a **quadratic** expression in x because the *highest* power of x is 2. An expression such as $7x^3 + 2x^2 + 1$ is called a **cubic** because the highest power of x is 3. An expression such as $4x^2 + 3xy + 2y^2$ is often called a **trinomial** (it has 3 terms).

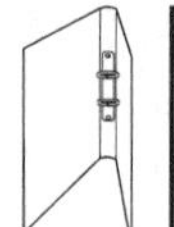

Exercise 10(c)

Remove the brackets and simplify where possible the following:

1 $(x+1)(x+2)$ **2** $(x+3)(x+5)$

3 $(2x+1)(x+2)$ **4** $(x-3)(x-2)$

5 $(2x-3)(x+1)$ **6** $(x-9)(x+9)$

7 $(a+b)(a+2b)$ **8** $(c-d)(c+2d)$

9 $(q+t)(q-2t)$ **10** $(3-x)(x+4)$

11 $(1-2x)(4+3x)$ **12** $(a+b)(c+2d)$

13 $(t-t^2)(1+t)$ **14** $(x-3)(2x+5)$

15 $(a-7)(a+4)$ **16** $(p-3q)(2p+q)$

17 $(3t+1)(2t+5)$ **18** $(q-3t)(q+3t)$

19 $(p+2q)^2$ **20** $(q-3p)^2$

10.4 Number patterns

The series of numbers 1, 3, 6, 10, 15, ... are known as **triangle numbers**. The reason for this is that if we draw a dot pattern for these numbers, they will be in the shape of a triangle (see Figure 10.2).

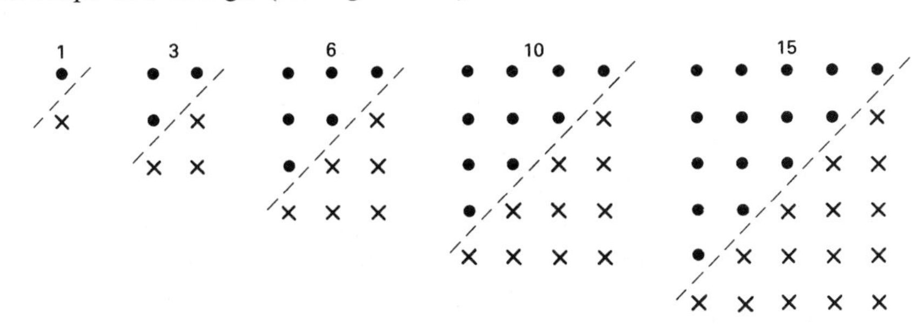

Figure 10.2

Also added to the diagram is the same number of crosses each time, making up a rectangle. The total number of crosses and dots in each diagram is

$$1 \times 2,\ 2 \times 3,\ 3 \times 4,\ 4 \times 5,\ 5 \times 6,\ \ldots$$

Each triangle number is half of this, and so we can write the triangle numbers as:

$$\frac{1 \times 2}{2},\ \frac{2 \times 3}{2},\ \frac{3 \times 4}{2},\ \frac{4 \times 5}{2},\ \frac{5 \times 6}{2},\ \ldots$$

Continuing this pattern, the formula for the nth triangle number is $\frac{n(n+1)}{2}$.

Example 10.4

Look at the sequence of numbers 2, 5, 10, 17, ...

(a) Find the difference between successive pairs of numbers. What do you notice?

(b) Find the eighth number in the sequence.

(c) By subtracting the triangle numbers from each of these numbers, can you see a pattern that would help you to find a formula for the nth number?

Solution

(a) and (b)

2 5 10 17 [26] [37] [50] [65]

differences → 3 5 7 [9] [11] [13] [15]

The differences increase by 2 each time.

If you continue this pattern shown in the boxes, the eighth number is 65.

(c) Subtract triangle numbers

	2	5	10	17	26	37
	−1	−3	−6	−10	−15	−21
to give	1	2	4	7	11	17

The differences are still not constant, so

Subtract triangle numbers again	−1	−3	−6	−10	−15	−21
	0	−1	−2	−3	−4	−5

The differences now decrease by 1 each time and so the nth number in the final row is $-(n-1)$

To get back to the original series, we must add the triangle numbers twice

The nth triangle number is $\dfrac{n(n+1)}{2}$, therefore add $2 \times \dfrac{n(n+1)}{2}$

$= n(n+1)$

the nth number in the series

$$= -(n-1) + n(n+1) = -n + 1 + n^2 + n$$

$$= n^2 + 1$$

[square numbers + 1].

You will see that this type of number pattern involves a quadratic expression. You should now try one of the next two Investigations to see how the ideas of Investigation VIII can be extended.

Investigation XIX: Handshakes

There are three people in a room, X, Y and Z.

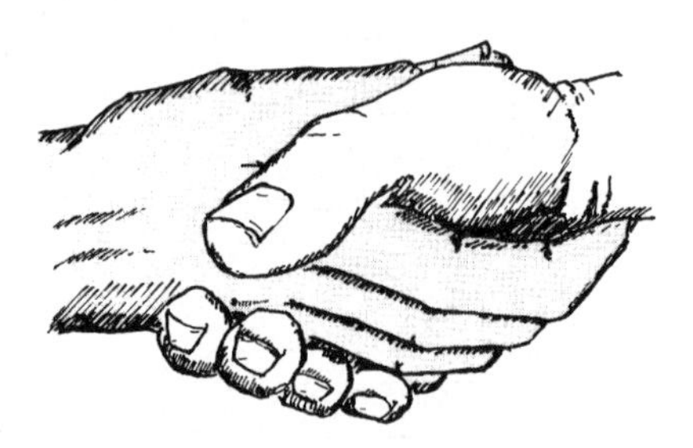

X shakes hands with Y

Y shakes hands with Z

X shakes hands with Z

Each person has greeted all of the others, and there have been 3 handshakes in all.

(i) Try to complete the following table.

No. of people	2	3	4	5	6
No. of handshakes	1	3			

Can you see how this table extends?

(ii) Try to look for a formula that predicts the numbers in the bottom row. Have you seen this number pattern anywhere else?

(iii) If you are given the number of handshakes, can you find how many people there were? For example, if there were 11 175 handshakes, how many people were there?

What problem arises?

(iv) Can you find other situations which are similar to this problem?

Investigation XX: Cutting a cake

If a square cake is cut with a knife, what is the *maximum* number of pieces you can get with a given number of cuts?

(i) Here are the first two examples.

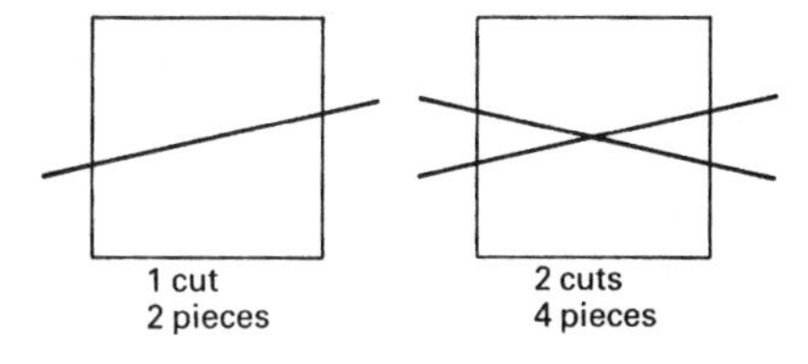

Try to continue. Remember, you are trying to find the *maximum* number of pieces.

(ii) Now try to fill in and extend this table.

No. of cuts	1	2	3	4	5	6
No. of pieces	2	4				

Can you find a formula for N cuts?

(iii) How many cuts produce 121 pieces?

(iv) How do the results alter, if you try and find the *smallest* number of pieces with a given number of cuts? How do the cuts have to be arranged?

(v) If the cuts are not straight, how does this affect results?

What about overlapping circles?

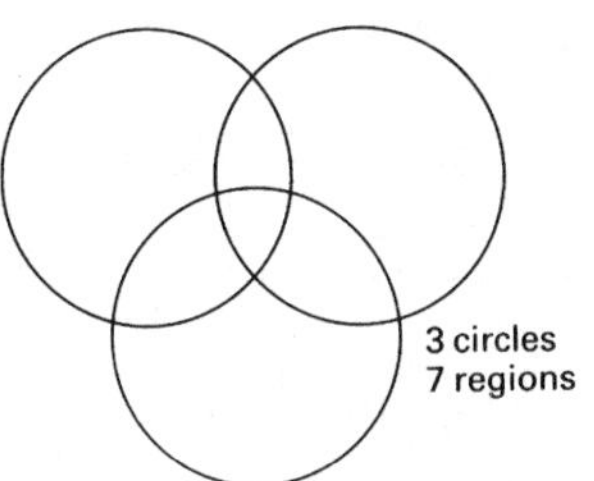

10.5 Factorisation

(i) General

In Section 3.4, we looked at how to remove a bracket in an expression such as $4(2x + 3t)$ which would become $8x + 12t$. The reverse process of this is called **factorisation**.

Consider the expression $16x - 24y$.

Since 8 is the *highest common factor* (HCF) of $16x$ and $24y$, the expression can be written $8(2x - 3y)$.

i.e. $16x - 24y = 8(2x - 3y)$ when factorised.

You should note that it could also be written as $4(4x - 6y)$, but the expression is then not completely factorised, because 2 is a factor of $4x$ and $6y$. You must always find the HCF of each term.

Now look at the following example.

Example 10.5

Factorise completely:

(a) $x^2 + x^2y$

(b) $3xy^2 - 12xy$

(c) $100 - 25x^3$.

Solution

(a) The HCF of each term is x^2

Hence $x^2 + x^2y = x^2(1 + y)$.

↑ note this becomes 1.

(b) The HCF of each term is $3xy$,

hence $3xy^2 - 12xy = 3xy(y - 4)$.

(c) 25 is the highest common factor of each term, and so

$$100 - 25x^3 = 25(4 - x^3).$$

The expression we have just considered contained two terms. If the expression contains four terms, they need to be grouped in pairs in the following way:

Example 10.6

Factorise completely:

(a) $ac + bc + bd + ad$

(b) $3p^2 + 2qt + 6qp + pt$

(c) $ap - 2bp - 4bq + 2aq$.

Solution

(a) $ac + bc + bd + ad = (ac + bc) + (bd + ad)$

$$= c(a + b) + d(b + a)$$

Since $a + b$ and $b + a$ are the same, we have a common factor of $a + b$, which can be taken outside.

Hence the expression $= (a + b)(c + d)$.

Always multiply out to check it is correct

REMEMBER

(b) You do not *always* group in pairs as the expression is written. Since $3p^2$ and $2qt$ have nothing in common, swap $2qt$ and $6qp$ over.

Hence

$$3p^2 + 2qt + 6qp + pt = (3p^2 + 6qp) + (2qt + pt)$$

$$= 3p(p + 2q) + t(2q + p)$$

$$= (p + 2q)(3p + t).$$

(c) $ap - 2bp - 4bq + 2aq = (ap - 2bp) - (4bq - 2aq)$

Notice that since the second bracket has a minus sign in front, the sign in front of $+2aq$ is changed to $-2aq$.

This factorises to

$$p(a - 2b) - 2q(2b - a)$$

Since $(a - 2b) = -1 \times (2b - a)$, the expression can be rewritten as

$$p(a - 2b) + 2q(a - 2b) = (a - 2b)(p + 2q)$$

Before moving on to expressions with three terms, make sure you have understood this section by trying the next exercise.

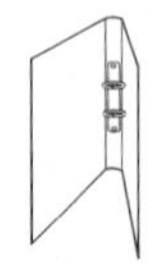

Exercise 10(d)

Factorise if possible the following expressions:

1	$4x + 8y$	**2**	$12t + 3$
3	$ab + ac$	**4**	$4ax + 6x^2$
5	$4pq - p^2$	**6**	$24a - 16a^2$
7	$12bt - 4b^2t$	**8**	$x^2 + x^3$
9	$x^2 + y^2$	**10**	$4abc - 2ab$
11	$xy^2 + yx^2$	**12**	$100x - 100x^2$
13	$xq + xt + yq + yt$	**14**	$ac + ad + bc + 2bd$
15	$ax - bx + by - ay$	**16**	$ax - bx + ay - by$
17	$pt - 3pc + 2qt - 6qc$	**18**	$4qc - 3tc + 8qd - 6td$
19	$20px + 15py - 8qx - 6qy$	**20**	$x^3 - x^2y^2 - y^3 + xy.$

In Section 10.3, we looked at how to multiply out two brackets and simplify the answer to give a quadratic expression.

The process of **factorising a quadratic** is not quite so easy and will be considered in stages.

(i) $$\begin{aligned} x^2 + 5x + 6 &= x^2 + 2x + 3x + 6 \\ &= x(x + 2) + 3(x + 2) \\ &= (x + 3)(x + 2). \end{aligned}$$

(ii) $$\begin{aligned} x^2 + x - 6 &= x^2 + 3x - 2x - 6 \\ &= x(x + 3) - 2(x + 3) \\ &= (x - 2)(x + 3). \end{aligned}$$

(iii) $$\begin{aligned} x^2 - 7x + 10 &= x^2 - 5x - 2x + 10 \\ &= x(x - 5) - 2(x - 5) \\ &= (x - 2)(x - 5). \end{aligned}$$

(iv) $$\begin{aligned} 2x^2 - x - 6 &= 2x^2 + 3x - 4x - 6 \\ &= x(2x + 3) - 2(2x + 3) \\ &= (x - 2)(2x + 3). \end{aligned}$$

(v) $$\begin{aligned} 18x^2 + 45x - 8 &= 18x^2 + 48x - 3x - 8 \\ &= 6x(3x + 8) - (3x + 8) \\ &= (6x - 1)(3x + 8). \end{aligned}$$

You should notice that when splitting the middle term into two parts, if you *multiply these together*, it is the same as *multiplying* the *first* and *third* term.

So

(i) $2x \times 3x = x^2 \times 6 = 6x^2$

(ii) $3x \times -2x = x^2 \times -6 = -6x^2$

(iii) $-5x \times -2x = x^2 \times 10 = 10x^2$

(iv) $3x \times -4x = 2x^2 \times -6 = -12x^2$

(v) $48x \times -3x = 18x^2 \times -8 = -144x^2$.

It is also worth trying to remember the following rule which may save you a lot of work. It is not always possible to put a quadratic expression into brackets; the following rule can save you a lot of time.

Rule

For the expression $ax^2 + bx + c$, work out $b^2 - 4ac$.
If it is a perfect square, it factorises, if not it doesn't.

For example:

(i) $x^2 + 6x + 8$: $a = 1$, $b = 6$, $c = 8$

$b^2 - 4ac = 36 - 32 = 4$, which *is* a perfect square, hence it factorises

In fact $x^2 + 6x + 8 = (x + 4)(x + 2)$.

(ii) $x^2 - 4x - 12$: $a = 1$, $b = -4$, $c = -12$

$b^2 - 4ac = 16 + 48 = 64$, which is a perfect square

This also factorises to $(x - 6)(x + 2)$.

(iii) $x^2 + 11x + 8$: $a = 1$, $b = 11$, $c = 8$

$b^2 - 4ac = 121 - 32 = 89$

This is not a perfect square and so it doesn't factorise.

(ii) Special cases

(1) $x^2 - y^2$ is known as a **difference of two squares**.

It can be put into brackets as follows:

Rule

$$x^2 - y^2 = (x - y)(x + y)$$

(check by multiplying out the right hand side)

Hence $4x^2 - 25t^2$ is really $(2x)^2 - (5t)^2$
$= (2x - 5t)(2x + 5t)$.

(2) $x^2 + 2xy + y^2$ is a **perfect square**.

It can be written as:

Rule

$$x^2 + 2xy + y^2 = (x + y)(x + y) = (x + y)^2$$

(check by multiplying out)

Also $x^2 - 2xy + y^2 = (x - y)^2$

For example $(3t - 4q)^2 = (3t)^2 + 2(3t)(-4q) + (-4q)^2$
$= 9t^2 - 24tq + 16q^2$.

10.6 Quadratic equations

The equation $x^2 - 5x + 6 = 0$ is referred to as a **quadratic equation**. We consider here two methods of solution, (i) by factors; (ii) by formula.

(i) Using factors

First factorise the left hand side, using the techniques developed in Section 10.5.

Hence $(x - 3)(x - 2) = 0$

If two numbers multiply to give zero, then one of them must be zero.

Hence either $x - 3 = 0$ i.e. $x = 3$
or $x - 2 = 0$ i.e. $x = 2$

A quadratic equation *usually* has *two* solutions.

Consider another example: $2x^2 + 5x - 3 = 0$

This involves more work to factorise the left hand side, but you should be able to get it to

$(2x - 1)(x + 3) = 0$

Therefore either $2x - 1 = 0$ and $x = \frac{1}{2}$
or $x + 3 = 0$ and $x = -3$.

(ii) By formula

It can be shown that for the general quadratic equation

$$ax^2 + bx + c = 0,$$

the two solutions are given by the formula

$$x = \frac{-b \pm \sqrt{b^2 - 4ac}}{2a}.$$

[*Note*: $\pm$ means you get two answers, one using the $+$ sign, another using the $-$ sign.]

You do not need to understand the following proof of the formula, but it is given for completeness.

$$ax^2 + bx + c = 0$$

$$x^2 + \frac{bx}{a} + \frac{c}{a} = 0 \qquad \text{(dividing by } a\text{)}$$

Hence $$x^2 + \frac{bx}{a} = -\frac{c}{a}$$

$$x^2 + \frac{bx}{a} + \frac{b^2}{4a^2} = \frac{b^2}{4a^2} - \frac{c}{a} \qquad \left(\text{adding } \frac{b^2}{4a^2} \text{ to each side}\right).$$

The left hand side is now a perfect square. (This technique is called **completing the square**.)

So $$\left(x + \frac{b}{2a}\right)^2 = \frac{b^2}{4a^2} - \frac{c}{a}$$

$$= \frac{b^2 - 4ac}{4a^2}$$

Take the square root of each side

$$x + \frac{b}{2a} = \pm\frac{\sqrt{b^2 - 4ac}}{2a}$$

So $$x = -\frac{b}{2a} \pm \frac{\sqrt{b^2 - 4ac}}{2a}$$

$$= \frac{-b \pm \sqrt{b^2 - 4ac}}{2a}.$$

Example 10.7

Use the formula to solve the equations if possible.

(a) $2x^2 - 9x - 18 = 0$ (b) $t^2 + t - 4 = 0$

(c) $y^2 + y + 1 = 0$ (d) $m^2 - 6m + 9 = 0$.

Solution

(a) Always write down the values of a, b and c at the beginning, making sure you have the correct sign.

Hence, $a = 2$, $b = -9$, $c = -18$

$$\text{So} \quad x = \frac{-(-9) \pm \sqrt{(-9)^2 - 4 \times 2 \times -18}}{4}$$

$$= \frac{9 \pm \sqrt{81 + 144}}{4} = \frac{9 \pm \sqrt{225}}{4} = \frac{9 \pm 15}{4}$$

$$\text{So} \quad x = \frac{9 + 15}{4} = 6 \quad \text{or} \quad \frac{9 - 15}{4} = -1.5.$$

(b) $a = 1$, $b = 1$, $c = -4$

$$\text{So} \quad t = \frac{-1 \pm \sqrt{1 - 4 \times 1 \times -4}}{2} = \frac{-1 \pm \sqrt{1 + 16}}{2} = \frac{-1 \pm \sqrt{17}}{2}.$$

The $\sqrt{17}$ is not exact, and so we will have to give the answer to, say, 2 decimal places.

$$t = \frac{-1 \pm 4.123}{2}$$

$$\text{Hence} \quad t = \frac{-5.123}{2} = -2.56 \quad \text{or} \quad \frac{3.123}{2} = 1.56.$$

(c) $a = 1$, $b = 1$, $c = 1$

$$y = \frac{-1 \pm \sqrt{1 - 4 \times 1 \times 1}}{2} = \frac{-1 \pm \sqrt{-3}}{2}.$$

We cannot find the square root of a negative number, and so no solution exists.

(d) $a = 1$, $b = -6$, $c = 9$

$$\text{Therefore} \quad m = \frac{6 \pm \sqrt{(-6)^2 - 4 \times 1 \times 9}}{2} = \frac{6 \pm 0}{2}$$

Hence there is only one solution, $m = 3$.

10.7 Trial and improvement

The equation $x^3 = 4 - x$ can be solved by a guessing process. First, rewrite the equation so that the right hand side is zero

Therefore $x^3 - 4 + x = 0$.

Now look for possible values of x.

If

$x = 0$

$0 - 4 + 0 = -4$ Should be zero, so too low

$x = 1$

$1 - 4 + 1 = -2$ too low

$x = 2$

$8 - 4 + 2 = 6$ too high

the solution lies between $x = 1$ and $x = 2$ (The right hand side has changed sign)

$x = 1.5$

$1.5^3 - 4 + 1.5 = 0.875$ too high

$x = 1.4$

$1.4^3 - 4 + 1.4 = 0.144$ too high

$x = 1.3$

$1.3^3 - 4 + 1.3 = -0.503$ too low

the solution lies between $x = 1.3$ and 1.4

$x = 1.35$

$1.35^3 - 4 + 1.35 = -0.19$ too low

$x = 1.36$

$1.36^3 - 4 + 1.36 = -0.12$ too low

$x = 1.37$

$1.37^3 - 4 + 1.37 = -0.06$ too low

$x = 1.38$

$1.38^3 - 4 + 1.38 = 0.008$ too high

Therefore $x = 1.38$ (2 d.p.).

10.8 Problem solving

In a similar way to Section 3.6, certain problems give rise to quadratic equations. Look carefully at the next two examples, and then try Exercise 10(e).

Example 10.8

A solid cuboid has a square cross-section, and length 3 cm greater than the side of the square. If the total surface area of the cuboid is 210 cm^2, find the measurements of the solid.

Solution

If we let the square be of side x cm, then the length must be $(x+3)$ cm (see Figure 10.3). The total surface area

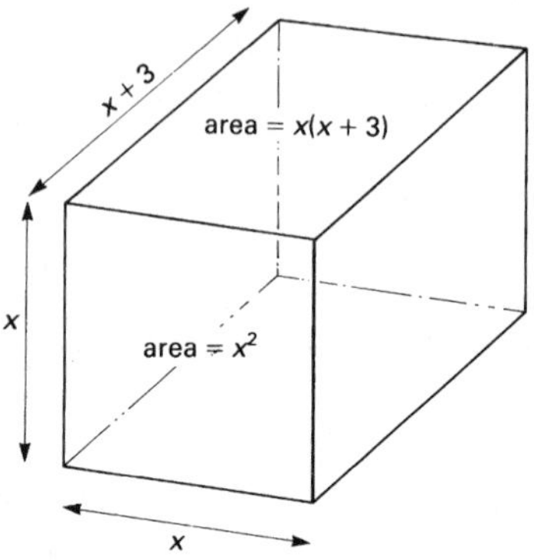

$$= 2 \times (x^2) + 4 \times (x \times (x+3))$$
$$= 2x^2 + 4x^2 + 12x$$
$$= 6x^2 + 12x$$

This must equal 210, and so

$6x^2 + 12x = 210$ (This is not useful in this form)

$6x^2 + 12x - 210 = 0$ (Subtract 210 from each side)

The equation can be slightly simplified by dividing by 6

$$x^2 + 2x - 35 = 0$$

This factorises to $(x-5)(x+7) = 0$

Hence $x = 5$ or -7.

Clearly x cannot be -7, and so $x = 5$, hence the measurements of the solid are 5 cm by 5 cm by 8 cm.

Example 10.9

A speed boat travelled downstream a distance of 10 km at an average speed of v km/h. On the return journey, the average speed was increased by 5 km/h, and the journey took 6 minutes less. Find v.

Solution

For the first journey, using time $= \dfrac{\text{distance}}{\text{speed}}$, the time for this part is $\dfrac{10}{v}$ hours.

On the return journey, the time is $\dfrac{10}{v+5}$ hours.

Now $\dfrac{10}{v} - \dfrac{10}{v+5}$ must be 6 minutes $= \dfrac{1}{10}$ hours

$$\text{i.e.} \quad \frac{10}{v} - \frac{10}{v+5} = \frac{1}{10}.$$

The common denominator of all three fractions is $10v(v+5)$.

Multiplying each fraction by this we get

$$\frac{100(v+5)}{10v(v+5)} - \frac{100v}{10v(v+5)} = \frac{v(v+5)}{10v(v+5)}$$

Now they all have the same denominator, the numerators give

$$100(v+5) - 100v = v(v+5)$$
$$100v + 500 - 100v = v^2 + 5v$$
$$0 = v^2 + 5v - 500 = (v-20)(v+25)$$
$$\text{So } v = 20 \text{ or } -25$$

Hence $v = 20$ (since v cannot be negative).

Exercise 10(e)

1 Factorise if possible:

(i) $x^2 - 11x + 24$ (ii) $x^2 + 2x - 3$

(iii) $x^2 + 11x - 26$ (iv) $q^2 - 7q + 8$

(v) $2p^2 + 13p + 15$ (vi) $4x^2 + 8x + 3$

(vii) $6x^2 - 5x + 1$ (viii) $9t^2 - 7t - 2$

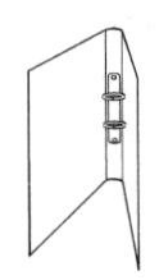

(ix) $4y^4 - 3y^2 - 1$ (x) $4x^2 - 25$

(xi) $1 - x^4$ (xii) $\pi R^2 - \pi r^2$

(xiii) $121x^2y^2 - 4$ (xiv) $4x^2 - 12xy + 9y^2$

(xv) $x^2 + 6x + 9$ (xvi) $4x^2 - 20x + 25$.

2 Solve where possible the following equations:

(i) $x^2 - 2x - 3 = 0$ (ii) $t^2 - 10t + 16 = 0$

(iii) $2p^2 + p - 1 = 0$ (iv) $3x^2 - 10x + 3 = 0$

(v) $a^2 - 9a = 0$ (vi) $3x^2 - x + 3 = 0$

(vii) $4y^2 = 9y + 1$ (viii) $n + \dfrac{1}{3n} = 2$.

3 Solve by trial and improvement:

(i) $2x^3 = 8 - x$ (ii) $x^4 = x + 1$

4 A rectangular lawn measuring 12 m by 9 m is surrounded by a path of width x m (see Figure 10.4). If the total area of the path is 72 m^2, find x.

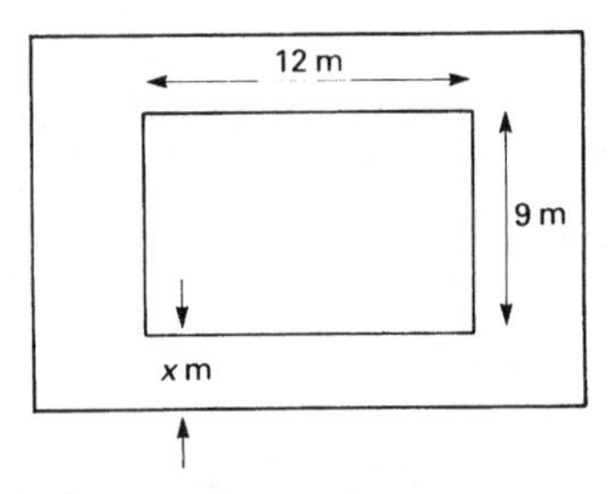

Figure 10.4

5 A target on a dartboard has a centre region of radius R cm, and a concentric circle surrounding it of radius $(R + 1)$ cm. If the area between the two circles is one-tenth of the area of the centre region, find R.

6 An aeroplane travelled a distance of 400 km at an average speed of x km/h. Write down an expression for the time taken. On the return journey, the speed was increased by 40 km/h. Write down an expression for the time for the return journey. If the return journey took 30 minutes less than the outward journey, write down an equation in x and solve it.

7 For the following sequences, find the tenth number, and a formula for the nth number.

(i) 1, 4, 9, 16, 25, ... (ii) 1, 2, 4, 7, 11, ...

(iii) 2, 4, 6, 8, ... (iv) 5, 9, 15, 23, ...

8 Figure 10.5 shows two equilateral triangles drawn on an isometric grid of dots. The dots are 1 centimetre apart. The length of each side of the smaller triangle is 1 cm and that of the larger is 3 cm.

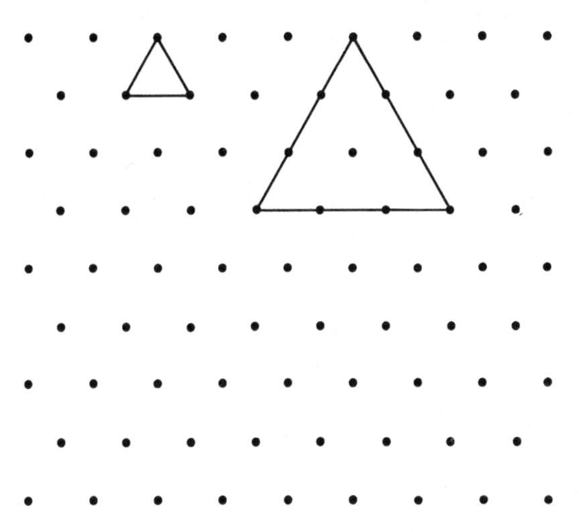

Figure 10.5

(a) How many tiles of the shape and size of the smaller triangle would be required to tessellate the larger triangle?

(b) The larger triangle has 9 dots on the perimeter and 1 dot inside it. There are 10 dots on and in the triangle. A triangle of side length 4 cm is drawn in a similar way on the grid. Find

(i) the number of dots on the perimeter of the triangle

(ii) the number of dots inside the triangle

(iii) the total number of dots in and on the triangle;

(c) Copy and complete the following table of various sizes.

Length of side (L cm)	1	2	3	4	5
No. of dots on perimeter (P)	3		9		
No. of dots inside triangle (Q)	0		1		
Total no. of dots (T)	3		10		
No. of 1 cm triangles required to tessellate (A)	1		9		

(d) Write down a formula for

  (i) P in terms of L

  (ii) A in terms of L.

(e) Verify, for $L = 5$, that $T = \dfrac{A+P}{2} + 1$.

(f) Assuming that the result in part (e) is true for all values of L, or otherwise, show that

$$T = \frac{(L+1)(L+2)}{2}.$$

9 A gardener wished to protect his seeds from the birds.

He placed stakes in the ground and joined each stake to each of the others using pieces of string. He found that if he had 2 stakes there was 1 join. 3 stakes would need 3 joins and 4 stakes would need 6 joins.

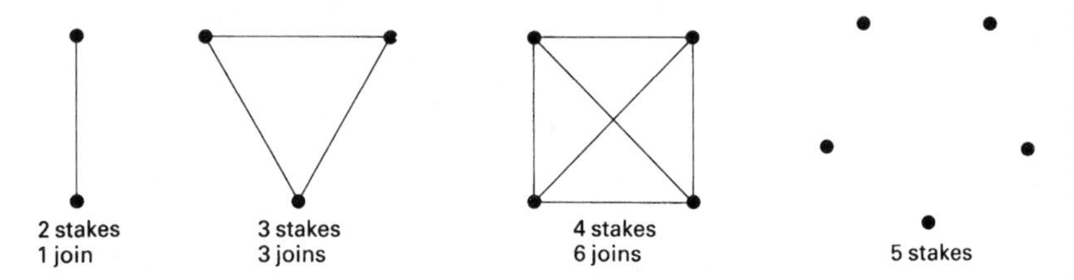

Figure 10.6

(i) Copy and complete the diagram for 5 stakes, showing the joins.

(ii) Find how many joins there would be if he used 6 stakes.

(iii) Find how many joins there would be if he used 7 stakes.

(iv) Write down the first eight terms of the sequence which begins 1, 3, 6, ... and explain how the sequence is formed.

(v) Given that the number of stakes is n, find how many joins there will be.

[SEG]

10.9 Further rearrangement of formulae

In Section 3.7, we looked at the basic ideas of changing the subject of a formula. It is important that you have understood those ideas before working through the following examples. The golden rule throughout this work is not to take short cuts, and always do the same thing to each side.

Example 10.10

Make x the subject of the following formulae.

(a) $t = ax + b$

(b) $t = a(x + b)$

(c) $t = 2a\sqrt{x + y}$

(d) $y = \frac{1}{t}(ax - k)$

(e) $y = \frac{2t}{(1 + x)}$

(f) $p = a\sqrt{x} + \frac{q}{t}$

(g) $y = \frac{ax + b}{cx + d}$

(h) $y = \frac{t}{1 - x^2}$.

Solution

It is important to realise that the following solutions are not the only ones, but are an attempt at a systematic approach.

(a) $$t = ax + b$$

Subtract b from each side: $$t - b = ax$$

Divide each side by a: $$\frac{t - b}{a} = x$$

or, better: $$x = \frac{(t - b)}{a}$$

(b) $$t = a(x + b)$$

Remove brackets: $$t = ax + ab$$

Subtract ab from each side: $$t - ab = ax$$

Divide each side by a: $$\frac{t - ab}{a} = x$$

i.e. $$x = \frac{t - ab}{a}$$

(c) $$t = 2a\sqrt{x + y}$$

Square both sides (remember to square $2a$): $$t^2 = 4a^2(x + y)$$

Remove brackets: $$t^2 = 4a^2x + 4a^2y$$

Subtract $4a^2y$ from each side $$t^2 - 4a^2y = 4a^2x$$

Divide each side by $4a^2$: $$x = \frac{(t^2 - 4a^2y)}{4a^2}$$

(d) $$y = \frac{1}{t}(ax - k)$$

Remove brackets: $$y = \frac{ax}{t} - \frac{k}{t}$$

Multiply each term by t: $$yt = ax - k$$

Add k to each side: $$yt + k = ax$$

Divide each side by a: $$x = \frac{(yt + k)}{a}$$

(e) $$y = \frac{2t}{1 + x}$$

Multiply each side by the common denominator $1 + x$: $$y(1 + x) = 2t$$

Remove brackets: $$y + xy = 2t$$

Subtract y from each side: $$xy = 2t - y$$

Divide each side by y: $$x = \frac{(2t - y)}{y}$$

(f) $$p = a\sqrt{x} + \frac{q}{t}$$

Subtract $\frac{q}{t}$ from each side: $$p - \frac{q}{t} = a\sqrt{x}$$

(The term containing a square root should always be on its own before you can square each side.)

Square both side: $$\left(p - \frac{q}{t}\right)^2 = a^2 x$$

Divide each side by a^2: $$x = \frac{\left(p - \frac{q}{t}\right)^2}{a^2}$$

$$\text{or} \quad \frac{1}{a^2}\left(p - \frac{q}{t}\right)^2$$

(g) $$y = \frac{ax + b}{cx + d}$$

Multiply both sides by the common denominator $(cx + d)$: $$y(cx + d) = ax + b$$

Remove brackets: $$ycx + yd = ax + b$$

Subtract ax from each side and then subtract yd from each side: $$ycx - ax = b - yd$$

You will notice here that the new subject appears **twice**.

Factorise the side containing new subject x: $x(yc - a) = b - yd$

Divide each side by $(yc - a)$: $x = \dfrac{(b - yd)}{(yc - a)}$

(h) $y = \dfrac{t}{1 - x^2}$

Multiply each side by the common denominator $1 + x$: $y(1 - x^2) = t$

Remove brackets: $y - yx^2 = t$

Add yx^2 to both sides: $y = yx^2 + t$

Subtract t from both sides: $y - t = yx^2$

Divide each side by y: $x^2 = \dfrac{y - t}{y}$

Take square root of each side: $x = \sqrt{\dfrac{y - t}{y}}$.

Exercise 10(f)

1 In each of the following examples, rearrange the formula so that x becomes the new subject.

(i) $y = \dfrac{2x}{t}$

(ii) $z = x(a + b)$

(iii) $H = ax^2 + b$

(iv) $v^2 = u^2 + 2ax$

(v) $\dfrac{1}{f} = \dfrac{1}{u} + \dfrac{1}{x}$

(vi) $T = 2\pi\sqrt{\dfrac{x}{g}}$

(vii) $y = \dfrac{q}{1 + 2x}$

(viii) $y = \dfrac{2}{t}(q - x)$

(ix) $p = k\sqrt{x} - 2$

(x) $y = \dfrac{a + x}{b + x}$

(xi) $t = \dfrac{a}{x} + \dfrac{b}{x}$

(xii) $p = \dfrac{ax}{b} + \dfrac{cx}{d}$

(xiii) $y = \dfrac{x^2 - y^2}{x^2}$

(xiv) $t = 4a\sqrt{x - y}$

(xv) $q = 4\sqrt{x - 1} + 1$

2 The formula $R = 231 - 0.45N$ gives a very good estimate of the world record in the men's 1500 metres race. R seconds is the record time and N is the number of years after 1930.

(i) (a) Use the formula to find the value of R when $N = 30$.

(b) The world record in 1960 was 3 minutes 35.6 seconds. Calculate the difference, in seconds, between this time and the time given by the formula.

(ii) (a) Use the formula to find the value of N when $R = 195$.

(b) Write down the year in which, according to the formula, the world record will be 3 minutes 15 seconds.

(iii) Explain why the formula must eventually go wrong.

[MEG]

3 The formula $v^2 = u^2 + 14d$, gives the final speed v m/s of a vehicle that starts with a speed of u m/s and travels a distance of d m.

(i) Find v if $u = 12$ and $d = 10$.

(ii) Rearrange the formula to make d the subject.

(iii) Find d if $v = 30$ and $u = 20$.

(iv) Find u if $v = 18$, $d = 10$.

10.10 Simultaneous equations (linear)

If there are two unknown letters, say p and q, and we are given two equations satisfied by them, such as:

$$3p + q = 7 \quad (1)$$

$$2p + 5q = 9 \quad (2)$$

the equations are said to be true simultaneously. They are **simultaneous equations**. One of the letters is eliminated in the following way:

Multiply equation (1) by 5: $\quad 15p + 5q = 35 \quad (3)$

$$2p + 5q = 9 \quad (2)$$

Now find (3) – (2) $\quad 13p = 26, \quad p = 2$

Now substitute back in one of the equations, say (1):

$$3 \times 2 + q = 7, \quad q = 1$$

The solution of the equations is $p = 2$, $q = 1$.

A slightly harder example now.

$$2t = 5q + 6$$
$$4t + 3q = -1$$

First the equations must be written so that the two unknowns occur on the same side

$$\text{So} \qquad 2t - 5q = 6 \qquad (1)$$
$$4t + 3q = -1 \qquad (2)$$
$$(1) \times (3): \qquad 6t - 15q = 18 \qquad (3)$$
$$(2) \times 5: \qquad 20t + 15q = -5 \qquad (4)$$
$$\text{Now } (3) + (4): \qquad 26t \quad = 13, \quad t = \frac{13}{26} = \frac{1}{2}$$
$$\text{Substitute in } (1): \qquad 2 \times \tfrac{1}{2} - 5q = 6$$
$$-5q = 5, \quad q = -1$$

The solution is $q = -1$, $t = \frac{1}{2}$.

As with all types of equation, simultaneous equations can arise from problem situations.

Example 10.11

A bill of £355 was paid using £5 notes and £20 notes. If 35 notes were used altogether, find how many of each were used.

Solution

Let the number of £5 notes be f.

Let the number of £20 notes be t.

Then

$$f + t = 35 \qquad (1)$$
$$5f + 20t = 355 \qquad (2)$$
$$(1) \times 5: \qquad 5f + 5t = 175 \qquad (3)$$
$$(2) - (3): \qquad 15t = 180, \quad t = 12$$
$$\text{Substitute in equation } (1): \qquad f + 12 = 35, \quad f = 23$$

Hence 12 £20 notes, and 23 £5 notes were used.

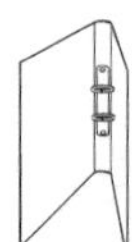

Exercise 10(g)

1 Solve the following simultaneous equations:

(i) $x + y = 7,\ 2x - y = 8$

(ii) $3t - 2q = 4,\ 5t - 2q = 0$

(iii) $4a - 3b = 5,\ 7a - 5b = 9$

(iv) $2p + q = 8,\ 5p - q = 6$

(v) $x + 3y = 2,\ 2x - 4y = 1$

(vi) $3x + 2y = 13,\ 2x + 3y = 12$

(vii) $4c - 5d = 21,\ 6c + 7d = -12$

(viii) $3h + 2j = 17,\ 4j - h = -8$

2 The cost of 8 pens and 12 pencils is £9.00, and the cost of 6 pens and 11 pencils is £7.05. Find the cost of a pen and a pencil.

3 The cost of £C of making n articles is given by the formula $C = p + qn$ where p and q are constants. The cost of making 7 articles is £29, and of making 5 articles is £21.

(i) Write down two equations in p and q.

(ii) Solve these equations to find the values of p and q.

[LEAG]

10.11 Variation and proportion

If two variables x and y are related by the formula $y = kx$, where k is a constant, then we say that y is **directly proportional** to x, or y **varies directly** as x. The symbol for proportion is a $\propto$, k is called the **constant of proportionality**.

Example 10.12

The time swing, T seconds, of a pendulum clock varies as the square root of its length L cm.

If $T = 1$ when $L = 25$, calculate

(a) T when $L = 36$

(b) L when T is increased to 1.5 seconds.

Solution

Since $t \propto \sqrt{L}$, then $T = k\sqrt{L}$

$T = 1$ when $L = 25$, so $1 = k\sqrt{25} = 5k$

Hence $k = \frac{1}{5}$.

So $T = \frac{1}{5}\sqrt{L}$.

(a) When $L = 36$, $T = \frac{1}{5}\sqrt{36} = 1.2$ seconds

(b) When $T = 1.5$, $1.5 = \frac{1}{5}\sqrt{L}$

So $7.5 = \sqrt{L}$

$L = 7.5^2 = 56.25$ cm.

The following example illustrates the idea of joint variation.

Example 10.13

The lift, L, produced by the wing of an aircraft, varies directly as its area, A, and as the square of the airspeed, V.

For a certain wing $A = 15$, and $L = 1200$ when $V = 200$.

(a) Find an equation connecting L, A and V.

(b) If, for the same wing, the airspeed is increased by 10%, find the corresponding percentage increase in lift.

Solution

(a) If L varies directly as A, and as the square of V, then $L = kAV^2$

$L = 1200$ when $A = 15$, $V = 200$

So $1200 = k \times 15 \times (200)^2$

Hence $k = \dfrac{1200}{15 \times 40\,000} = \dfrac{1}{500}$

The formula is $L = \dfrac{AV^2}{500}$.

(b) If V increases by 10%, then $V = 220$

Hence $L = \dfrac{15 \times (220)^2}{500} = 1452$

The percentage increase $= \dfrac{\text{Increase}}{\text{Original}} \times 100$

$\dfrac{252}{1200} \times 100 = 21\%$.

When one quantity y varies as the reciprocal of another x, then we say that y varies inversely as x.

Example 10.14

Assuming that the length (L) of paper on a roll of fixed dimensions varies inversely as the thickness (D) of the paper

(a) find a formula connecting L and D if $D = 0.75$ mm when $L = 100$ m

(b) find D when $L = 80$ m.

Solution

(a) $L = \dfrac{k}{D}$ so $100 = \dfrac{k}{0.75}$, $k = 75$

Hence $L = \dfrac{75}{D}$.

(b) $D = \dfrac{75}{L} = \dfrac{75}{80} = 0.9375$ mm.

Exercise 10(h)

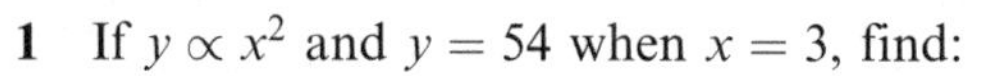

1 If $y \propto x^2$ and $y = 54$ when $x = 3$, find:

(i) y when $x = 5$

(ii) x when $y = 24$

2 If y is inversely proportional to x and when $x = 2.5$, $y = 20$, what is the value of x when $y = 4$?

3 P is inversely proportional to x^2. If $P = 32$ when $x = \frac{1}{2}$, find:

(i) the constant of proportion

(ii) The value of P when $x = -3$.

4 A quantity z varies directly as x and inversely as y^2. If $z = 3$ when $x = 12$ and $y = 2$, find the value of z when $x = 4$ and $y = 1/4$.

5 The number of revolutions made per minute by a wheel varies as the speed of the vehicle and inversely as the diameter of the wheel. If a wheel of diameter 60 cm makes 100 revolutions per minute at a speed of 11.3 km/h, calculate the number of revolutions made per minute by a wheel of 50 cm diameter when the speed is 15 km/h.

6 The area A of a triangle is directly proportional to the length of its base, b, and its height h. If $A = 90$ when $b = 18$ and $h = 10$, what is the height of a triangle with a base length 30 and area 115?

7 F is directly proportional to the square root of T, but varies inversely as the square root of M. Given that $F = 25$ when $T = 20$ and $M = 5$, calculate:

(i) the value of F when $T = 50$ and $M = 8$

(ii) the value of M when $F = 80$ and $T = 10$.

8 A cylinder of variable radius has constant height h. Its volume is denoted by V. Draw a sketch to show the relationship between V and r as r varies.

9 Given that y varies as x^n, write down the value of n in each of the following cases:

(i) y is the area of a circle of radius x

(ii) y is the volume of a cylinder of given base area and height x

(iii) y and x are the side of a rectangle of given area.

10 It is given that x is both directly proportional to y and inversely proportional to the square of z. Complete the table of values:

x	y	z
3	1	2
2		3
1	3	

10.23 Venn diagrams

A **set** is the name used in mathematics to define a collection of distinguishable objects. The members of the set are often referred to as **elements**. A set is usually denoted by a capital letter, and the members of the set are enclosed by brackets { }. Hence $R = \{$red, orange, yellow, green, blue, indigo, violet$\}$ would be the set of colours of the rainbow. $L = \{a, b, c, d, e\}$ simply stands for the set of letters a, b, c, d, e. If we want to say that something *belongs* to a set, we use the symbol $\in$. Since d belongs to L, we write $d \in L$ (d belongs to L, or d is a member of L). Because x does not belong to L, we write $x \notin L$ (x does *not* belong to L).

The number of members of set L would be written $n(L)$. In this case, $n(L) = 5$.

If $n(A) = 0$, the set A has no members, and is empty. For example, if $T = \{$people over ten feet tall$\}$, we would hope that $n(T) = 0$. There is a special symbol for the empty set, which is $\emptyset$, or just $\{\}$. A set where the objects can be counted is **finite**. A set such as {all integers} cannot be counted as it continues indefinitely. This is an **infinite** set.

The **universal set** $\mathscr{E}$ is a convenient mathematical symbol to denote the set of objects that we are interested in for a particular problem. A' is the set of elements that are in the universal set, but are not in A. A' is called the **complement** of A.

Hence if $\mathscr{E} = \{1, 2, 3, 4, 5, 6\}$ and $A = \{3, 4, 5\}$ then $A' = \{1, 2, 6\}$. Clearly if we did not specify the universal set, there would be rather a lot of numbers not in A.

There are two symbols namely $\cup$ and $\cap$ which are commonly used in working with sets.

If $A = \{a, b, c, d, e\}$ and $B = \{b, c, d, e, f, g\}$, then $A \cup B$ (which reads A **union** B is obtained by listing all possible members from either A or B (or both).

So $A \cup B = \{a, b, c, d, e, f, g\}$,

Notice that by convention, you do not put down any member **twice** if it appears in both sets.

$A \cap B$ (which reads A **intersection** B) is obtained by listing all possible elements that are common to both A and B.

So $A \cap B = \{b, c, d, e\}$.

If $X \cap Y = \emptyset$, then X and Y have no elements in common, and they are called **disjoint** sets.

For example, if $X = \{1, 3, 5\}$ and $Y = \{2, 6\}$, there are no common elements, and so $X \cap Y = \emptyset$.

If you look at the sets $P = \{\text{red, blue, green, yellow}\}$ and $Q = \{\text{red, blue}\}$, all of the elements of Q belong to P. We say that Q is a **subset** of P, and write this $Q \subset P$. A strange piece of mathematical notation associated with this idea is that we allow any set to be called a subset of itself!

So $\{1, 4, 6\} \subset \{1, 4, 6\}$ even though they are equal.

This idea is only of use in higher mathematics.

In order to help solve problems involving sets, we use a diagram known as a **Venn diagram**. In these, each set is represented by an oval (or circle) and the universal set is represented by a rectangle. The diagrams help to give you a clearer picture of the relationship between one set and another. Figure 10.7 shows how a Venn diagram can be used in six different examples. In each case, the shaded part represents where the elements are.

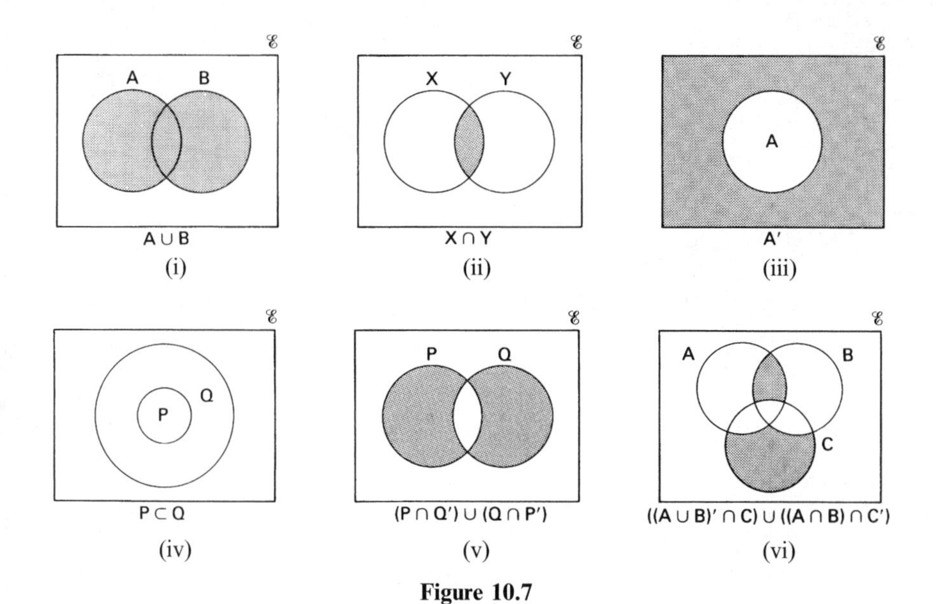

Figure 10.7

Parts (v) and (vi) are not easy to understand, and you should not worry too much if you cannot see how they arise.

Sometimes, the diagrams can be used to represent the **number of members** of various sets, from which deductions can be made.

Example 10.15

Figure 10.8 shows the results of interviewing some women students at the local polytechnic.

$B = \{\text{women who play badminton}\}$
$H = \{\text{women who play hockey}\}$
$T = \{\text{women who play tennis}\}$

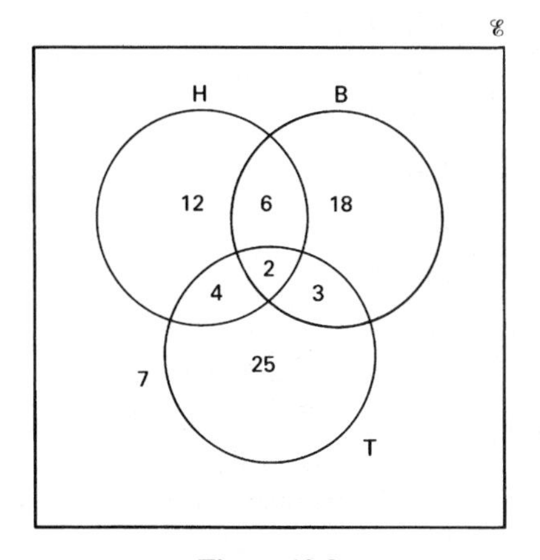

Figure 10.8

(i) How many women were interviewed?
(ii) How many women played tennis?
(iii) How many women play just two of the sports?

Solution

(i) We need to find the total of all numbers in the diagram.

$$7 + 12 + 6 + 2 + 4 + 3 + 25 + 18 = 77.$$

(ii) Here all numbers within the T circle are required.

i.e. $4 + 2 + 3 + 25 = 34.$

(iii) The centre region must not be included, because these play all 3 sports, and it is the number who play exactly 2 sports we need.

i.e. $4 + 6 + 3 = 13.$

Exercise 10(i)

1 Draw Venn diagrams to illustrate the following sets:

(i) $A \cap B'$ (ii) $(A \cup B)'$

(iii) $A \cap (B \cup C)$ (iv) $A' \cup B$.

2 If $n(A) = 20$, $n(B) = 18$, and $n(A \cup B) = 30$, find $n(A \cap B)$.

3 Newspapers are delivered to 40 houses. Each house receives either one copy of the *Times* or one copy of the *Gazette* or one copy of each. In all, 26 copies of the *Times* and 24 copies of the *Gazette* are delivered. Find the number of houses which receive the *Times* only.

4 The universal set $\mathscr{E}$ is defined as

$\mathscr{E} = \{\text{whole numbers from 2 to 25 inclusive}\}$,

and P, Q and R are subsets of $\mathscr{E}$ such that

$P = \{\text{multiples of 3}\}$

$Q = \{\text{square numbers}\}$

$R = \{\text{numbers, the sum of whose digits is 8}\}$.

Find:

(i) $P \cap Q$

(ii) $P \cap Q'$

(iii) $(R \cup Q) \cap P'$.

5 A student interviewed 70 people to ask them which channels they had watched on TV the night before. He discovered that:

52 watched BBC1
24 watched BBC2
22 watched Channel 4
14 watched BBC1 and BBC2 only
8 watched BBC2 and Channel 4
13 watched BBC1 and Channel 4
2 watched all three channels
2 watched BBC2 only

Draw a Venn diagram, letting

$B = \{$people who watched BBC1$\}$,

$C = \{$people who watched BBC2$\}$,

$F = \{$people who watched Channel 4$\}$.

Complete the Venn diagram with the information given and use it to find, for the 70 people interviewed:

(i) the number who did not watch TV at all

(ii) the number who watched Channel 4 only.

6 A total of 90 copies of either the *Times*, the *Mail* or the *Express* was delivered to the 60 houses in a certain street.

Each house took at least one paper.

Five houses each took a copy of each paper.

No house took only the *Mail*.

Seven houses took both the *Times* and the *Mail*, but not the *Express*.

Nine houses took both the *Mail* and the *Express*, but not the *Times*.

Draw a Venn diagram to illustrate this information and find

(i) the number of houses which took both the *Times* and the *Express*, but not the *Mail*

(ii) the least and the greatest number of copies of the *Times* that might be required.

[UCLES]

7 (i) In a survey of shoppers it was found that 60% were pensioners and 30% were male. If 60% of the pensioners were female, what percentage of the shoppers were female non-pensioners? Show the information on a Venn diagram.

(ii) A survey of cars in a car park revealed that the most popular optional extras were radios, clocks and sunroofs.

6 of the cars had radios and sunroofs, 14 had radios and clocks and 5 had clocks and sunroofs but did not have radios.

20 had sunroofs, 60 had radios and 22 had clocks but neither radios nor sunroofs.

(a) If x cars had clocks, radios and sunroofs, complete a Venn diagram showing all the above information.

(b) If there were 120 cars in the car park, write down, on the Venn diagram, how many cars had none of these optional extras (answer in terms of x).

(c) If 19 cars had exactly 2 of these optional extras, write down an equation and solve it to find x.

11 Probability

11.1 Definition

Probability is the study of **events** which may or may not happen. The value given to the *likelihood* of an event happening is called the probability of the event, and is usually expressed as a fraction or decimal in the range 0 (impossible) to 1 (certainty). [Sometimes the fraction is changed to a *percentage*.] In order to calculate this numerical value, we need to consider possible outcomes of the situation under consideration.

If a fair die is rolled, then there are *six* possible outcomes.

Figure 11.1

If asked what is the probability that the die lands showing 4, then we say that since there are 6 possible outcomes, and 1 is *favourable* to the required event, then the probability is defined as:

$$\text{P(score is 4)} = \frac{\text{number of favourable outcomes}}{\text{total number of outcomes}} = \frac{1}{6}$$

↑
this stands for
the probability
that the score is 4

A more involved example will show that it is not always quite so straightforward to calculate the probability.

Suppose a red die and a blue die are rolled together, and the total score obtained by adding the score on each die is found. How would we find the probability that the total score is 6?

The set of all possible outcomes is sometimes referred to as the possibility space. Since each die can land in six ways, the *total* number of outcomes is $6 \times 6 = 36$. These can be represented in a diagram similar to that in Figure 11.2.

red die							
	1	2	3	4	5	⑥	7
	2	3	4	5	⑥	7	8
	3	4	5	⑥	7	8	9
	4	5	⑥	7	8	9	10
	5	⑥	7	8	9	10	11
	6	7	8	9	10	11	12
		1	2	3	4	5	6
				blue die			

Figure 11.2

It can be seen that there are 5 occasions when the score is six.

We say that

$$P(\text{score is } 6) = \frac{5}{36}.$$

These examples are both calculated from theory, and give us a **theoretical probability**.

In reality, if we rolled two dice a number of times, the score would not be six exactly 5/36 of the time. You would have to roll them for a very large number of times to get close to this value. Very often, we have to state an **empirical probability** based on experience (or past records). For example, if a woman is expecting a child, and we wanted to know the probability that it was a girl, then if a survey of hospitals showed that of 500 children born, 280 were girls, the empirical probability of it being a girl is:

$$P(\text{girl}) = \frac{280}{500} = \frac{28}{50} = \frac{14}{25} \text{ or } 0.56.$$

The following Investigation emphasises the connection between theoretical and empirical probability.

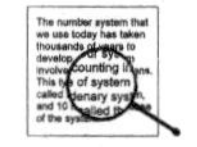

Investigation XXI: Rolling dice

This investigation is about the laws of probability.

Imagine you are throwing two dice (1 white, 1 red) from a shaker.

(i) Complete Table 11.1 below to show the possible scores you can obtain.

blue die

red die	1	2	3	4	5	6
1						
2					7	
3						
4		6				
5				9		
6						

Table 11.1

(ii) When you have completed Table 11.1, you will be able to see the relative frequency with which each score occurs. Hence you can complete Table 11.2 to show these frequencies.

Score	2	3	4	5	6	7	8	9	10	11	12
Frequency	1							4			
Probability											

Table 11.2

Since the number of possible outcomes is 36, the probabilities can also be completed.

(iii) Now experiment by rolling the dice 36 times. Do you get the same frequencies as in part (ii)? If not, why not?

11.2 Probability from graphs

Example 11.1

Figure 11.3 shows a cumulative frequency curve of the average weekly wages earned in a factory by 2500 employees. Estimate:

(a) the probability that an employee selected at random earned not more than £120

(b) the value of x, if the probability than an employee earns more than £x is 0.4.

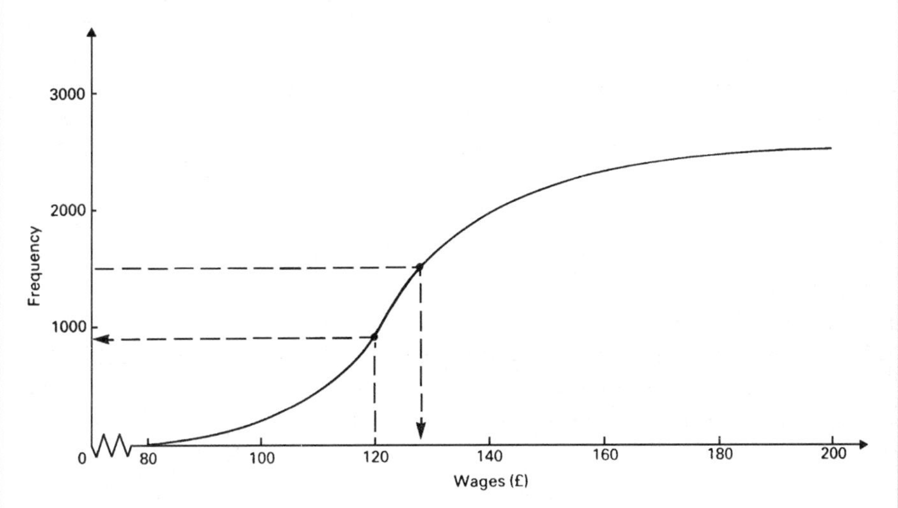

Figure 11.3

Solution

(a) Reading £120 on the graph gives a frequency of approximately 900. Hence the required probability is

$$\frac{900}{2500} = \frac{9}{25} = 0.36.$$

(b) If the probability of earning more than £x is 0.4, then the probability of earning less than £x is

$$1 - 0.4 = 0.6$$

Now $0.6 \times 2500 = 1500$, and so 1500 must earn less than £x.

If we read the frequency of 1500 on the graph, then we get a wage of £128.

Exercise 11(a)

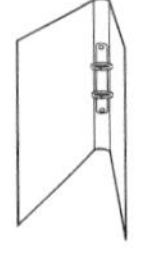

1 What is the probability that a single roll of a die results in a score of 2 or more?

2 A card is chosen at random from a well-shuffled pack of 52 playing cards. What is the probability that it is: (i) a spade (ii) a five (iii) the ace of clubs (iv) a picture card?

3 The following table shows the marks out of 8 scored by a group of students on an assignment.

Marks	0	1	2	3	4	5	6	7	8
Number of students	1	2	4	6	8	5	2	2	1

Calculate:

(i) the mean mark

(ii) the probability of selecting at random a student who obtained a mark

(a) greater than the mean

(b) 5 or 6.

4 Two unbiased coins are spun together. What is the probability of obtaining: (i) two heads (ii) at least one head. If the coins are spun again, what is the probability that two heads were obtained on both occasions?

5 A bag contains twenty discs numbered 1–20. A disc is drawn from the bag. What is the probability that the number on it: (i) is a multiple of 3 (ii) contains a 5.

6 A bag contains 8 red discs, and x blue discs. If the probability of drawing a blue disc is 5/7, what is the value of x?

7 Two dice are rolled. Draw a diagram to represent all possible outcomes. What is the most likely total that you will get by adding the value on each die?

11.3 Using Venn diagrams

If the possible outcomes in a situation are regarded as the **outcome space** (denoted here by S), and various events as subsets of the outcome space, then it is possible to use a Venn diagram together with set notation, to represent the probabilities.

[*Note*: It is possible to omit this section and the next, and go straight to Section 11.5.]

Suppose the outcome space is the 52 possible outcomes when a card is chosen at random from a pack of cards.

Let D be the event of selecting a diamond and K the event of selecting a king.

D contains 13 possible outcomes and so we can write $n(D) = 13$.

K contains 4 possible outcomes and so $n(K) = 4$.

However $D \cap K$ is the king of diamonds, so the diagram would be as in Figure 11.4.

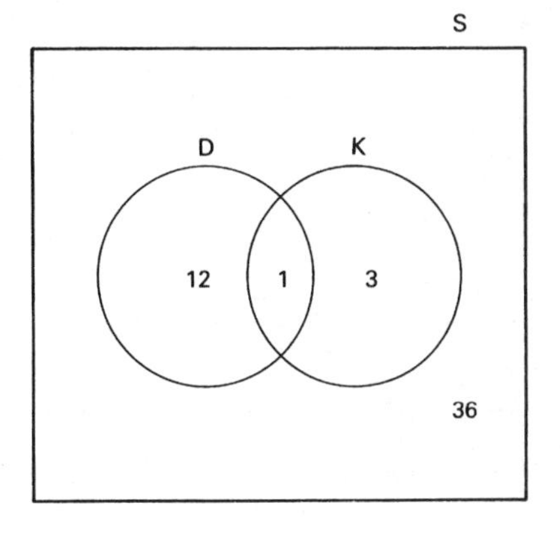

Figure 11.4

We have

$$\text{P(drawing a diamond)} = \frac{n(D)}{n(S)} = \frac{13}{52} = \frac{1}{4}$$

$$\text{P(drawing a king)} = \frac{n(K)}{n(S)} = \frac{4}{52} = \frac{1}{13}$$

$$\text{P(drawing the king of diamonds)} = \mathrm{P}(K \cap D)$$

$$= \frac{n(K \cap D)}{n(S)} = \frac{1}{52}$$

$$\text{P(Not drawing a diamond)} = P(D')$$

$$= \frac{n(D')}{n(S)} = \frac{39}{52} = \frac{3}{4}.$$

Example 11.2

A card is drawn from an ordinary pack of 52 playing cards. Find the probability of drawing a red picture card, illustrating your answer with a Venn diagram.

Solution

Let R be the event of drawing a red card and P the event of drawing a picture card.

The Venn diagram in Figure 11.5 shows the number of possibilities in each region. Note that 6 picture cards will be red.

$$\begin{aligned}\text{P(red picture card)} &= \text{P}(R \cap P)\\ &= \frac{n(R \cap P)}{n(S)}\\ &= \frac{6}{52} = \frac{3}{26}.\end{aligned}$$

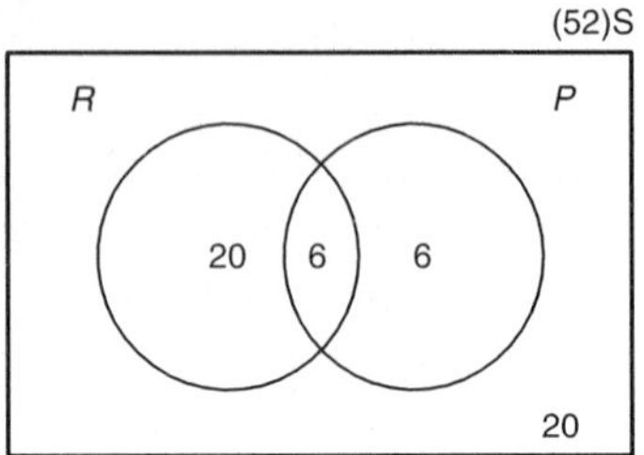

Figure 11.5

Exercise 11(b)

1 Determine the probabilities of the following outcomes, illustrating your answer with a Venn diagram.

(i) A red ace appears in drawing a card from an ordinary pack of playing cards.

(ii) A white marble appears in drawing a single marble from a bag containing 4 white, 3 red and 5 blue marbles.

(iii) Choosing either a b or an i from the word probability.

(iv) An even number or a square number is drawn from a pack of cards numbered 1 to 25.

(v) If two dice are rolled, at least one of them shows a six.

2 $S = \{1, 2, 3, \ldots, 30\}$, $A = \{\text{multiples of } 7\}$, $B = \{\text{multiples of } 5\}$, $C = \{\text{multiples of } 3\}$.

If a member x of S is chosen at random, find the probability that:

(i) $x \in A \cup B$ (ii) $x \in B \cap C$ (iii) $x \in A \cap C'$

3 When a group of children were asked about their pets, 15% said they did not have one. Of the rest, 48% said they had a cat, 50% said they had a dog, but 20% had both a cat and a dog. Using a Venn diagram, find the probability that a child chosen at random from this group: (i) only has a dog (ii) doesn't have a cat.

4 Figure 11.6 represents the likes in outdoor pursuits of 100 students. If a student is chosen at random from this group, find the probability (expressed as a fraction in its lowest term) that

(i) they like one pursuit only

(ii) they do not like jogging

(iii) they like either swimming or fishing.

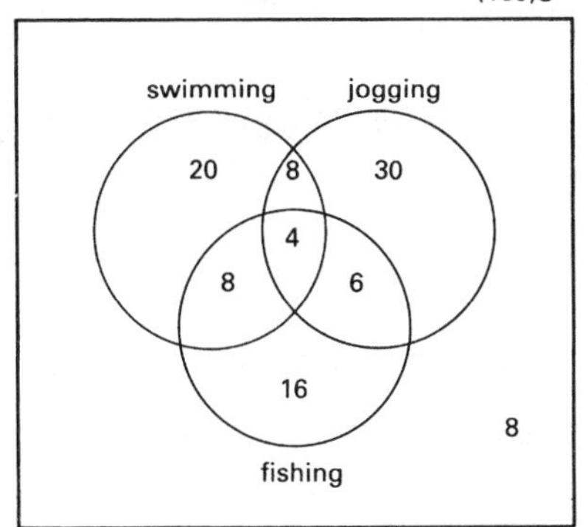

Figure 11.6

5 Anne and Jason are members of the local tennis club. There are 15 men (including Jason) in the club and 12 women (including Anne). A mixed doubles pair is selected at random by choosing one man and one woman. Calculate the probability that: (i) Anne is in the team but Jason is not (ii) at least one of them is in the team.

Illustrate your answers with a Venn diagram.

11.4 Sum and product rules

If you add the number of elements in A and B together, then clearly the elements in common would be included *twice*. Hence to find the number of elements in the union $A \cup B$, we use the formula

$$n(A \cup B) = n(A) + n(B) - n(A \cap B)$$

Divide each term by $n(S)$ to give

$$\frac{n(A \cup B)}{n(S)} = \frac{n(A)}{n(S)} + \frac{n(B)}{n(S)} - \frac{n(A \cap B)}{n(S)}$$

These can be expressed as probabilities, to give

Rule

$$\mathrm{P}(A \cup B) = \mathrm{P}(A) + \mathrm{P}(B) - \mathrm{P}(\mathrm{A} \cap \mathrm{B})$$

If A and B do not overlap, the events cannot happen together, and we say that events A and B are **mutually exclusive**, so $\mathrm{P}(A \cap B) = 0$.

Hence $\mathrm{P}(A \cup B) = \mathrm{P}(A) + \mathrm{P}(B)$ if A and B are **mutually exclusive**. This is the **addition rule** of probabilities.

$P(A \cup B)$ is the probability that *at least one* of events A and B occur.

Referring to the Venn diagram in Figure 11.4, we have that

$$\text{P(drawing either a diamond or a king or both)} = P(D) + P(K) - P(D \cap K)$$
$$= \frac{1}{4} + \frac{1}{13} - \frac{1}{52} = \frac{4}{13}$$

In this example, it can be seen that

Rule

$$P(D \cap K) = \frac{1}{52} = \frac{1}{13} \times \frac{1}{4} = P(D) \times P(K)$$

This is known as the **multiplication rule** for probabilities, and is only true if A and B are independent events. In other words, if the outcome of A does not affect the outcome of B.

11.5 Complement

If a die is rolled, the probability that the score is 4 is given by

$$P(4) = \frac{1}{6}$$

If $F = \{4\}$, then $F' = \{1, 2, 3, 5, 6\}$, the complement of F would be the event of *not* scoring 4.

We have $P(F') = \frac{5}{6}$

Hence $P(F) + P(F') = 1$, or

Rule

$$P(F') = 1 - P(F)$$

This statement is true for all events and their complements.

Example 11.3

The probability that an event A occurs is $\frac{1}{5}$ and the probability that event B occurs is $\frac{1}{3}$. Assuming that A and B are independent events, find the probability that:

(a) A and B both occur
(b) neither event occurs
(c) at least one event occurs
(d) only one of the events occurs.

Solution

(a) $P(A \cap B) = P(A) \times P(B)$ since A and B are independent.

$$= \frac{1}{5} \times \frac{1}{3} = \frac{1}{15}$$

(b) The probability that an event doesn't occur is $P(A')$ where A' is the complement of A

But $P(A') = 1 - P(A) = 1 - \frac{1}{5} = \frac{4}{5}$

Also $P(B') = 1 - P(B) = 1 - \frac{1}{3} = \frac{2}{3}$

And so $P(A' \cap B') = P(A') \times P(B')$

$$= \frac{4}{5} \times \frac{2}{3} = \frac{8}{15}.$$

A' and B' are independent since A and B are independent.

(c) $A \cup B$ is the event of at least one occurring.

Therefore

$$P(A \cup B) = P(A) + P(B) - P(A \cap B)$$
$$= \frac{1}{5} + \frac{1}{3} - \frac{1}{15} = \frac{7}{15}.$$

(d) Exactly one occurrence excludes the situations where both occur or where neither occur.

P(both occur) $= \frac{1}{15}$ (part (a))

P(neither occurs) $= \frac{8}{15}$ (part (b))

Hence P(exactly one occurs) $= 1 - \frac{1}{15} - \frac{8}{15} = \frac{2}{5}$.

Investigation XXII: Expectation

If three coins are tossed, the possible outcomes are HHH, HHT, HTH, HTT, THH, THT, TTH, TTT.

If we look at the number of heads obtained, we can tabulate the **probability distribution** of the number of heads as follows:

No. of heads	0	1	2	3
Probability	$\frac{1}{8}$	$\frac{3}{8}$	$\frac{3}{8}$	$\frac{1}{8}$

If we find

$$\frac{1}{8}\times 0+\frac{3}{8}\times 1+\frac{3}{8}\times 2+\frac{1}{8}\times 3=\frac{12}{8}=1\tfrac{1}{2},$$

we find the expected or average number of heads each time we spin the three coins.

10	■	3	2
■	2	5	■
1	■	3	1
5	1	2	■

At the local scout fete, a game consists of throwing a dart at a board like that in the diagram. You pay 2p for one throw. If it lands on a black square, you lose, otherwise you get the amount of money back shown in the square on which the dart lands. Find the expected return for each throw. Is the game worth playing?

Extension: Design your own game, and analyse the expected winnings.

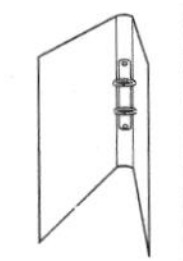

Exercise 11(c)

1 The probability that United will win next Saturday is $\frac{1}{4}$, and the probability that Rovers will win is $\frac{2}{3}$. Find the probability that: (i) both teams win (ii) at least one of them wins.

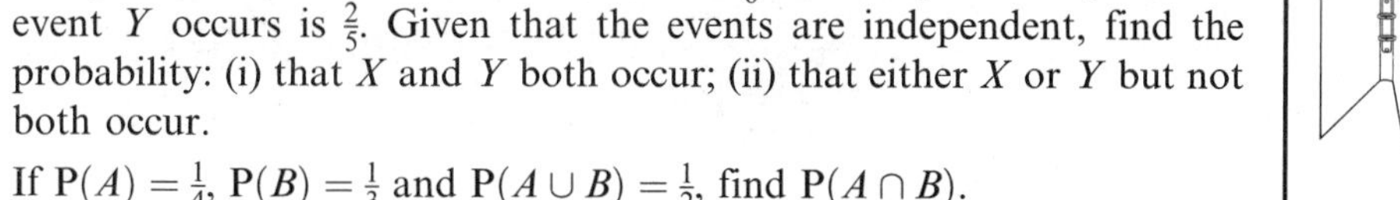

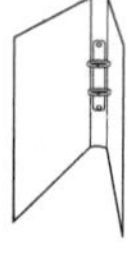

2 The probability that an event X occurs is $\frac{1}{6}$ and the probability that an event Y occurs is $\frac{2}{5}$. Given that the events are independent, find the probability: (i) that X and Y both occur; (ii) that either X or Y but not both occur.

3 If $P(A) = \frac{1}{4}$, $P(B) = \frac{1}{3}$ and $P(A \cup B) = \frac{1}{2}$, find $P(A \cap B)$.

4 If $P(X) = \frac{1}{2}$ and $P(Y) = \frac{1}{3}$ where X and Y are independent events, find:
(i) $P(X \cup Y)$ (ii) $P(X \cap Y')$ (iii) $P(X' \cup Y')$.

5 If $P(A) = \frac{1}{3}$, $P(A \cap B') = \frac{1}{4}$ and A and B are independent, find:
(i) $P(B)$ (ii) $P(B \cap A')$
Illustrate your answers with a Venn diagram.

6 In a set of 100 students, 37 studied mathematics, 32 physics and 31 chemistry. Nine students took both mathematics and physics, 12 took mathematics and chemistry, 11 physics and chemistry, and 28 none of these subjects. Assuming that the selection is random and with replacement, so that each choice is made from 100 students, calculate:

(i) the probability that the first student chosen was studying all three subjects;
(ii) the probability that the first student took mathematics only;
(iii) the probability that of the first two chosen, at least one was studying other subjects.

11.6 Tree diagrams

Example 11.4

Tim and Jason are playing darts. The probability that Tim throws a double is $\frac{1}{3}$, and the probability that Jason throws a double is $\frac{2}{5}$. If Tim and Jason each throw one dart, what is the probability that:

(a) neither scores a double;
(b) exactly one double is scored.

Solution

Although this can be solved using the methods of Section 11.4, we will now look at the use of a tree diagram.

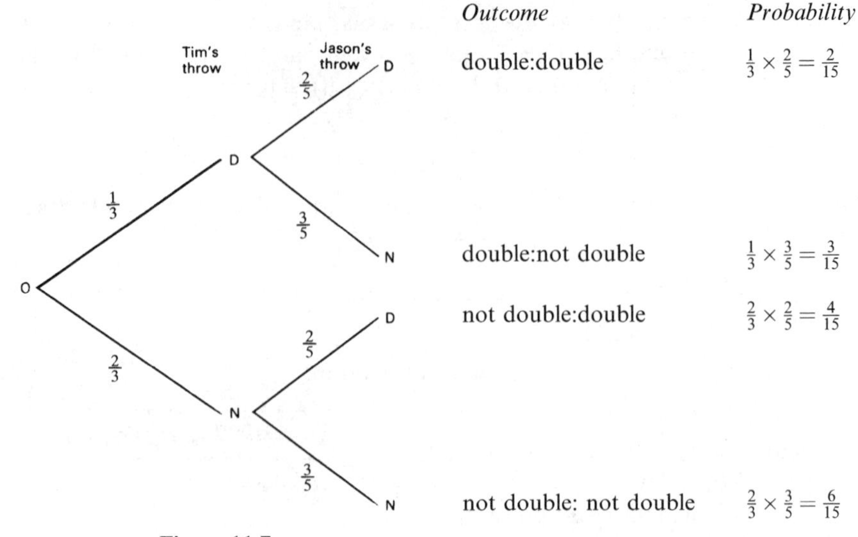

Figure 11.7

In Figure 11.7, D stands for double, N for not a double. The probability of each outcome is written alongside a **branch** of the tree. There are four different routes of travelling along the tree from 0, each giving a different outcome. The probability for each of these outcomes is obtained by *multiplying* the probabilities along that route.

(a) $\text{P(neither scores a double)} = \frac{2}{3} \times \frac{3}{5} = \frac{2}{5}.$

(b) The event 'exactly one double' means Tim throws a double and Jason does not, or Tim does not throw and double and Jason does. There are, therefore, two possible outcomes on the tree, and the probability of each outcome is added together.

$$\text{Hence P(exactly one double)} = \frac{1}{3} \times \frac{3}{5} \times \frac{2}{3} \times \frac{2}{5}$$
$$= \frac{3}{15} + \frac{4}{15} = \frac{7}{15}.$$

Example 11.5

A bag contains 5 red discs and 4 blue discs. If 3 discs are drawn from the bag without replacement, find the probability that:

(a) all three are blue

(b) there are two red and one blue

(c) the second disc is red.

Solution

The tree is shown in Figure 11.8. Note that because the discs are not replaced, the probability of drawing, say, a blue changes on each branch.

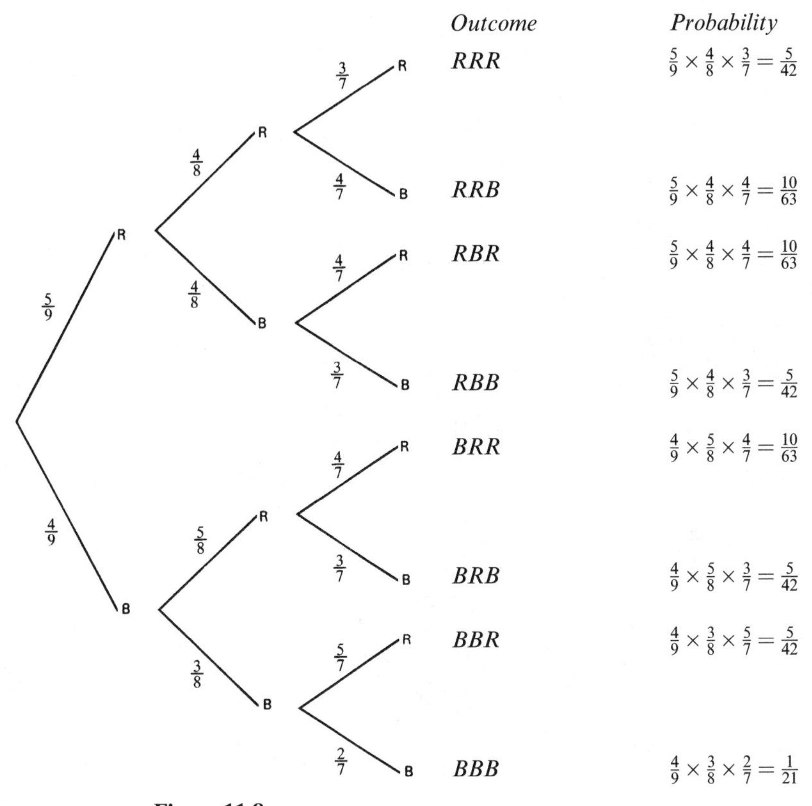

Figure 11.8

(a) all blue (*BBB*) $= \dfrac{4}{9} \times \dfrac{3}{8} \times \dfrac{2}{7} = \dfrac{1}{21}$

(b) two red and one blue is *RRB* or *RBR* or *BRR* $= \dfrac{10}{63} + \dfrac{10}{63} + \dfrac{10}{63} = \dfrac{10}{21}$

(c) the second disc red is *RRR* or *RRB* or *BRR* or *BRB* $= \dfrac{5}{42} + \dfrac{10}{63} + \dfrac{10}{63} + \dfrac{5}{42} = \dfrac{5}{9}.$

Example 11.6

A game consists of spinning a coin. If a head shows you roll a fair die once and note the score. If a tail shows, you select a card from a shuffled pack of cards. If it is a heart, you score 1, a diamond you score 2, a club you score 3, and a spade you score 4. A person wins on the game if his score is 3 or 4. Do you think this is a reasonably fair game to place a bet on?

Solution

The tree for this problem is shown in Figure 11.9

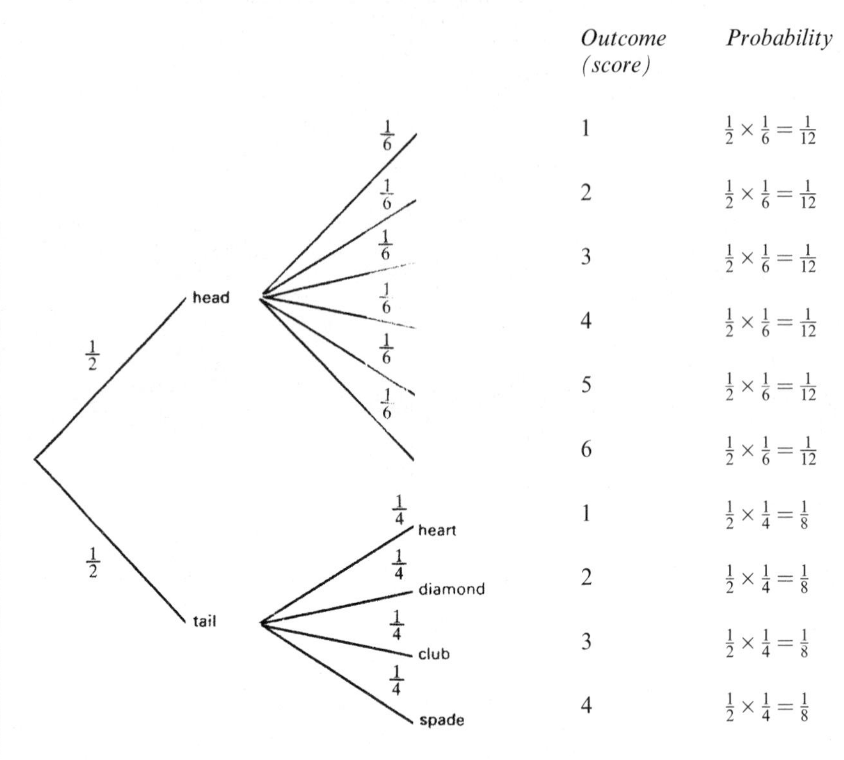

Figure 11.9

$$\text{P(score is 3 or 4)} = \frac{1}{12}+\frac{1}{12}+\frac{1}{8}+\frac{1}{8}=\frac{5}{12}=0.42$$

Since this is less than 0.5, you are *more likely to lose* than win.

Exercise 11(d)

1 The probability of City hockey team winning at home is 0.6 and of winning away is 0.1.

(i) (a) What is the probability of City NOT winning at home?

(b) What is the probability of City NOT winning away?

(ii) Complete the following tree diagram for two consecutive matches, one home and one away.

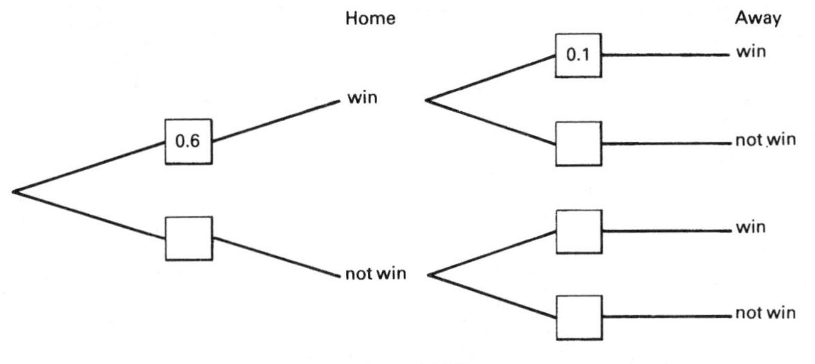

Figure 11.10

(iii) (a) What is the probability that City wins both matches?

(b) What is the probability that City wins at least one of the matches?

2 Each letter of the word

CAERONNEN

is written on identical squares of cardboard, one letter per card, and the cards placed in a bag.

(i) The cards are well shuffled and a card is taken out of the bag.

(a) What is the probability that the letter on the card is R?

(b) Which letter is most likely to be chosen?
What is the probability of this letter being chosen?

(ii) A card is taken out of the bag and not replaced.
Another card is then taken out of the bag.
What is the probability that the two cards have the letter E on them?

[WJEC]

3 A die numbered 1, 1, 2, 3, 4, 4 is rolled twice. Find the probability that the total score on the two goes is: (i) odd (ii) greater than 6.

4 A coin is biased in such a way that the probability of a head appearing if it is spun is $\frac{2}{3}$. The coin is spun three times.

What is the probability that you get: (i) three heads (ii) one head (iii) at least two heads?

5 The probability of a male birth is 0.52. If a woman has three children, what is the probability that:

(i) all three are girls
(ii) at least two are boys.

6

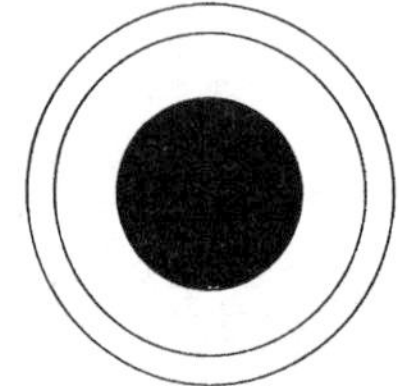

A target at a rifle range consists of a centre of radius 3 cm which scores 100 and two concentric circles of radius 5 cm and 6 cm, which score 50 and 20 respectively.

Assuming that the target is always hit, and that the probability of hitting a particular score is proportional to the area for that score, find the probabilities of scoring 100, 50 and 20 respectively. What is the expected total score for 10 shots?

7 Ann and Jane play a game against each other which starts with Ann aiming to throw a bean bag into a circle marked on the ground.

(i) The probability that the bean bag lands entirely inside the circle is $\frac{1}{2}$, and the probability that it lands on the rim of the circle is $\frac{1}{3}$. Show that the probability that the bag lands entirely outside the circle is $\frac{1}{6}$. What are the probabilities that two successive throws land:

(a) both outside the circle

(b) the first on the rim of the circle and the second inside the circle?

(ii) Jane then shoots at a target on which she can score 10, 5, or 0. With any one shot, the probability that she scores 10 is $\frac{2}{5}$, the probability that she scores 5 is $\frac{1}{10}$, and the probability that she scores 0 is $\frac{1}{2}$. With exactly two shots, what are the probabilities that she scores:

(a) 20 (b) 10?

8 When John and Henry play darts, the probability that John wins a game is 0.7 and that Henry wins a game is 0.3. They play a match to be decided by the first player to win 3 games (the match is finished as soon as this happens).

Calculate the probabilities that:

(i) John wins the first three games

(ii) John loses the first game but wins the next

(iii) John wins by three games to one

(iv) John wins the match in not more than four games.

9 Four cards numbered 1 to 4 are placed face downwards on the table.

(i) Two cards are picked up at the same time. Find the number of ways that this can be done. Hence find the probability that if two cards are picked up at random from the table, one of them is numbered 4.

(ii) In a second experiment, one card is picked up from the table and marked on the face. The card is then replaced and a second card is picked up and marked. Find the probability that: (a) the same card is then marked twice (b) that the numbers of the two marked cards add up to 5.

12 Coordinates and straight line graphs

12.1 Using straight line graphs

(i) Conversion

A straight line graph illustrates the **relationship** between two quantities. We have already looked at scatter diagrams in Section 9.5 which involve straight lines. In this chapter a variety of topics will be looked at which involve the use of straight lines. A very simple application is in the conversion from one quantity to another. Most people are aware that in measuring temperature, 0°C is the same as 32°F and 100°C is the same as 212°F. If these two relationships are plotted on a diagram similar to Figure 12.1, and the points are joined up with a straight line then we obtain a **conversion graph** between temperature in °C and temperature in °F. This graph can then be used to convert any temperature in °C to °F or °F to °C.

(i) 100°F is the same as 38°C

(ii) 80°C is the same as 175°F

(iii) On the horizontal axis, it can be seen that 0°F is the same as −18°C.

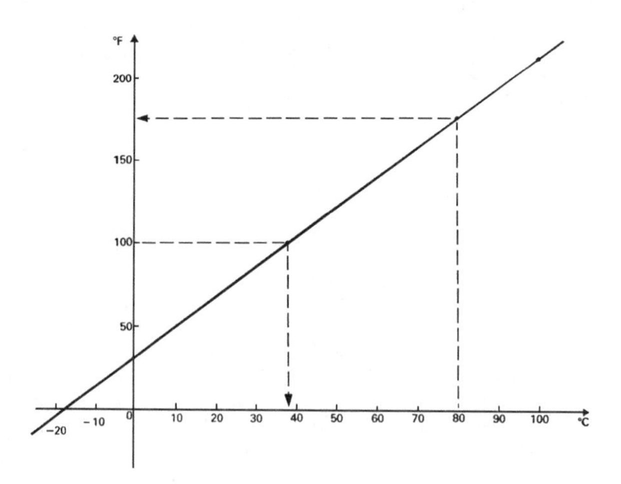

Figure 12.1

(ii) Proportion

If two quantities are **proportional** to each other, then a graph of their relationship will be a straight line through the origin. For example, if material can be bought at a cost of £4.50 per metre, then the cost (C) is proportional to the length (L) purchased.

In Figure 12.2 a graph has been drawn to show the relationship. The mathematical symbol for proportion is $\propto$. Hence $C \propto L$. In fact, it can be seen that $C = 4.5L$.

This would be the **equation** of the line, and in the next section, this idea will be expanded.

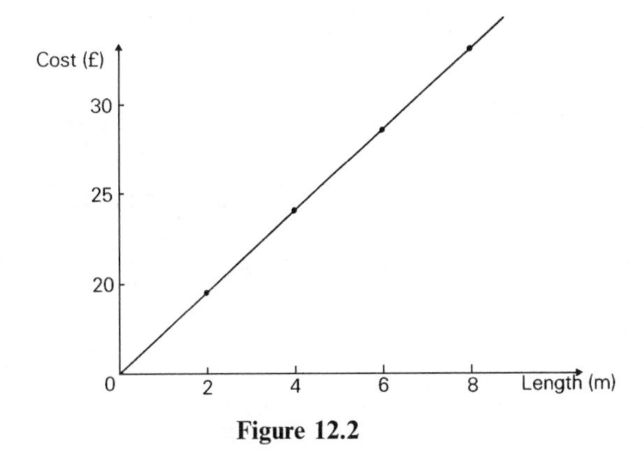

Figure 12.2

12.2 Coordinates and the equation of a straight line

The *position* of a point is defined by its **coordinates**, referred to **axes**. The *horizontal* axis is called the x-axis, and the *vertical* axis is called the y-axis. In Figure 12.3 A has coordinates $(3, 4)$. The x-coordinate is 3, and the y-coordinate is 4. The point B has coordinates $(-2, -3)$. The x-coordinate is -2, and the y-coordinate is -3.

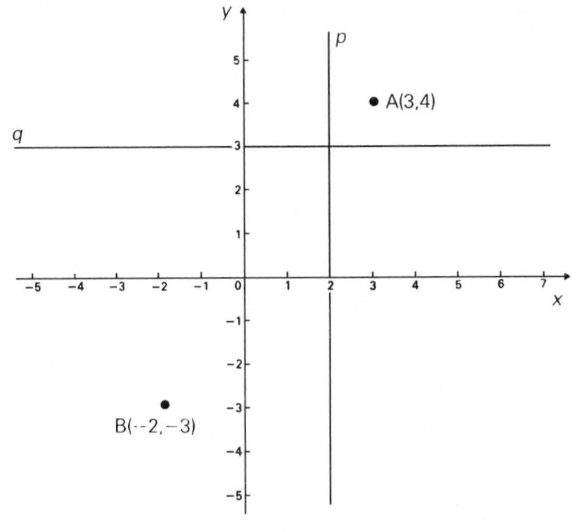

Figure 12.3

If you look at line p in Figure 12.3 you will see that every point on p has an x-coordinate of 2. We say that the **equation** of p is $x = 2$. Similarly, on line q, the y-coordinate is always 3, hence its equation is $y = 3$.

It should be noted that the x-axis can be referred to as the line $y = 0$, and the y-axis can be referred to as $x = 0$. If the lines are not parallel to the x or y axis, then the problem becomes slightly more involved.

Rules

$x = k$ is parallel to the y-axis

$y = k$ is parallel to the x-axis

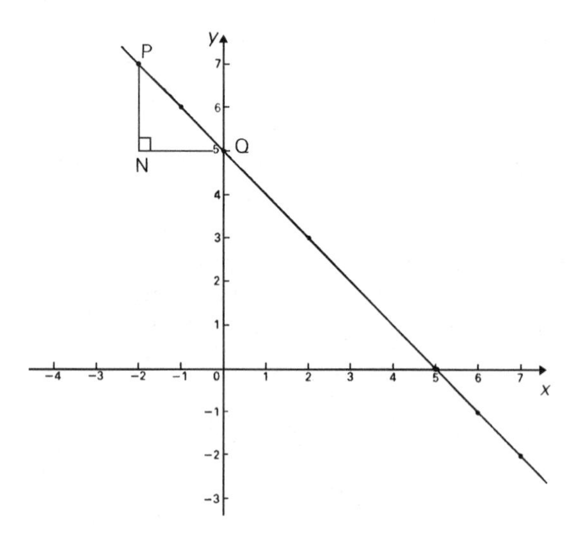

Figure 12.4

Look at the points shown in Figure 12.4. The coordinates of the points are $(-2, 7)$, $(-1, 6)$, $(0, 5)$, $(2, 3)$, $(5, 0)$, $(6, -1)$ and $(7, -2)$. Clearly, both the x- and y-coordinates are changing, but careful inspection shows that if you add the x- and y-coordinates, you always get 5.

Hence $\quad x + y = 5$

or $\qquad\quad y = -x + 5$

This would be referred to as the **equation of the given line**.

If you are given the equation of the line it can be **plotted** by working out a **table of values** as we now show.

For the equation $y = 3x + 2$ for values of x between -3 and 3, a table is set up as follows:

x	-3	-2	-1	0	1	2	3
$y = 3x + 2$	-7	-4	-1	2	5	8	11

The values of y are found by substituting each value of x into the equation $y = 3x + 2$, so for $x = -2$,

$$y = 3 \times -2 + 2$$
$$= -6 + 2 = -4, \text{ etc.}$$

The points $(-3, -7)$, $(-2, -4)$, $(-1, -1)$, ... can now be plotted, choosing suitable scales to fit the graph on the piece of graph paper you are using. The points should lie on a straight line which can then be drawn as in Figure 12.5.

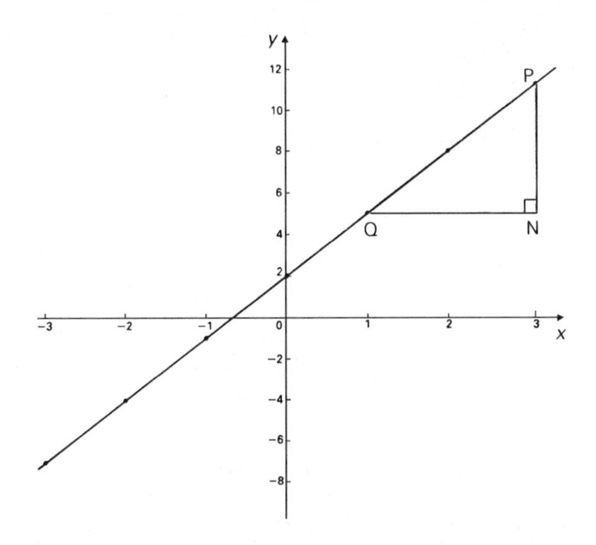

Figure 12.5

The lines drawn in Figure 12.4 and Figure 12.5 are sloping at different angles. The idea of *steepness* is measured by calculating the **gradient** of the line.

In Figure 12.5 the gradient is measured by

$$\tan \angle PQN = \frac{PN}{NQ} = \frac{6}{2} = 3.$$

Care should be taken to read the scales correctly when stating the values of PN and NQ. Notice here that the scales used on the x-axis and y-axis are different.

In Figure 12.4 the line slopes **down** towards the right. To distinguish this from the previous example the gradient is **negative**.

Although

$$\tan \angle PQN = \frac{PN}{NQ} = \frac{2}{2} = 1$$

the gradient of the line is -1.

12.3 $y = mx + c$

> **Rule**
>
> The equation of a straight line can be written $y = mx + c$, where m = gradient of the line, and c is the value of y where the graph cuts the y-axis (this is often called the y-intercept)

If the equation is written in a different way, it must be rearranged if you want to find m and c.

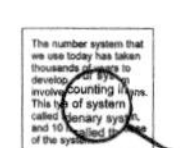

Investigation XXIII: $y = mx + c$

Choose sensible values for x and y and try the following.

(i) On one sheet of graph paper, plot the lines $y = x + 3$, $y = x + 1$, $y = x - 2$. What do you notice about the lines?

(ii) On one sheet of graph paper, plot the lines $y = x + 3$, $y = 2x + 3$, $y = 3x + 3$, $y = -2x + 3$. What do you notice about these lines?

The graphs you have plotted should enable you to see how changing m and c affect the position and gradient of the line.

Example 12.1

Find the gradient and y-intercept for the following lines, and give a sketch to show the position of the line:

(a) $2y = 3x - 1$

(b) $3x + 4y = 12$

(c) $2x - 5y = 8$.

Solution

(a) $2y = 3x - 1$

Divide by 2 to give: $y = \frac{3}{2}x - \frac{1}{2}$

The gradient is $\frac{3}{2}$, the y-intercept is $-\frac{1}{2}$.

(b) $3x + 4y = 12$, so $4y = -3x + 12$

Divide by 4: $y = -\frac{3}{4}x + 3$

The gradient is $-\frac{3}{4}$, the y-intercept is 3.

(c) $2x - 5y = 8$, so $-5y = -2x + 8$

Divide by -5: $y = \frac{2}{5}x - \frac{8}{5}$

The gradient is $\frac{2}{5}$, the y-intercept is $-\frac{8}{5}$.

The graphs can now be sketched as in Figure 12.6.

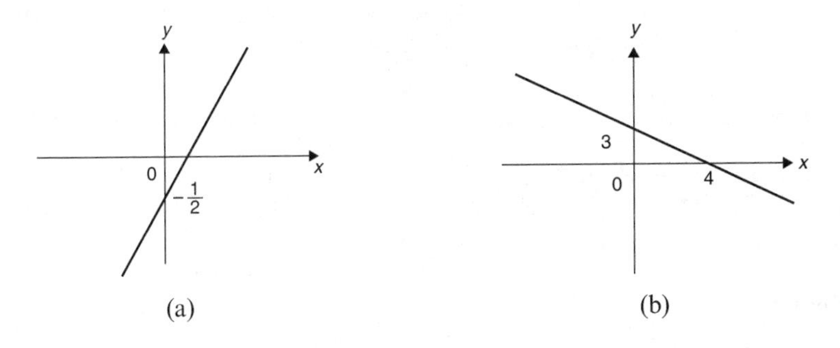

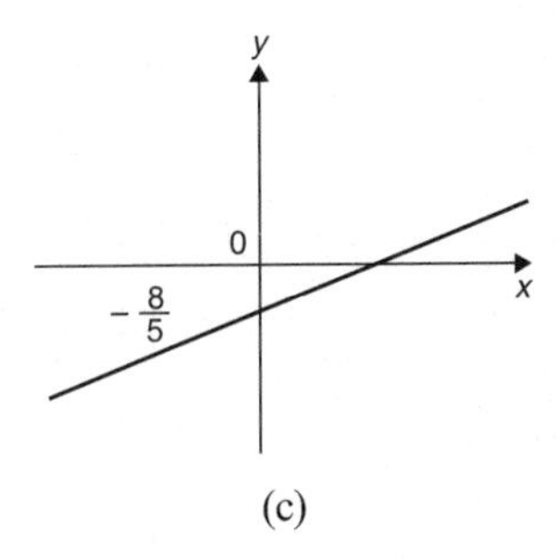

Figure 12.6

Example 12.2

Find the equation of the line that passes through the point $A(1, 2)$ and $B(3, 5)$.

Solution

Let the equation be $y = mx + c$

Since it passes through $(1, 2)$, we can put $y = 2$ when $x = 1$

So $2 = m + c$ (1)

Similarly, $x = 3$, $y = 5$

So $5 = 3m + c$ (2)

We now have two simultaneous equations to solve.

(2) – (1): $3 = 2m$, hence $m = \frac{3}{2}$

substituting in (1), $2 = \frac{3}{2} + c$, hence $c = \frac{1}{2}$

The equation of the line is $y = \frac{3}{2}x + \frac{1}{2}$.

12.4 Simultaneous equations

It is also possible to solve **simultaneous equations** by using graphs and the equation $y = mx + c$. The following example illustrates clearly the method used.

Example 12.3

Use a graphical method to solve the simultaneous equations $2x + 3y = 12$, and $4x - 3y = 8$. You should label the horizontal axis for x between -2 and 5, and the vertical axis for y between -1 and 7.

Solution

For the line $2x + 3y = 12$,

if $x = 0$; $0 + 3y = 12$, so $y = 4$
if $y = 0$; $2x + 0 = 12$, and $x = 6$

Hence the points $(0, 4)$ and $(6, 0)$ are on the line.

Find where the lines cross the axes

Also, for $4x - 3y = 8$,

if $x = 0$; $0 - 3y = 8$, so $y = -2\frac{2}{3}$
if $y = 0$; $4x - 0 = 8$, and $x = 2$

Hence the points $(0, -2\frac{2}{3})$ and $(2, 0)$ are on this line.

Plot each pair of points as shown in Figure 12.7 and join up each line. The point of intersection A gives the solution of the equations. It is $x = 3.3$, $y = 1.8$.

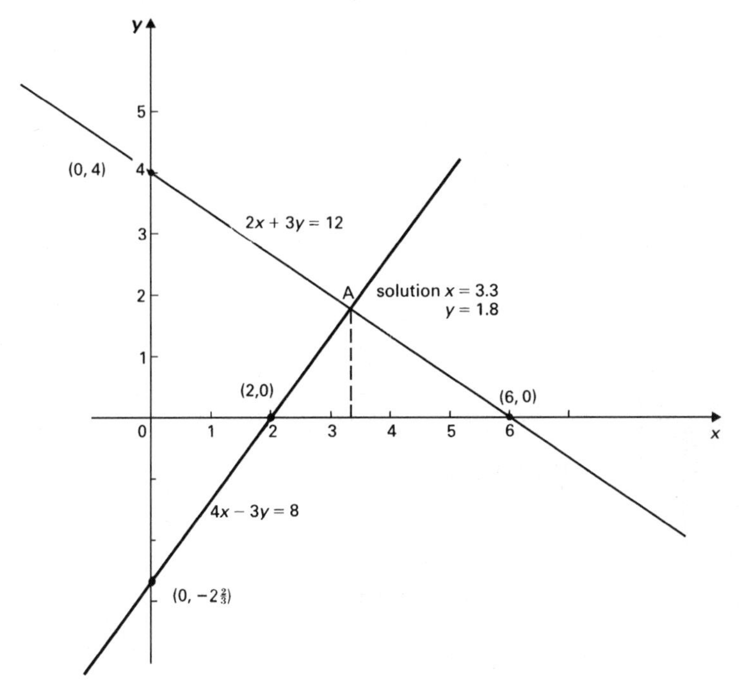

Figure 12.7

Exercise 12(a)

1 The total charge which the firm 'Self Motoring Providers' makes for hiring a car is made up of two parts: (a) a fixed basic charge (b) a charge for each mile travelled.

The total charges for three distances travelled are shown in the table.

Distance travelled (in miles)	200	400	700
Total charge (£)	47	59	77

(i) On graph paper, draw a pair of axes. Using a scale of 2 cm to 100 miles, mark your distance axis across the paper from 0 to 800 miles. Using a scale of 2 cm to £10, mark the charge axis vertically.

(ii) Mark on your graph the three points from the table.

(iii) Draw the straight line graph passing through the three plotted points.

(iv) Use your graph to find:

  (a) the distance travelled when the total charge is £56;

  (b) the fixed basic charge.

A second firm 'Motoring Everywhere Group' charges £65 for hiring a car with no extra charge for the distance travelled.

(v) Draw a straight line on your graph to represent the charge of the 'Motoring Everywhere Group'.

(vi) A motorist has to make a journey of 650 miles. Which firm is the cheaper and by how much?

(vii) What is the shortest journey for which a 'MEG' car is cheaper than a 'SMP' car?

[MEG]

2

Huw observes a bird flying directly away from a bird box. He starts his watch and finds out how far the bird is from the box at different times.

This graph in Figure 12.8 is drawn from his results.

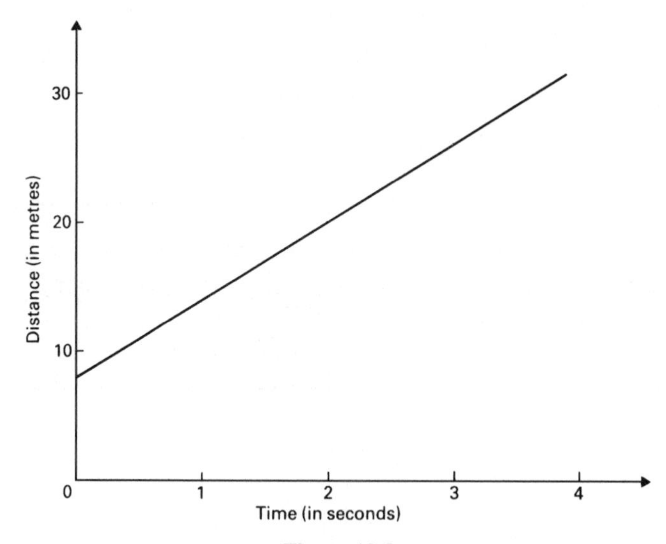

Figure 12.8

(i) How far is the bird from the box when Huw starts his watch?

(ii) How fast is the bird flying?

(iii) Write down a formula for the distance, d, the bird is from the box in terms of time, t.

[WJEC]

3 Two old cookery books offer advice on making tea.

(i) Book A says use one teaspoonful of tea per person and one for the pot.

So the formula is $t = p + 1$

where t = number of teaspoonsful, p = number of people

(a) Fill in the table for this recipe.

p	2	6	8	10	20	40
t						

(b) On graph paper, plot these points and join them up in a straight line. Label the line Recipe A.

(ii) Book B has a recipe that gives the formula

$$t = \frac{3p}{4} + 3.$$

(a) Complete the table for this recipe.

p	2	6	8	10	20	40
t		$7\frac{1}{2}$		$10\frac{1}{2}$		

(b) On the same graph, plot these points and join them up in a straight line. Label the line Recipe B.

(iii) The two recipes use the same number of teaspoonsful for a certain number of people. How many people is that?

(iv) 16 people need tea.

(a) Which recipe uses less tea for 16 people?

(b) How much less does it use?

[MEG]

4 For the following lines, state (a) the gradient of the line; (b) the point where the line cuts the y-axis.

(i) $y = 4x - 7$ (ii) $y = 1 - x$ (iii) $2y = x - 4$
(iv) $y = 4 - 3x$ (v) $2y + 4x = 1$ (vi) $4x - 3y = 24$
(vii) $y = x - 3$ (viii) $2y + 7x = 11$

5 Find the equation of the line through the point $(0, 4)$ with gradient 3.

6 Find the equation of the line which passes through the points $A(1, 2)$ and $B(3, 6)$. If the point $C(2, k)$ is also on the line, find k.

7 Find the equation of the line through the point $(2, 1)$ which is parallel to the line $y = 3x + 2$.

8 Solve the following simultaneous equations using a graphical method.

(i) $2y + 3x = 6$; $4y - 3x = 12$

(ii) $6p + q = 18$; $2p - 5q = 10$

(iii) $2m + 7n = 14$; $6m - 3n = 8$.

12.5 Travel graphs

The speed of an object is defined as the distance it travels in a certain time. So 10 km/h (or 10 kmh^{-1}) means that in 1 hour it would travel 10 km.

Since

$$\text{speed} = \frac{\text{distance}}{\text{time}},$$

it follows that

$$\text{distance} = \text{speed} \times \text{time},$$

and

$$\text{time} = \frac{\text{distance}}{\text{speed}}$$

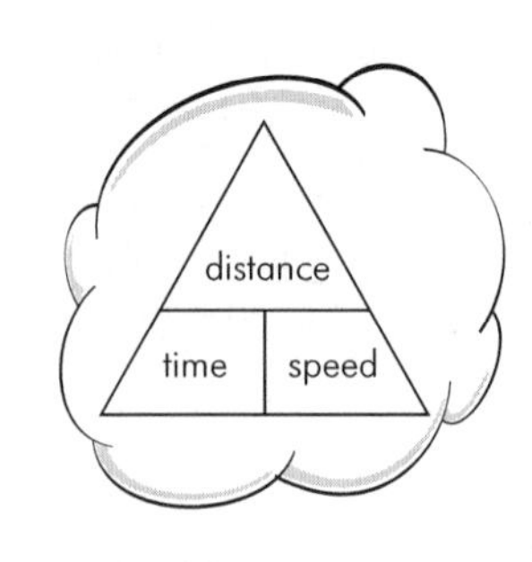

All of these quantities can be represented on a travel graph of distance (or more correctly **displacement**) plotted against time. The word displacement means distance in a certain direction.

You can see that if distance is plotted on the vertical axis, and time is plotted on the horizontal axis, then since gradient is the steepness of the line:

Rule

The **gradient** of a **distance–time graph** gives the **speed** of the object.

If the travel graph is not a straight line, then this means the speed is changing. Hence the object is either **accelerating** or **decelerating**. If you want to find the speed at any particular time, draw a tangent to the curve at that time, and find the gradient of the tangent. This gives the speed at the time.

Example 12.4

Susan and her mother visited friends who live in Clifton, 50 km away. Their journeys are shown in Figure 12.9. Susan left home at 9.30 am and cycled. Her mother left home at 11 am and travelled by car.

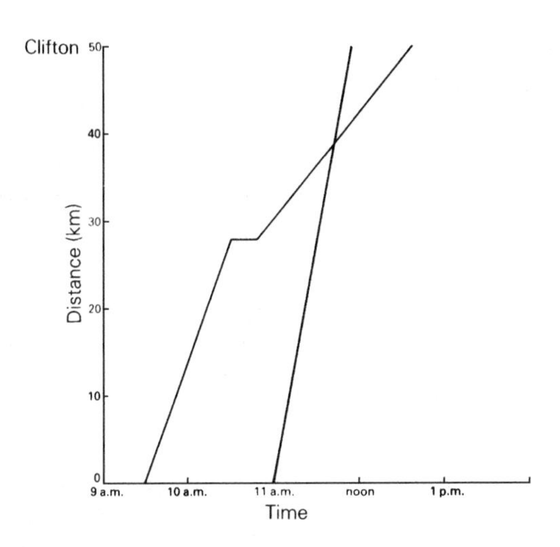

Figure 12.9

(a) At what time did Susan stop for a rest?

(b) How far had they both travelled from home when Susan was passed by her mother?

(c) Her mother took 54 minutes. How much longer did Susan take?

[MEG]

Solution

(a) Susan stops when her speed is zero, hence the gradient of the line is zero. This occurs at 10.30 am.

(b) They pass each other at the point where the graphs meet. Reading across on the vertical axis, the distance is 39 km.

(c) Reading across from Clifton at 50 km, we find that Susan arrived at 12.36 pm. Since she left home at 9.30 am, her journey took 3 hours and 6 minutes. Hence Susan takes 2 hours 12 minutes longer than her mother.

Example 12.5

A miniature racing car was timed over a distance of $\frac{1}{2}$ km at intervals of 100 m.

The results are shown in the following table:

Distance (100 m)	1	2	3	4	5
Time (secs)	12	25	40	50	62

Plot a graph of these results, and estimate the speed of the car after 400 m.

Solution

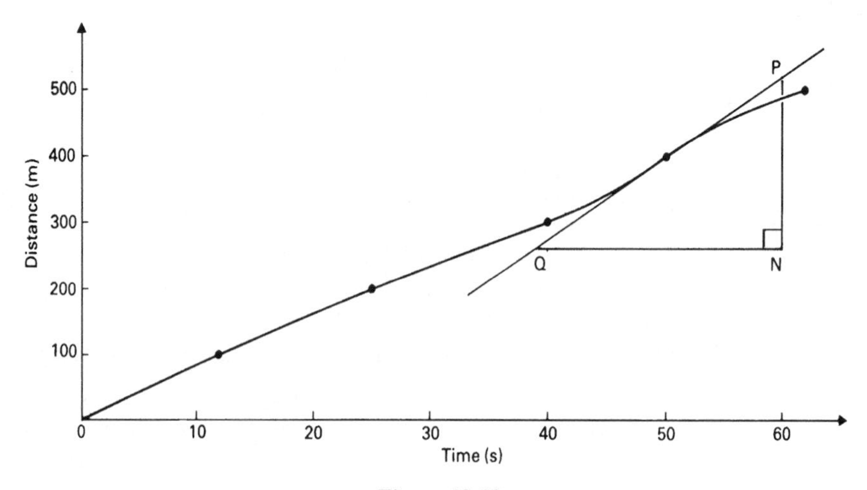

Figure 12.10

The results have been plotted in Figure 12.10. PQ is tangent to the curve where the distance is 400 m. The triangle PQN will give the gradient at 400 m.

$$\text{Speed} = \frac{PN}{QN}$$

$$= \frac{260}{21} = 12.4\text{ms}^{-1}$$

In a similar fashion, we would plot a graph of the speed (or **velocity**, i.e. speed in a given direction) of a body against time, then:

Rule

The **gradient** of a **speed–time** graph gives us the **acceleration**

Acceleration is measured in m/s^2 (or ms^{-2}) or similar units. However, probably the most important fact that can be calculated from a speed–time graph is the *area* under the graph which gives the *distance travelled.*

Example 12.6

Figure 12.11 gives the speed–time graph of a tube train travelling between two stations *A* and *B*.

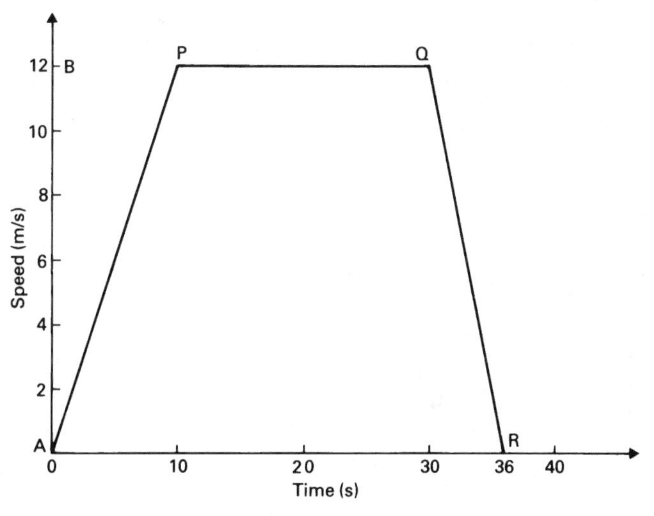

Figure 12.11

(a) What is the acceleration of the train at the beginning?

(b) Describe what happens between *P* and *Q* on the graph.

(c) Find the distance between the two stations.

Solution

(a) The gradient of the first part $= \dfrac{12}{10} = 1.2$

The acceleration is 1.2 ms^{-2}

(b) Between P and Q the train is travelling with a constant speed of $12\ \text{ms}^{-1}$.

(c) The shape can be treated as a trapezium.

$$\text{Area} = \tfrac{1}{2} \times 12 \times (PQ + AR) = \tfrac{1}{2} \times 12 \times (20 + 36)$$
$$= 6 \times 56 = 336$$

The distance is 336 m.

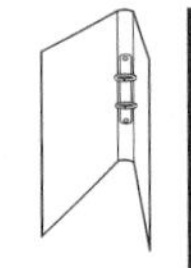

Exercise 12(b)

1

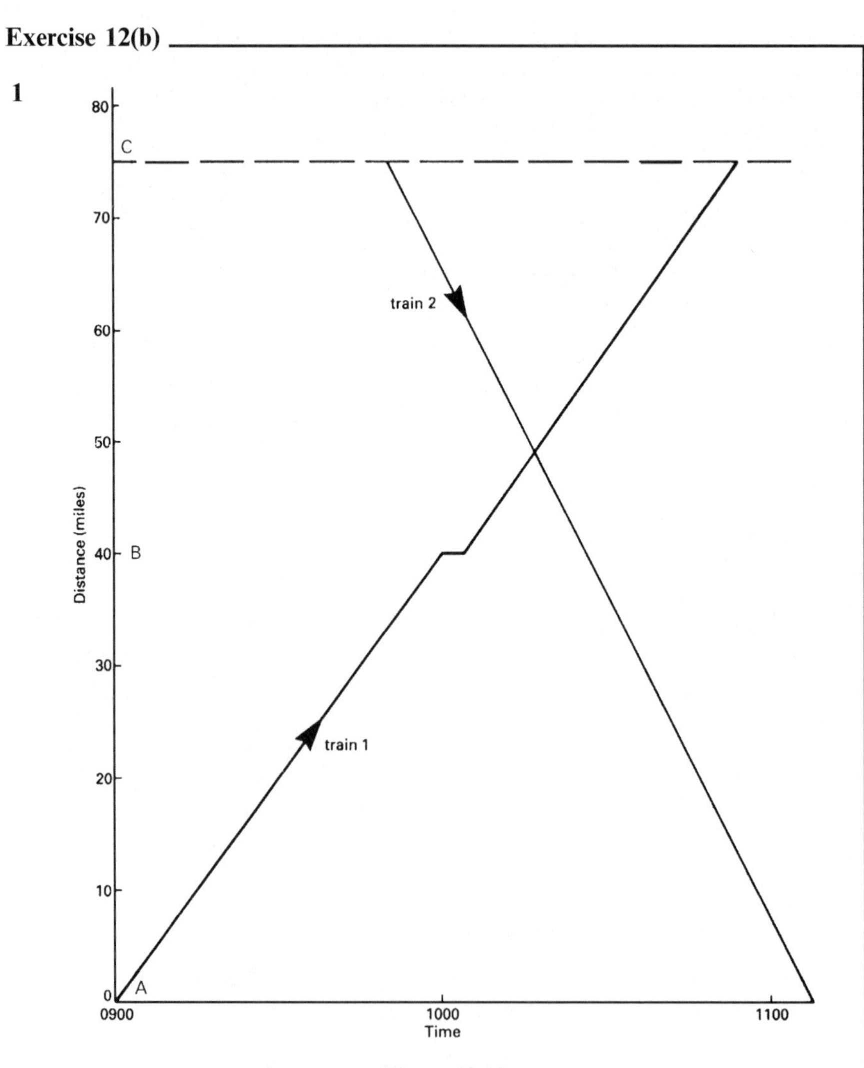

Figure 12.12

The graphs in Figure 12.12 represent the journeys of two trains.

Train 1 travels from A to C, stopping at B.

Train 2 travels from C to A, without stopping at B.

(i) (a) How far is it from A to C?

(b) At what time does Train 1 arrive at C?

(c) How long does Train 2 take for the journey from C to A?

Train 3 leaves A at 0930.

It takes 40 minutes to travel to B where it stops for 6 minutes.

It then travels on to C at an average speed of 60 miles per hour.

(ii) Copy Figure 12.12 on graph paper and add the travel graph for the journey of Train 3.

[LEAG]

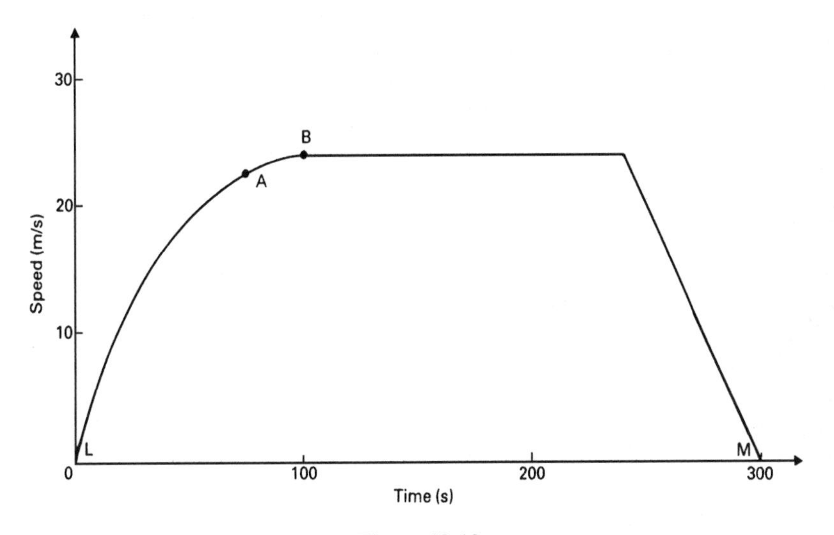

Figure 12.13

Figure 12.13 shows the speed–time graph of an electric train for its journey from the station at Lym (L) to the station at Man (M).

(i) (a) Write down the maximum speed, in m/s, of the train on the journey.

(b) Express this speed in km/h.

(ii) For how many seconds was the train travelling at this maximum speed?

(iii) Calculate the uniform deceleration, in m/s^2, of the train in the last stage of the journey.

(iv) Copy Figure 12.13 on graph paper and then draw a tangent to the curve at A.

Hence calculate an approximation for the acceleration, in m/s^2, of the train 75 s after it left Lym.

[LEAG]

3 Nathan pulled away from a petrol station on the M1. He accelerated for 20 seconds up to a speed of 70 km/h. He then cruised at this speed for a further 5 minutes, before reducing speed for 15 seconds to 30 km/h in order to leave the motorway. Taking particular care over the labelling of the axes, draw the speed–time graph to represent this journey. Calculate the distance that Nathan travelled along the motorway.

4 The graph in Figure 12.14 shows the speed of a train on a narrow-gauge railway as it travelled between two stations.

(i) How long did the train take to reach its maximum speed?

(ii) What was the acceleration of the train over this period? Give your answer in ms^{-2} (metres per second per second or m/s^2).

(iii) How far, in kilometres, did the train travel between the two stations?

(iv) What was the average speed of the train between the two stations? Give your answer in kilometres per hour.

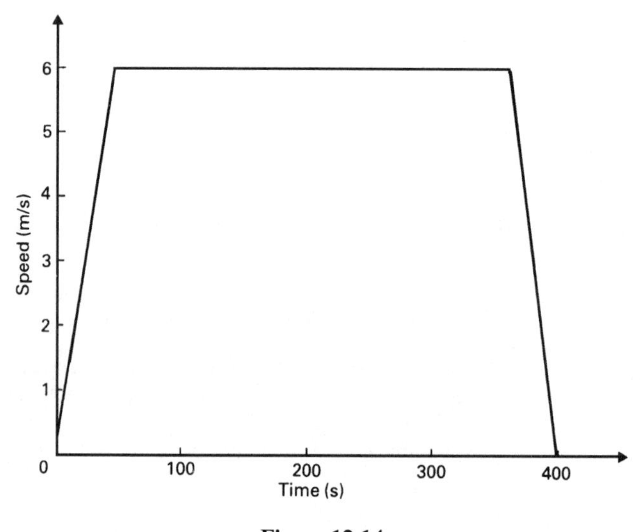

Figure 12.14

12.4 Experimental laws

Very often, a set of points which has been found by experiment, when plotted on a graph, appears to be roughly in a straight line. The line of best fit is drawn, from which its equation can be found.

Example 12.7

In an experiment to investigate how a spring stretches, the following results were obtained for the amount a spring stretched for given weights attached.

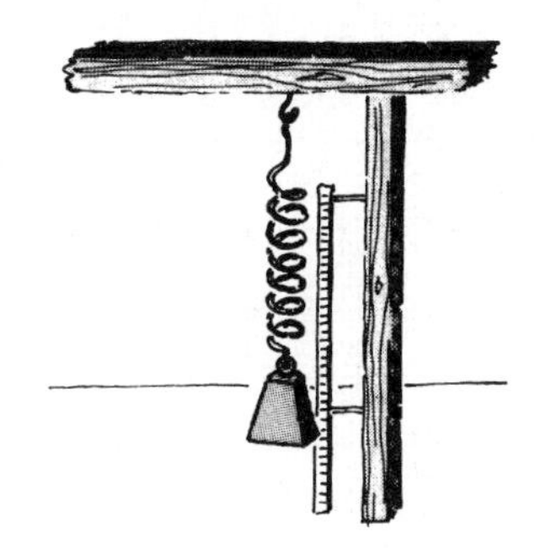

Mass (M) (kg)	0	0.5	1.5	3.0	4.0	5.1
Extension (E) (mm)	0	0.9	3.1	7.0	8.2	9.8

Plot these points on graph paper, draw the line of best fit, and hence find the equation that connects mass (M) and extension (E).

Use your equation to find what mass would cause an extension of

(i) 4 mm (ii) 11 mm.

Solution

The points are shown plotted in Figure 12.15 together with the line of best fit. It must go through the origin, because zero mass produces zero extension.

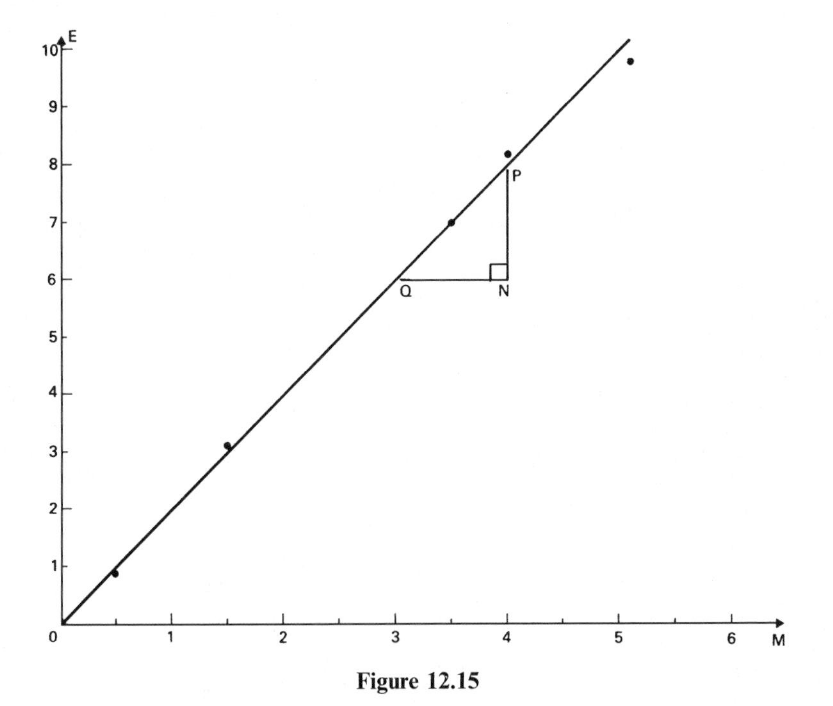

Figure 12.15

The gradient of the line $= \frac{PN}{QN} = \frac{2}{1} = 2$

In the standard equation $y = mx + c$, y is represented by E, x is represented by M, $m = 2$ and $c = 0$.

The equation is $E = 2M$.

(i) If $E = 4$, $4 = 2M$, $M = 2$. Hence the mass required is 2 kg.

(ii) If $E = 11$, this would be outside the original range of points, and so we must *assume the equation works outside this range.*

Hence $11 = 2M$, $M = 5.5$. Hence the mass required is 5.5 kg.

Where a point is calculated with the given range of points, this is called **interpolation** and where we are making a deduction outside the given points, this is called **extrapolation**. It should be noted that in general interpolation is usually more reliable than extrapolation.

Sometimes it may be necessary to adapt an equation before it is plotted as shown in the following example.

Example 12.8

The following values are thought to satisfy the equation $y = \frac{A}{x}$. By plotting y against $\frac{1}{x}$, estimate the value of A, and find y when $x = 1.8$.

x	2.5	2.0	1.5	1.2	0.9
y	1.00	1.25	1.67	2.08	2.78

Solution

Let $z = \frac{1}{x}$, then $y = Az$

Calculate z first, and we have the following table:

z	0.40	0.50	0.67	0.83	1.11
y	1.00	1.25	1.67	2.08	2.78

The graph can now be plotted, see Figure 12.16.

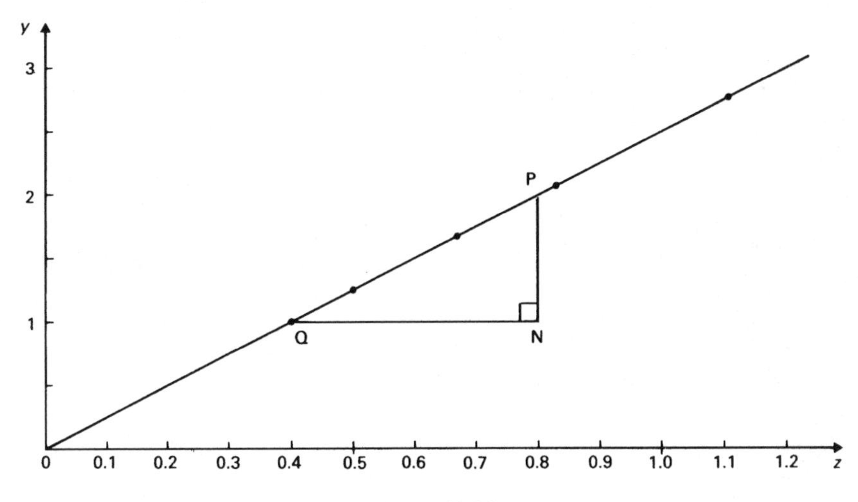

Figure 12.16

The gradient of the line is

$$A = \frac{PN}{QN} = \frac{1}{0.4} = 2.5$$

Hence

$$y = \frac{2.5}{x}$$

If $x = 1.8$, $y = \dfrac{2.5}{1.8} = 1.4$ (1 d.p.)

Exercise 12(c)

1 The resistance (R) of a wire was measured at various temperatures (T) and the following results were obtained:

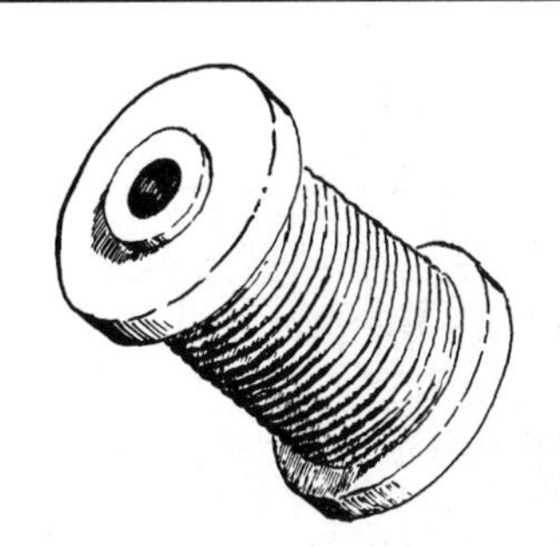

T	10	20	40	60	100
R	4.8	5.4	6.9	8.3	12.2

Plot a graph of T against R, and find the equation connecting T and R. Find (i) R when $T = 30$ (ii) T when $R = 7.5$.

2 When gas expands, the pressure and volume are related by a formula $P = \frac{k}{V}$, where k is a constant.

For the following values of P and V, draw a graph of P against $\frac{1}{V}$.

From the graph, estimate the value of k.

P	96	63	42	35	32
V	4	6	9	11	15

3 It is known that two variables r and s are connected by the formula

$$s = kr^2 + t,$$

where k and t are constants.

In an experiment, the following values of s and r were found:

s	5.25	6.0	7.3	9.0	11.2
r	1	2	3	4	5

Plot a graph of s on the vertical axis, against r^2 on the horizontal axis.

What do you notice?

Use your graph to find the values of k and t.

What would you expect s to equal when $r = 10$?

12.7 Inequalities

The inequality signs allow us to arrange numbers in order. The statement $6 < 8$ means 6 is *less than* 8, or alternatively 8 is *greater than* 6. The statement $x \leqslant -4$, means x is *less than or equal to* -4, or alternatively -4 is *greater than or equal to* x. We often use $x > 0$ to mean x is positive, and $x < 0$ to mean x is negative. If $-3 < t < 2$, this means that t is greater than -3 and less than 2. In other words, t lies between -3 and 2.

The equation $3x + 2 = 8$ has the simple solution $x = 2$. However, $3x + 2 < 8$ has rather a lot of solutions. In fact, any number less than 2. In many ways, inequalities can be rearranged like equations, provided you do not *multiply* or *divide* by a *negative number*:

For example: $3 < 5$
Multiply by -2: $-6 < -10$ is not true,
but: $-6 > -10$ is true.

Rule

If you multiply or divide by a negative number, the inequality is reversed

Example 12.9

Solve the inequalities:

(a) $2x - 3 < 3x + 4$
(b) $-15 < 2x - 3 < 7$
(c) $2x + 1 \leqslant 3x + 6 \leqslant x + 8$ (x is an integer)

Solution

(a) $2x - 3 < 3x + 4$

Subtract $2x$ from each side: $-3 < x + 4$

Subtract 4 from each side: $-7 < x$

The solution is: $x > -7$

(b) $-15 < 2x - 3 < 7$

Add 3 to each part: $-12 < 2x < 10$

Divide each part by 2: $-6 < x < 5$

(c) This type of inequality must be considered in two parts:

$2x + 1 \leqslant 3x + 6$

So: $1 - 6 \leqslant 3x - 2x$

$-5 \leqslant x$

Also: $3x + 6 \leqslant x + 8$

$3x - x \leqslant 8 - 6$

$2x \leqslant 2$

So $x \leqslant 1$

To satisfy both of these statements,

$$-5 \leqslant x \leqslant 1$$

Since x is an integer, the solution is

$$\{-5, -4, -3, -2, -1, 0, 1\}.$$

12.8 Errors

Inequalities occur in a number of ways. When measurements are made, there is always a degree of **error** made. The following example shows how small errors in quantities can have quite a big effect in substituting into a formula.

Example 12.10

It is given that $H = \dfrac{4a}{2+4t}$, with a and t measured in centimetres to the nearest centimetre. If $a = 6$ and $t = 4$, find the range of possible values of H.

Solution

Since a and t are measured to the nearest whole number, a can be anywhere between 5.5 and 6.5,

i.e. $\quad 5.5 \leqslant a \leqslant 6.5$

Similarly, $\quad 3.5 \leqslant t \leqslant 4.5$

Strictly speaking, we should exclude 6.5 and 4.5, as these would normally be rounded up. But this leads to complications, and does not really affect the result. Since H is a fraction, the *largest* value of H will have the *largest* numerator, and the *smallest* denominator. The *smallest* value of H will have the *smallest* numerator and the largest *denominator*.

This means

$$\frac{4 \times 5.5}{2 + 4 \times 4.5} \leqslant H \leqslant \frac{4 \times 6.5}{2 + 4 \times 3.5}$$

and so

$$1.1 \leqslant H \leqslant 1.625.$$

12.9 Regions and linear programming

In Figure 12.17, the line $y = \frac{3}{4}x + 1$ has been drawn. If this line is continued indefinitely in each direction, it divides the plane into two regions A and B.

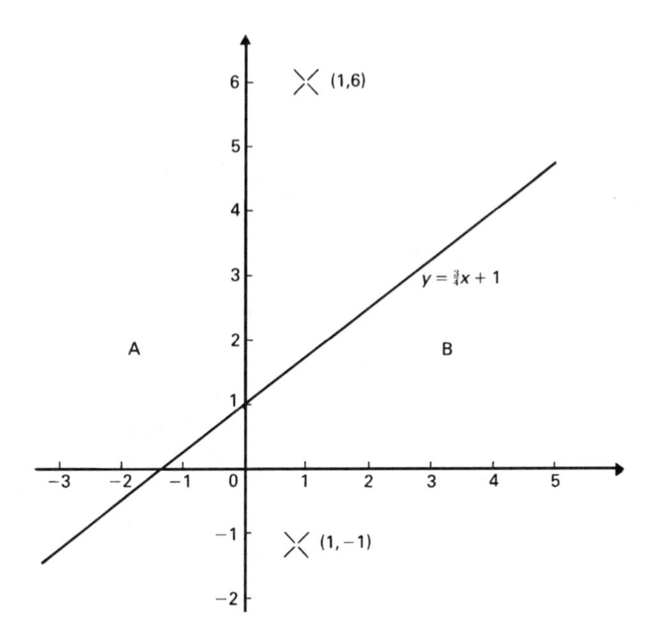

Figure 12.17

Everywhere *on* the line, $y = \frac{3}{4}x + 1$

The point $(1, 6)$ is in region A

$$6 > \tfrac{3}{4} \times 1 + 1 = 1.75 \qquad \text{i.e.} \quad y > \tfrac{3}{4}x + 1$$

The point $(1, -1)$ is in region B

$$-1 < \tfrac{3}{4} \times 1 + 1 = 1.75 \qquad \text{i.e.} \quad y < \tfrac{3}{4}x + 1$$

In general

$y > \frac{3}{4}x + 1$ in region A (above the line)

$y < \frac{3}{4}x + 1$ in region B (below the line)

Combining inequalities and the equation of a straight line enables us to define various regions in the plane.

Example 12.11

Sketch the region defined by the inequalities

$x \geqslant 1$; $y \geqslant x - 3$; $x + 2y \leqslant 6$.

Find the maximum value of $x + 3y$ for any point in the region:

(a) if x and y are unrestricted

(b) if x and y are whole numbers.

Solution

First draw the triangle formed by these lines.

$x \geqslant 1$ means the region to the right of $x = 1$

$y \geqslant x - 3$ means the region above the line $y = x - 3$

$x + 2y \leqslant 6$ means the region below the line $x + 2y = 6$.

Hence the region satisfied by these inequalities must be the triangle shown shaded in Figure 12.18.

The coordinates of the vertices are $A(1, 2.5)$, $B(4, 1)$, $C(1, -2)$.

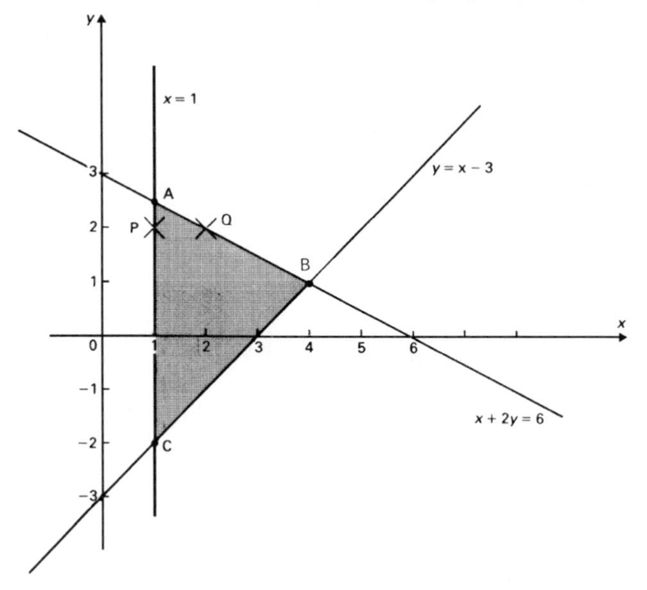

Figure 12.18

(a) and (b)

At A: $x + 3y = 1 + 7.5 = 8.5$

B: $x + 3y = 4 + 3 = 7$

C: $x + 3y = 1 - 6 = -5$

Hence if x and y are unrestricted, the largest value of $x + 3y$ is 8.5, and it occurs at A. If however, x and y are integers, then there are two points $P(1, 2)$ and $Q(2, 2)$ near to A.

At P: $x + 3y = 1 + 6 = 7$

Q: $x + 3y = 2 + 6 = 8$

Hence the maximum value is 8.

We can now apply this techniqe to a variety of planning problems known as **linear programming** problems.

Example 12.12

An island tour company has two types of plane available, A and B. Type A holds 120 passengers, and type B 300 passengers. On Saturday 1200 tourists have to be taken to a carnival, but only 3 planes of type B are available, and a total of 8 pilots are on duty. (1 plane needs 1 pilot.) If x type A planes are used, and y type B planes are used, write down inequalities satisfied by x and y, and construct the region in which they lie. If running type B planes is 50% more expensive than type A planes, find the values of x and y to minimise the cost.

Solution

$$y \leqslant 3 \quad \text{(obvious)}$$

$$x + y \leqslant 8 \quad \text{(pilots)}$$

$$120x + 300y \geqslant 1200 \quad \text{(passengers)}$$

i.e. $$2x + 5y \geqslant 20 \quad (\div 60)$$

The region (shown shaded in Figure 12.19) can now be constructed, using the ideas of Example 12.11.

The vertices of the region are $P(2.5, 3)$, $Q(5, 3)$, $R(6.7, 1.3)$.

Since type B is 50% more expensive, the total cost is proportional to $x + 1.5y$ (known as the cost line).

At P: $x + 1.5y = 7$

Q: $x + 1.5y = 9.5$

R: $x + 1.5y = 8.65$

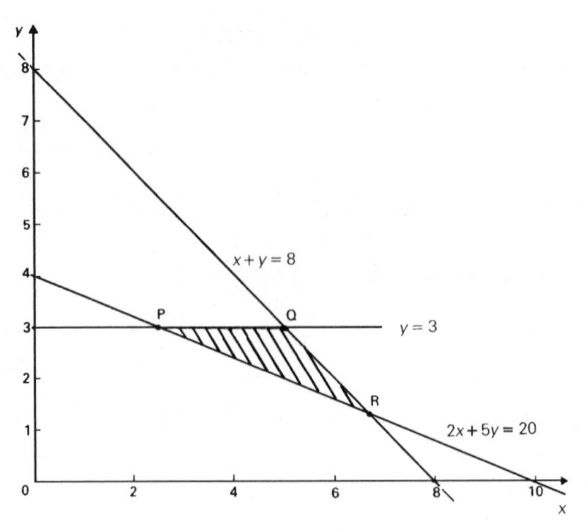

Figure 12.19

The minimum value is at P. However, x and y must be whole numbers, which is not true at P. Looking at points near to P, we have at $(3, 3)$:

$$x + 1.5y = 7.5.$$

This is clearly less than at Q and R, so the answer would be $x = 3$, $y = 3$.

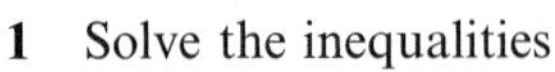

Exercise 12(d)

1 Solve the inequalities

(i) $2x + 5 < 19$

(ii) $3(x - 1) < 8$

(iii) $3 - 2x < 4 - 3x$

(iv) $-2 \leqslant 5 - x \leqslant 3$ (and x is a whole number)

(v) $1 - 2x > 4 - 3x$.

2 v, u, a and t are related by the formula $v = u + at$. If an error of 0.1 can be made in the measurements of u, a and t, find the possible range of values of v if $u = 12$, $a = 2$ and $t = 4$.

3 H, p and q are related by the formula $H = \dfrac{2p}{4+q}$. The values of p and q are correct to the nearest whole number. If $p = 3$ and $q = 2$, find the possible range of values of H.

4 Draw a diagram and shade the region given by the inequalities:

$$y \leqslant x + 4;\ y \geqslant 2x - 3;\ x + y + 2 \geqslant 0$$

Find the maximum value of $2x + 3y$ in this region:

(i) if x and y can take any value

(ii) if x and y are whole numbers.

5 Josie is baking two kinds of cake for the local charity bazaar. Part of the ingredients are given in the following table.

	fat (g)	*flour (g)*	*sugar (g)*
Cake X	120	150	60
Cake Y	75	160	100

On looking in the cupboard, she has the following quantities available: 4.5 kg flour, 3.6 kg fat and 2.1 kg of sugar.

Let x be the number of cake X baked, and y the number of cake Y baked. Write down 5 inequalities that apply to x and y. On graph paper, draw a graph to show the region defined by these inequalities.

She is told that cake X will make a profit of 35p, and cake Y will make a profit of 20p. Write down an expression in terms of x and y for the total profit made. Find x and y so that this profit is a maximum.

6 Jonah's Stores sell two types of canned drinks, cola and shandy. Jonah is about to order new supplies, and reckons he can stock 1000 cans. He always orders at least twice as many cans of cola as of shandy.

He needs at least 100 cans of shandy, and never stocks more than 800 cans of cola. Let x be the number of cans of cola ordered, and y the number of cans of shandy ordered. Write down four inequalities that are satisfied by x and y.

On a graph where 1 cm represents 100 cans, construct the region satisfied by these inequalities. If he makes 8p profit on a can of cola, and 6p profit on a can of shandy, determine the number of cans of each type that Jonah should order to make the greatest profit.

13 Curves, functions and equations

13.1 Plotting curves

In the same way that we were able to write down the equation of a straight line in Section 12.2, we can also write down the equation of a curved line. The techniques of plotting these curves are very similar to plotting a straight line, but there are some precautions that need to be taken. If we want to know what the curve $y = x^2 - 2x - 6$ looks like for values of x between -3 and -5, then it is necessary to draw up a table of values. In order to avoid mistakes, the table is expanded in the following way:

x	-3	-2	-1	0	1	2	3	4	5
x^2	9	4	1	0	1	4	9	16	25
$-2x$	6	4	2	0	-2	-4	-6	-8	-10
-6	-6	-6	-6	-6	-6	-6	-6	-6	-6
$y = x^2 - 2x - 6$	9	2	-3	-6	-7	-6	-3	2	9

The values can now be plotted as shown in Figure 13.1. This method should be used if y consists of *two or more* terms added together.

If we wish to find the gradient of a curve, say at the point $(2, -6)$, it is necessary to draw a tangent at this point, and find the gradient of the tangent.

[*Note*: The tangent just touches the curve at one point.]

$$\text{Hence the gradient} = \frac{PN}{PQ} = \frac{4}{2} = 2.$$

[*Note*: The scale is different on the x and y axes.]

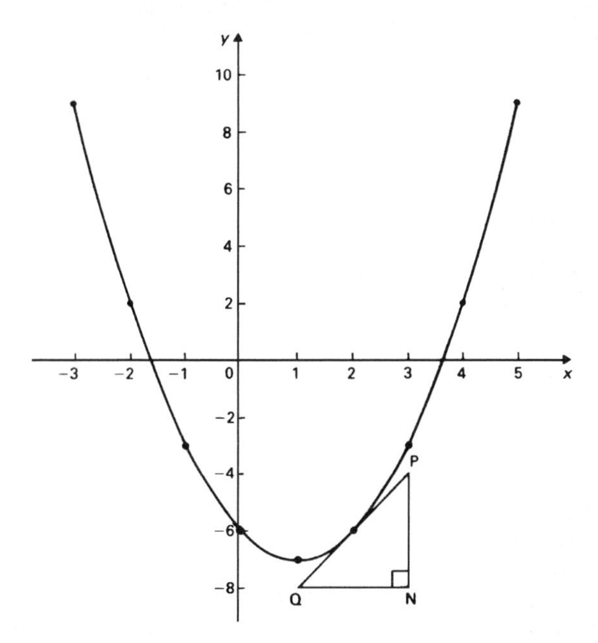

Figure 13.1

Example 13.1

The temperature of a liquid after heating was recorded at intervals of 10 seconds. The results are shown in the following table:

Time (s)	10	20	30	40	50	60
Temperature (°C)	85	75	66	59	55	52

Plot these results on a graph (cooling curve) and find the rate of cooling (the gradient) in °C/s after 25 seconds.

Solution

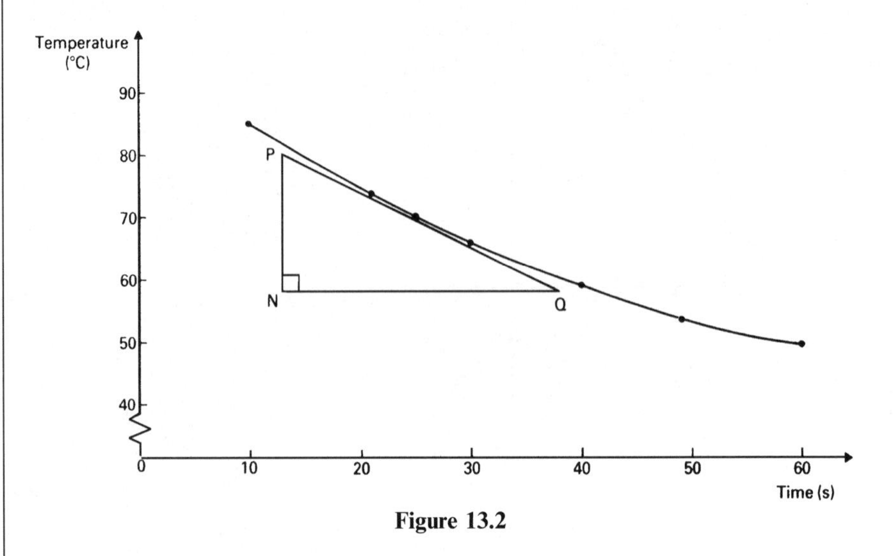

Figure 13.2

The results are plotted in Figure 13.2. You should notice that the vertical scale has been shortened. To find the rate of cooling, a tangent is drawn at 25 seconds, and the triangle PQN will give us the gradient.

$$\text{gradient} = \frac{PQ}{NQ} = \frac{22}{25}$$

$$= 0.88 \text{ (careful with the scales)}$$

The rate of cooling is 0.88°C/second.

13.2 Solving equations

In Section 12.2, graphical techniques involving straight lines were used to solve simultaneous equations. We can now extend these ideas to questions involving non-linear equations.

In Figure 13.3 two graphs have been plotted.

The parabola

$$y = x^2 + x - 2$$

and the straight line

$$y = \tfrac{3}{4}x + 2.$$

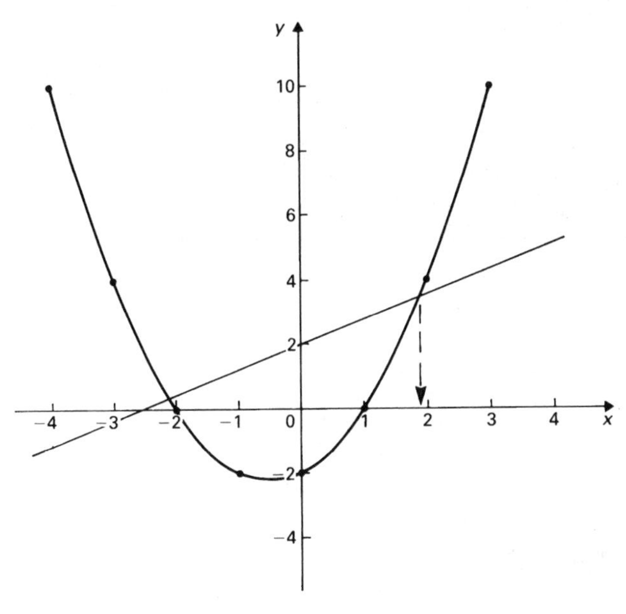

Figure 13.3

When the two graphs meet, then clearly the y and x values for each equation are the *same*.

Hence $x^2 + x - 2 = \frac{3}{4}x + 2$ (1)

From the graph $x = -2.1$ and 1.9.

If we simplify (1)

$$x^2 + \tfrac{1}{4}x - 4 = 0$$

Multiply each term by 4: $4x^2 + x - 16 = 0$

Solve the quadratic to check your answers

The solutions of this equation are:

$x = -2.1$ and 1.9.

Example 13.2

Plot the graph of $y = 2 + \dfrac{4}{x^2}$ for values of x between 1 and 6. In order to solve the equation $x^3 - 3x^2 - 12 = 0$, a suitable straight line must be drawn. Find the equation of this line, plot the line and hence solve the equation.

Solution

The table for $y = 2 + \frac{4}{x^2}$ gives the following values of y:

x	1	2	3	4	5	6
y	6	3	2.44	2.25	2.16	2.11

The graph is plotted in Figure 13.4

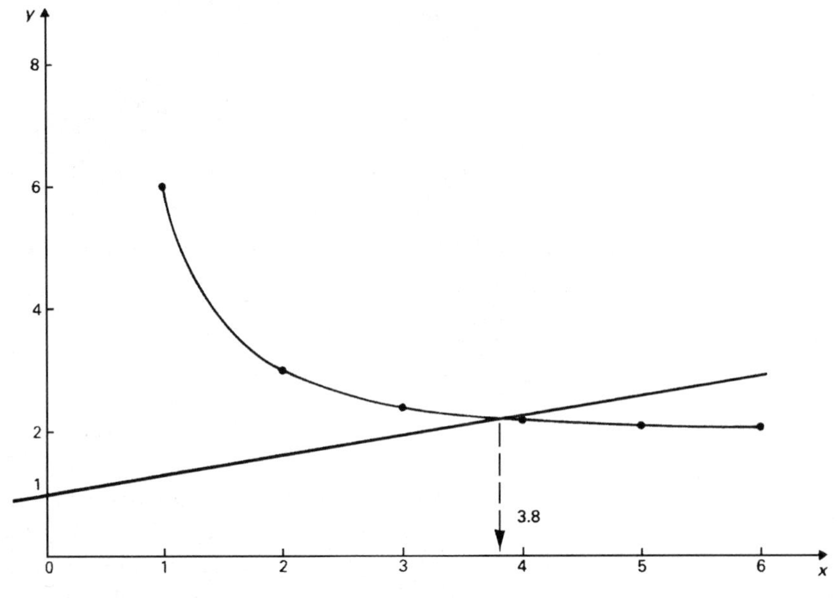

Figure 13.4

The equation now needs to be rearranged.

$$x^3 - 3x^2 - 12 = 0$$

So:

$$x^3 = 3x^2 + 12$$

Divide by $3x^2$:

$$\frac{x}{3} = 1 + \frac{4}{x^2}$$

Add 1 to both sides:

$$\frac{x}{3} + 1 = 2 + \frac{4}{x^2}$$

The right hand side is now the same as for the equation of the curve.

Hence the line $y = \frac{1}{3}x + 1$ must be drawn.

The solution where the graphs cross is $x = 3.8$.

13.3 The trapezium rule

Having drawn or plotted a curve, there are several mathematical situations that require us to find the area enclosed by a curved boundary. At this stage, we will look at a method that enables us to approximate to the answer.

The points in the following table have been plotted in Figure 13.5.

x	1	2	3	4	5
y	1.6	2.8	3.0	2.0	0.6

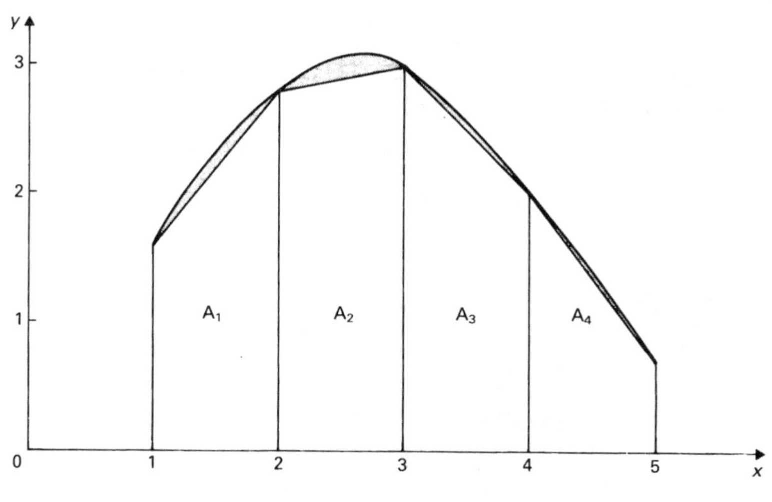

Figure 13.5

If we now want to find the area enclosed by the lines $x = 1$, $x = 5$, the x-axis and the curve, then a good approximation can be found by drawing trapezia of width 1 as shown.

$$\text{The area of} \quad A_1 = \tfrac{1}{2} \times 1 \times (1.6 + 2.8) = 2.2$$

$$A_2 = \tfrac{1}{2} \times 1 \times (2.8 + 3.0) = 2.9$$

$$A_3 = \tfrac{1}{2} \times 1 \times (3.0 + 2) = 2.5$$

$$A_4 = \tfrac{1}{2} \times 1 \times (2 + 0.6) = 1.3$$

$$\text{The total area} \quad = 2.2 + 2.9 + 2.5 + 1.3$$

$$= 8.9.$$

Note: This answer is *less* than exact value.

This can be summarised for a width of h, and n different y-values: $y_1, y_2, \ldots, y_n$, by the formula:

Rule

$$\text{Area} = \tfrac{1}{2}h(y_1 + 2y_2 + 2y_3 + \ldots + 2y_{n-1} + y_n)$$

You should note that you do not really need to draw a diagram, as the following example shows.

Example 13.3

The speed (v m/s) of a body t seconds after starting is given in the following table. Estimate the distance the body travels between $t = 1$ and $t = 6$.

Speed v (m/s)	4	8	10	7	6	2
Time t (s)	1	2	3	4	5	6

Solution

The y-values will be the different speeds, and the x-values the time. The width of each trapezium (the time interval) $h = 1$, so using the trapezium rule gives the approximate area (which is the distance travelled).

$$\text{Distance} = \tfrac{1}{2} \times 1 \times [4 + 2 \times (8 + 10 + 7 + 6) + 2] \text{ m}$$

$$= \tfrac{1}{2} \times [4 + 62 + 2] = 34 \text{ m}.$$

Investigation XXIV: The trapezium rule

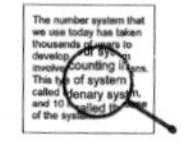

Draw axes on graph paper with x from 0 to 8, and y from 0 to 10.

(i) Plot the following points

x	0	1	2	3	4	5	6	7	8
y	0	2.4	4.6	6.4	7.8	8.8	6.5	3	0

(ii) Draw a smooth curve through these points and use the trapezium rule with these x- and y-values to estimate the area enclosed under the graph.

(iii) Now read off the y-coordinates for $x = 0.5$, 1.5 etc. You can now use the trapezium rule with 16 trapezia of width 0.5 to get another value for the area.

(iv) Repeat the process for $x = 0.25, 0.75, \ldots$ What do you notice about your answers?

Exercise 13(a)

1 The following table gives the depth of water in Mersea Marina at various times in the afternoon.

Time (p.m.)	1	2	3	4	5	6	7
Depth in metres	1.0	1.6	2.4	3.2	3.2	2.4	1.6

(i) Plot this information on graph paper.

(ii) Draw a smooth curve through the points.

(iii) From your graph, find the depth of water in the marina at 6.30 p.m.

(iv) Amanda needs a depth of 2 metres to sail her boat. Estimate the earliest time at which she will be able to sail into the marina.

(v) Estimate

(a) the greatest depth of water in the marina;

(b) the time at which this occurs.

2 An open rectangular tank with square base has a volume of 200 cm^3. The base has sides of length x cm and the height of the tank is h cm.

The surface area, S cm^2, of the tank, is given by $S = x^2 + \frac{800}{x}$.

x	2	3	4	6	8	10	12
x^2	4						
$\frac{800}{x}$	400	267					
S	404	276					

(i) Copy and complete the table above giving all figures to the nearest whole number.

(ii) Using a scale of 1 cm for 1 unit on the x-axis and 1 cm for 25 units on the S-axis, draw the graph of $S = x^2 + \frac{800}{x}$.

(iii) From your graph find the least value of the surface area of the tank and the length of the side of the tank, in this case.

[LEAG]

3 The table below shows the weight in kg of a new-born baby at various intervals during the first 6 weeks after its birth.

No. of days	0	2	5	10	21	26	35	42
Weight in kg	2.25	2.16	2.20	2.31	2.64	2.91	3.24	3.37

(i) Plot these points on graph paper and join them up with a smooth curve.

(ii) On which day was the baby gaining weight most quickly?

(iii) By drawing a tangent to your graph, estimate the rate at which the baby was gaining weight on this day.

[NEA]

4 $y = 5 + 3x - x^2$

(i) Copy and complete the following table:

x	−2	−1	0	1	2	3	4	5
5	5	5	5			5	5	
$3x$	−6		0			9	12	
$-x^2$	−4		0			−9	−16	
y	−5		5			5	1	

(ii) On graph paper, using a scale of 2 cm to represent 1 unit on the x-axis and 1 cm to represent 1 unit on the y-axis, draw the graph of the curve $y = 5 + 3x - x^2$ for $-2 \leqslant x \leqslant 5$.

(iii) Using your graph, find and write down the maximum value of y.

(iv) Using your graph write down, as accurately as possible, the values of x for which $5 + 3x - x^2 = 0$.

(v) Using the same axes and the same scales, draw the graph of the straight line $y = x + 3$.

(vi) Using your graphs write down, as accurately as possible, the values of x for which $x^2 - 2x - 2 = 0$.

5 The graph in Figure 13.6 shows the rate at which people of different ages and heights should be able to blow out. This is called the peak expiratory flow and is measured in litres per minute.

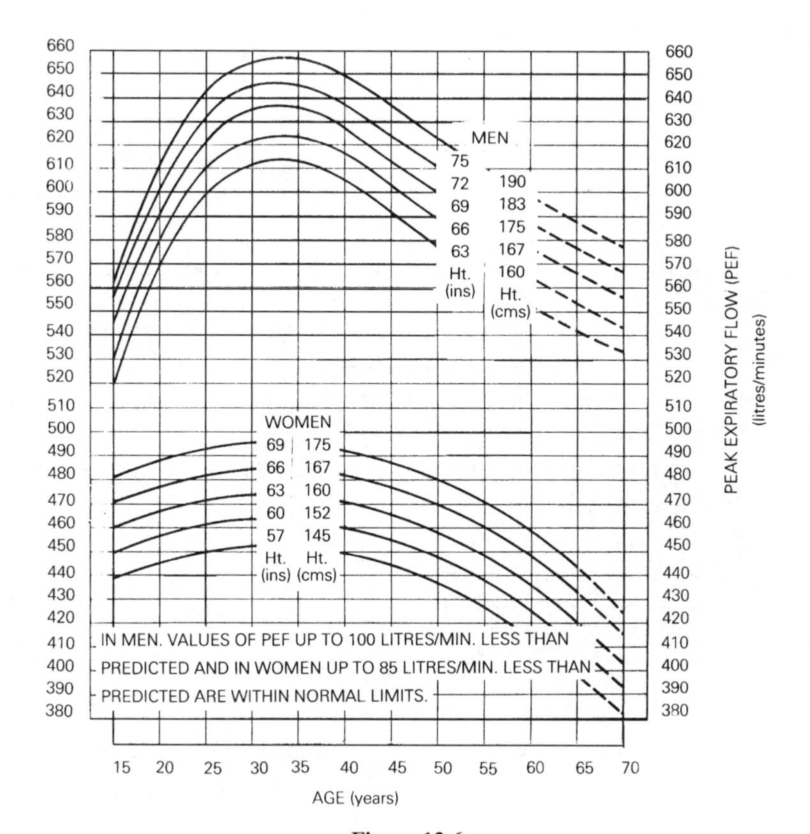

Figure 13.6

(i) Lisa is 25 years old and 145 cm tall. What should her peak expiratory flow be?

(ii) How old are men when their peak expiratory flow is greatest?

(iii) The note at the bottom of the graph says that men can be up to 100 litres/minute below the given rate and still be normal; the corresponding value for women is 85 litres/minute below the given rate.

Who is allowed the greater percentage difference, a 23-year-old man who is 183 cm tall or a 47-year-old woman who is 152 cm tall? Show your working to justify your answer.

[WJEC]

6 In a science class a pupil is attempting to find how the length of a pendulum (l) affects the time of the swing (t). The following results are obtained:

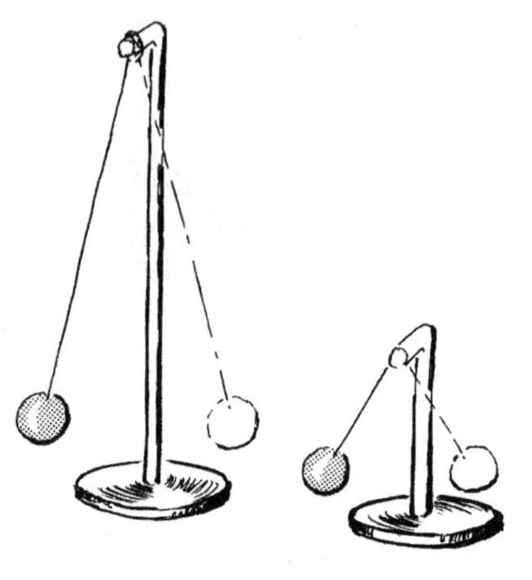

Length l (cm)	0	20	40	60	80	100
Time t (s)	0	0.89	1.27	1.55	1.80	2.00

(i) Explain how you can tell, from the table, that t is not directly proportional to l.

(ii) It is suspected that t is directly proportional to $\sqrt{l}$. Complete the table of values of t and $\sqrt{l}$.

$\sqrt{l}$	0	4.47		7.75		10
t	0	0.89	1.27	1.55	1.80	2.00

(iii) Plot a graph of t against $\sqrt{l}$.

(iv) Explain why the graph suggests that t is directly proportional to $\sqrt{l}$.

[NEA]

7 The distance d, in feet, in which it is possible to stop a car travelling at v miles per hour is given by the formula $d = 2v + \frac{v^2}{10}$ when the road surface is slippery.

(i) Complete the table below which shows stopping distances when travelling at speeds from 0 to 60 miles per hour.

v	0	10	20	30	40	50	60
$2v$							
$\frac{v^2}{10}$							
d							

(ii) Plot the graph of d against v on graph paper.

(iii) Use your graph to estimate:

(a) the stopping distance for a car travelling at 35 mph;

(b) the maximum speed at which it is safe to travel when visibility is down to 300 feet.

13.4 Function notation

Look at these two sets of numbers: $A = \{-2, -1, 0, 1, 2\}$ and $B = \{2, 3, 4, 5, 6\}$. Can you see any relationship between them?

Now look at the same numbers written in a different way:

x		y
−2	⟶	2
−1	⟶	3
0	⟶	4
1	⟶	5
2	⟶	6
set A		set B

Figure 13.7

We use the arrows to show that there is a simple rule for changing the value of x from set A (called the **domain**) into the value of y from set B (called the **codomain**). We call this rule together with the domain and codomain a **mapping**, or **function**. Our rule this time is that:

$$y = x + 4$$

Another way of writing this is $x \mapsto x + 4$, which is read as 'x maps on to $x + 4$'. In mathematics, we go on to write that the function which maps x on to $x + 2$ is written as:

$$\mathrm{f} : x \mapsto x + 2$$

Although it appears that the letter f is used to denote function, we can use any letter, so that $\mathrm{g} : x \mapsto x + 3$ denotes a different function which maps x on to $x + 3$. If we want to ask the question 'what does f map the number 1 on to?', we write this $\mathrm{f}(1)$, and say that $\mathrm{f}(1)$ is the **image** of 1 under f. Looking at Figure 13.7 or remembering what the rule here was, we can see that $\mathrm{f}(1) = 5$.

Another way of writing the function $\mathrm{f} : x \mapsto x^2 + 1$ would be $\mathrm{f}(x) = x^2 + 1$. Here $\mathrm{f}(x)$ gives the image of x using the function f. The letter x is called a dummy variable and does not affect the answer. We could write the above function as $\mathrm{f}(t) = t^2 + 1$. The rule here tells us that whatever number is inside the bracket, f will square that number and add 1 to the result. This notation has a wide range of uses as is illustrated by the following examples.

Example 13.4

If $\mathrm{f}(t)$ denotes the sum of the positive integers which are less than t, find
(a) $\mathrm{f}(1)$; (b) $\mathrm{f}(8)$; (c) $\mathrm{f}(-4)$.

Solution

(a) Since 1 is the first positive integer, there are no integers less than t, which in this case is equal to 1

So $\mathrm{f}(1) = 0$

(b) Here, $t = 8$, so the integers less than t are 1, 2, 3, 4, 5, 6, 7.

Hence $\mathrm{f}(8) = 1 + 2 + 3 + 4 + 5 + 6 + 7 = 28$

(c) Here $t = -4$

Although this is a question about positive integers, $\mathrm{f}(-4)$ still has a meaning, because there are no positive integers less than -4.

Hence $\mathrm{f}(-4) = 0$.

Example 13.5

If $\mathrm{r}(x)$ denotes the remainder when the whole number x is divided by 10, find
(a) $\mathrm{r}(73)$;
(b) $\mathrm{r}(10x)$;
(c) Under what conditions if any is $\mathrm{r}(x + y) = \mathrm{r}(x) + \mathrm{r}(y)$?

Solution

(a) 73 divided by 10 is 7 remainder 3

Hence $r(73) = 3$.

(b) Since $10x$ is a multiple of 10, there is no remainder when $10x$ is divided by 10, and so

$r(10x) = 0$ whatever the value of x.

(c) Since the remainder is a number less than 10, this can only be true if $r(x) + r(y) < 10$. This will be true when the last digit of x and of y add up to a number less than 10.

If we wish to plot the graph of $f : x \mapsto x^2$ say, then we need to know what values of x are needed. For example, if $-3 \leqslant x \leqslant 3$, then this is called the **domain** of the function f.

The set of values obtained is called the **range** or **image set** of the function, and if a value x is mapped on to y by a function f, then we say that y is the **image** of x.

Example 13.6

The function $f : x \mapsto x^2 - 2$ has domain $\{-2, -1, 0, 1, 2\}$.

Draw a mapping diagram to illustrate this function, and state the range.

Solution

The mapping diagram is drawn in Figure 13.8

The range $= \{-2, -1, 2\}$.

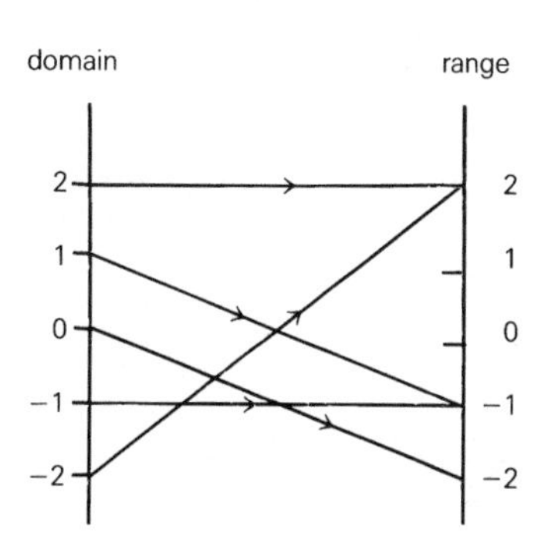

Figure 13.8

The idea of an **inverse** function can be considered by returning to Figure 13.7.

If we reverse the arrows, then we have as follows:

x		y
-2	$\longleftarrow$	2
-1	$\longleftarrow$	3
0	$\longleftarrow$	4
1	$\longleftarrow$	5
2	$\longleftarrow$	6

Since $y = x + 4$, $x = y - 4$

So $y \mapsto y - 4$

This is the inverse function, written f^{-1}

Hence $\mathrm{f}^{-1} : y \mapsto y - 4$

It is conventional to use the letter x for the variable, and so:

$\mathrm{f}^{-1}x : x \mapsto x - 4$

The idea of finding an inverse function is really just that of changing the subject of a formula.

If $\mathrm{f} : x \mapsto x + 1$, and $\mathrm{g} : x \mapsto 2x$, then fg is referred to as the **composition** of the functions f and g, and fg would be the composite function.

It is calculated by applying g *followed by* f.

Hence fg means $\mathrm{f}(\mathrm{g}(x)) = \mathrm{f}(2x) = 2x + 1$ (since f adds 1 to any value).

Now gf means $\mathrm{g}(\mathrm{f}(x)) = \mathrm{g}(x + 1) = 2x + 2$ (since g doubles any value).

You will notice that $fg \neq gf$.

Example 13.7

If $\mathrm{f} : x \mapsto 3x - 1$ and $\mathrm{g} : x \mapsto x + 1$, find

(a) $\mathrm{f}^{-1}\mathrm{g}^{-1}(3)$ (b) $\mathrm{f}^{-1}\mathrm{g}^{-1}(3x)$.

Solution

To find f^{-1}

$y = 3x - 1$

If we reverse the process

$$\frac{y+1}{3} = x$$

Hence $\mathrm{f}^{-1} : x \mapsto \dfrac{x+1}{3}$.

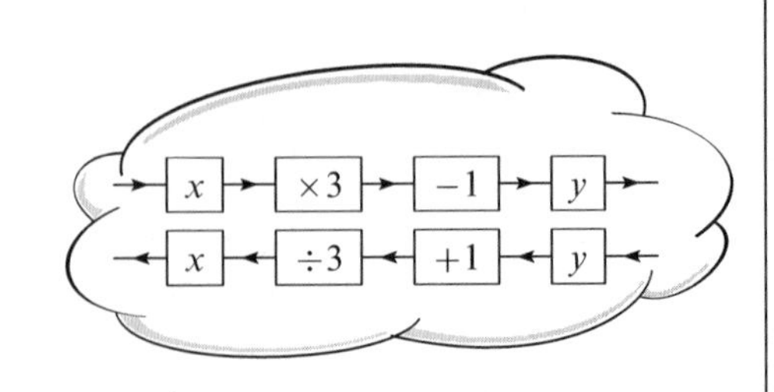

Similarly, $g^{-1} : x \mapsto x - 1$

(a) $g^{-1}(3) = 3 - 1 = 2$

We now apply the function f^{-1} to the number 2.

Hence $f^{-1}g^{-1}(3) = f^{-1}(2) = \frac{2+1}{3} = 1$.

(b) $g^{-1}(3x) = 3x - 1$

We now apply the function f^{-1} to the value $(3x - 1)$

Therefore $f^{-1}g^{-1}(3x) = \frac{(3x-1)+1}{3} = x$.

Exercise 13(b)

1 If $f(x) = \frac{x+1}{x}$, find: (i) $f(3)$ (ii) $f\left(\frac{1}{2}\right)$ (iii) $f(-4)$.

2 If $g(x)$ denotes the sum of all integers less than x which are perfect squares, find: (i) $g(25)$ (ii) $g(26)$.

3 If $g : x \mapsto 3x + 1$, find g^{-1}.

4 If $h : x \mapsto 2x - 1$, find $h^{-1}(-1)$.

5 If $f : x \mapsto 2x + 3$, find t such that $f(t) = 0$.

6 If $f(x)$ denotes the sum of all positive integers less than x which do not divide into x, find: (i) $f(8)$ (ii) $f(20)$.

7 $R(x)$ denotes the remainder when x is divided by 3. Find:

(i) $R(20)$ (ii) $R(20m)$ (iii) $R(21n)$ where m and n are integers.

8 Given that $m(x) = (x+1)(x-5)^2 - 28$, evaluate:

(i) $m(0)$ (ii) $m(1)$.

Hence estimate a value of x for which $m(x) = 0$.

9 Find the inverse of the following functions.

(i) $f : x \mapsto 1 - x$ (ii) $g : x \mapsto \frac{1}{x}$ (iii) $h : x \mapsto x^2$

(iv) $m : x \mapsto 1 - 3x$ (v) $n : x \mapsto \frac{1}{x+1}$ (vi) $q : x \mapsto \frac{ax+b}{cx+d}$.

10 Functions f and g map x on to $x + 1$ and $x^2 + 1$ respectively. Show that fg maps x on to $x^2 + 2$ and find gf. What does this tell you about composition of functions?

A third function n maps x on to $b - ax$. If fgn maps x on to $x^2 - 6x + 11$, find a and b.

11 The image of the set $\{-1, 0, 1\}$ under the function $ax^2 + b$ is $\{7, 5\}$. Find a and b.

12 The function f is defined on the domain $\{0, 1, 3, 5, 7, 9\}$ by $f : x \mapsto$ the units digit of x^3. Draw a mapping diagram for this function.

13 $[x]$ denotes the largest integer which is less than or equal to x, i.e. $[3\frac{1}{2}] = 3$, $[5] = 5$, $[-2\frac{1}{2}] = -3$. The function $f : x \mapsto [x]$ is defined on the domain $\{-2, -1\frac{1}{4}, -\frac{1}{2}, 0, 1, 1\frac{3}{4}, 1, 2, 2\frac{3}{4}, 3\}$. Draw a mapping diagram for this function.

14 If $f : x \mapsto 4x + 1$ and $g : x \mapsto x - 1$, find f^{-1} and g^{-1}. What is $g^{-1}f^{-1}(3)$?

15 Choose a suitable domain for the function $g : x \mapsto 2x^2 - 4$ so that the range should not be outside the limits of -6 to 10. Using this domain, plot the graph. What is the minimum value of $g(x)$ in this region?

16 Plot the graphs of the functions $f : x \mapsto x^2 - 5$ and $g : x \mapsto ax$ for $a = 1, 2, \frac{1}{3}$. Use the graphs you have drawn to solve the equations:

(i) $x^2 - x - 5 = 0$

(ii) $x^2 - 2x - 5 = 0$

(iii) $3x^2 - x - 15 = 0$.

Use a domain of $-4 \leqslant x \leqslant 4$.

13.5 Flow charts

A **flow chart** consists of a set of instructions which have to be followed in order. The instructions are written in rectangular-shaped boxes, and connected by an arrow showing the direction in which you move. If a decision has to be made at any stage, the instruction is put in a diamond-shaped box. Although here we are mainly concerned with flow charts that carry out mathematical processes, they can be used in a variety of planning situations or manufacturing processes.

Example 13.8

(a) Work through the flow diagram, writing down the values of x, y and z in a table, as shown, recording only the first three decimal places of your answers.

x	y	z
2	1.333	

(b) Cube the last value of x from your table and comment on the result.

(c) Give the last value of x correct to two decimal places.

[MEG]

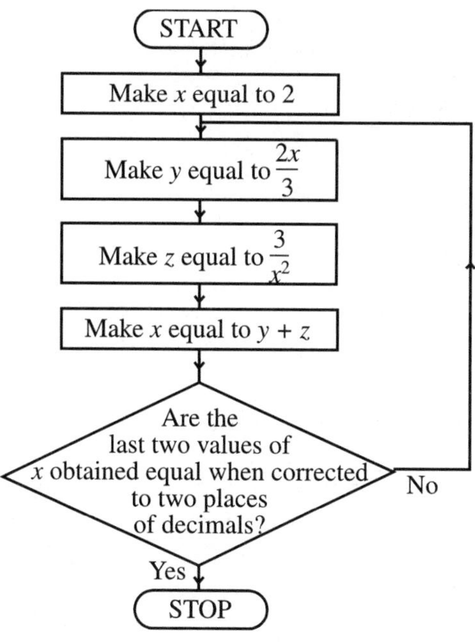

Figure 13.9

Solution

(a) $x = 2$

$$y = \frac{2 \times 2}{3} = \frac{4}{3}$$

$$z = 3 \div 2^2 = \frac{3}{4}$$

$$x = y + z = \frac{4}{3} + \frac{3}{4} = \frac{25}{12} = 2.083.$$

The last two values of x, namely 2 and 2.083 are not equal to two decimal places, you now follow the loop; the values are given in the following table:

x	y	z
2	1.333	0.750
2.083	1.388	0.691
2.079		

(b) $2.079^3 = 8.986$

We appear to have found the cube root of 9.

(c) $x = 2.08$ correct to 2 decimal places.

13.6 Iterative methods

Consider the formula $a = \dfrac{b}{2} + \dfrac{1}{b}$

$$\text{if } b = 1, \qquad a = \frac{1}{2} + \frac{1}{1} \qquad = 1.5$$

$$\text{if } b = 1.5, \qquad a = \frac{1.5}{2} + \frac{1}{1.5} \qquad = 1.417$$

$$\text{if } b = 1.417, \quad a = \frac{1.417}{2} + \frac{1}{1.417} = 1.414$$

$$\text{if } b = 1.414, \quad a = \frac{1.414}{2} + \frac{1}{1.414} = 1.414.$$

You can see that repeated use of the formula, replacing b with a each time eventually reaches an answer that stays unaltered. We say that the formula *converges* to the value of 1.414.

This method is summarised by the flow chart in Figure 13.10. A method of this type is called an **iterative formula**. Normally, the formula would be written

$$x_{n+1} = \frac{x_n}{2} + \frac{1}{x_n}$$

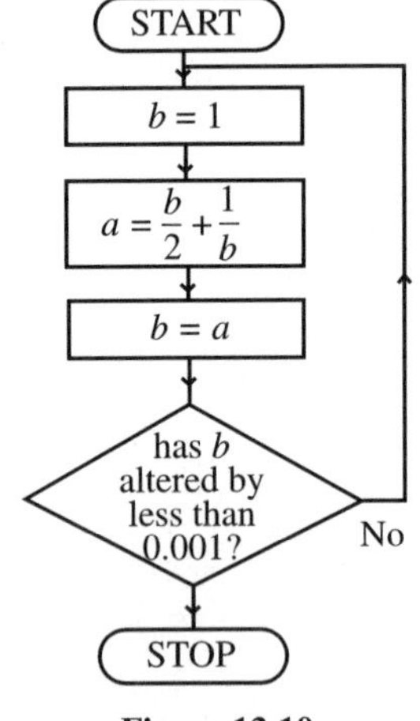

Figure 13.10

The starting value is called $x_0 = 1$

Hence $x_1 = 1.5$

$x_2 = 1.417$

$x_3 = 1.414.$

This type of repeating process is sometimes called an **iteration**.

Since eventually $x_{n+1} = x_n$, suppose the answer is X,

then $X = \dfrac{X}{2} + \dfrac{1}{X}$

Replace x_{n+1} *and* x_n by X

Multiply each term by $2X$: $2X^2 = X^2 + 2$

i.e. $X^2 = 2$

Hence $X = \sqrt{2} = 1.414$

The method enables us to find $\sqrt{2}$ to any accuracy.

The formula $\frac{x_{n+1}}{2} = \frac{x_n}{2} + \frac{N}{2x_n}$ leads to $\sqrt{N}$.

To find an iterative formula to carry out a particular purpose is not easy, and is really beyond the scope of this book. However, one example is given here, with a note on what must be avoided.

Example 13.9

The equation $x^3 - 2x - 1 = 0$ is to be solved by an iterative method.

Rearrange the formula to give $x = \sqrt{2 + \frac{1}{x}}$. Hence show that starting with $x_0 = 1$ and using the iteration formula $x_{n+1} = \sqrt{2 + \frac{1}{x_n}}$, a value is found (correct to 3 decimal places) that satisfies the equation $x^3 - 2x - 1 = 0$.

Solution

$x^3 - 2x - 1 = 0$, hence $\quad x^3 = 2x + 1$

$\div$ the equation by x: $\quad x^2 = 2 + \frac{1}{x}$

Take the square root: $\quad x = \sqrt{2 + \frac{1}{x}}$

Making this the iteration formula: $\quad x_{n+1} = \sqrt{2 + \frac{1}{x_n}}$

If $x_0 = 1$ $\quad x_1 = \sqrt{2 + \frac{1}{1}} = 1.732$

$$x_2 = \sqrt{2 + \frac{1}{1.732}} = 1.605$$

$$x_3 = \sqrt{2 + \frac{1}{1.605}} = 1.620$$

$$x_4 = \sqrt{2 + \frac{1}{1.620}} = 1.618$$

$$x_5 = \sqrt{2 + \frac{1}{1.618}} = 1.618$$

If we substitute $x = 1.618$ in the left hand side of the equation, we get

$$1.618^3 - 2 \times 1.618 - 1 = -0.0002.$$

Hence we have found the solution. You should note that there are several ways of rearranging the equation $x^3 - 2x - 1 = 0$, and hence several different iteration formulae. But some will *not* converge to an answer.

Exercise 13(c)

1

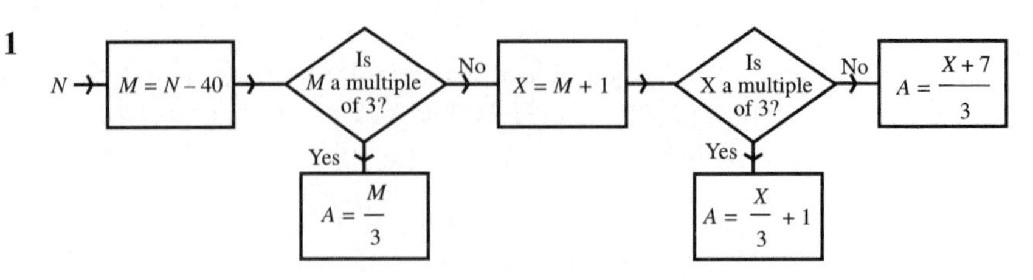

Figure 13.11

In the following question, N, M and A must be positive whole numbers less than 150 and greater than 40.

(i) Copy the following table, and complete it, using the flow chart above.

N	M	X	A
52			
90			
122			

(ii) Explain why the value for A is always possible.

(iii) (a) Find the largest possible value of A.

(b) Find the smallest possible value of A.

2 Jane is sorting a pile of buttons into four boxes. All the buttons have 2 or 4 holes. Some also have patterns on them.

She uses this flow chart to do the sorting.

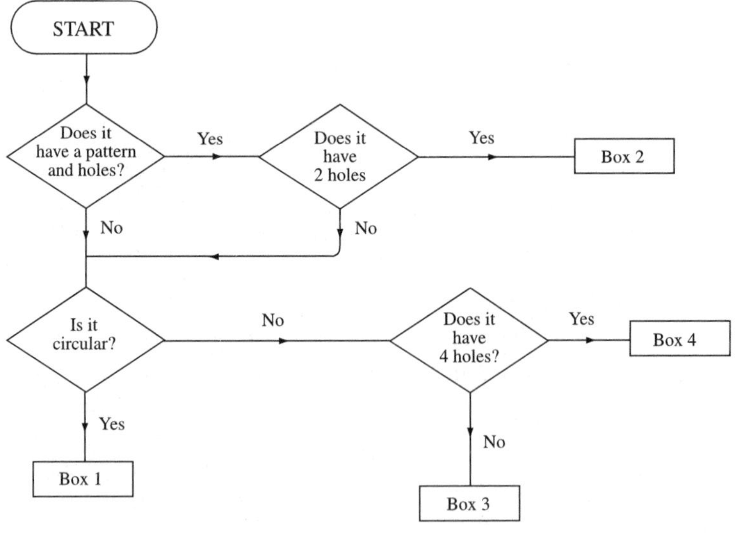

Figure 13.12

Use the flow chart to sort out the six buttons illustrated into box numbers.

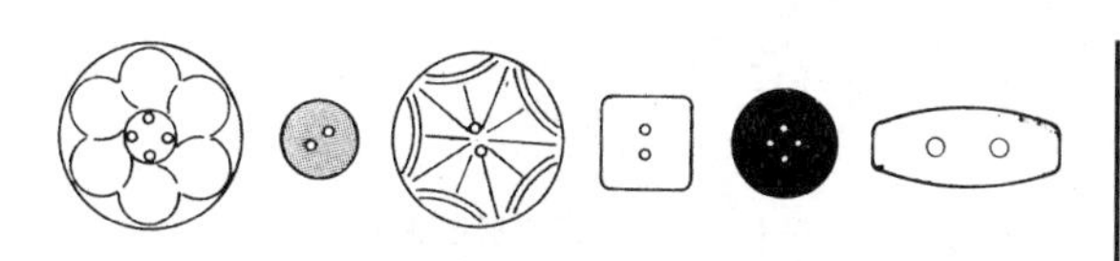

3 The flow chart in Figure 13.13 will find the $\sqrt{20}$. Work through and find the answer correct to 6 significant figures. Which box would you alter to find the square root of any whole number, and what would you alter it to?

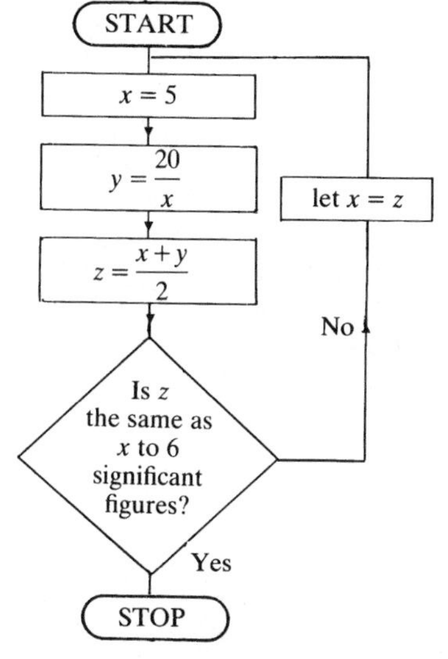

Figure 13.13

4

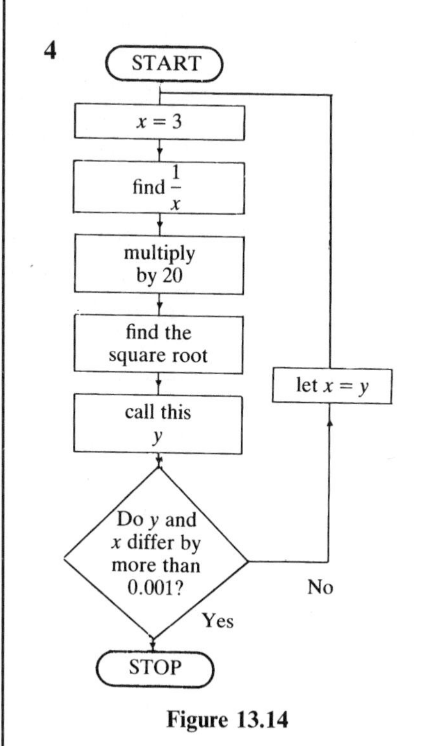

Figure 13.14

The flow chart in Figure 13.14 can be used to find the cube root of a number. Use this method, and state what number you have found the cube root of. Show that the iteration formula $x_{n+1} = \sqrt[4]{20x_n}$ will give the same result.

14 The circle and further trigonometry

14.1 Geometry of the circle

(i) Tangents, chords and segments

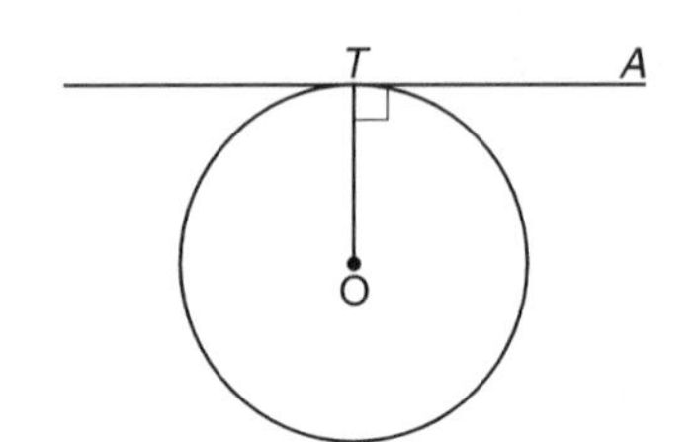

Figure 14.1

The angle between a tangent and radius at the point of contact is 90° (see Figure 14.1).

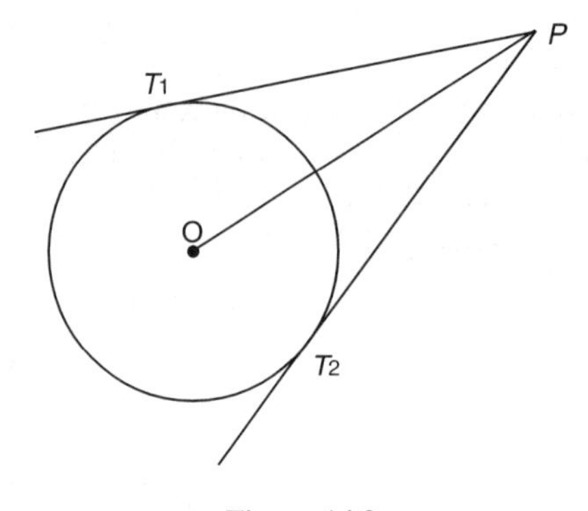

Figure 14.2

The two tangents drawn to a circle from a point outside the circle are equal in length (see Figure 14.2).

So $PT_1 = PT_2$.

To construct the position of T_1 and T_2 accurately, try the following investigation.

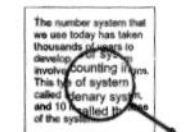

Investigation XXV: Tangents to a circle

(i) Draw a circle of radius about 3 cm, centre O.

(ii) Mark any point P outside the circle.

(iii) Join OP and find the midpoint of OP, marked M.

(iv) With M as centre, and MO as radius, draw a circle to cut the first circle in two places T_1 and T_2.

(v) Join PT_1 and PT_2. These will both be tangents to the circle.

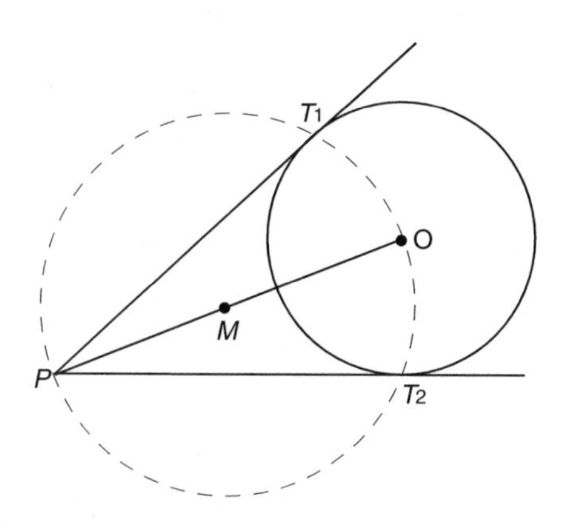

If AB is a chord of a circle, then since $OA = OB$ (radii of the circle), triangle OAB is **isosceles**. The angle AOB is referred to as the angle *subtended* at O, by the chord AB, or sometimes simply the **angle at the centre**. The chord divides the circle into two **segments**: the minor (smaller) segment shown shaded (Figure 14.3), and the major (larger) segment.

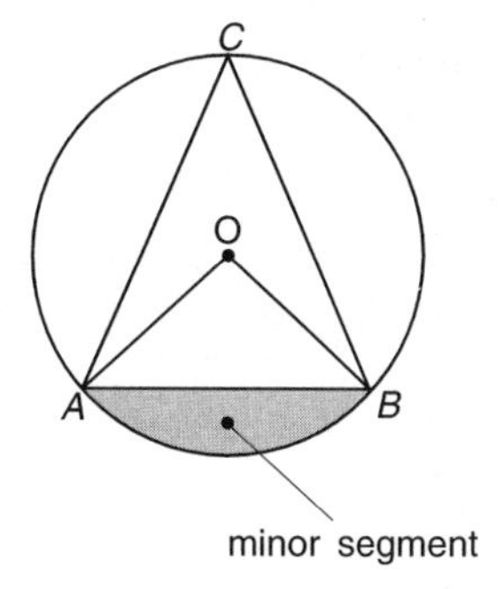

Figure 14.3

If C is a third point on the circumference, then $\angle ACB$ is referred to as an angle in the segment ACB.

If two chords are at equal distances from the centre of the circle, they subtend the same angle at the centre, and so must be equal in length. In Figure 14.4,
$\angle AOB = \angle DOC$ and so $AB = DC$.

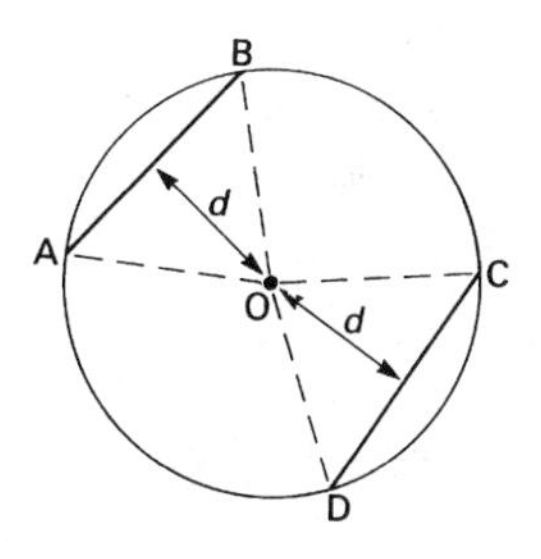

Figure 14.4

Conversely, if two chords of a circle are equal in length, they are at the same distance from the centre.

(ii) Angle at the centre

If A, B and P are three points on a circle as in Figure 14.5, then $\triangle AOP$ and $\triangle BOP$ are both isosceles. If we let $\angle APO = x^\circ$, then you can work out that $\angle AOQ = 2x^\circ$. Also, if $\angle OPB = y^\circ$, then $\angle QOB = 2y^\circ$.

Hence $\angle AOB = 2x^\circ + 2y^\circ = 2(x+y)^\circ$. Also $\angle APB = (x+y)^\circ$.

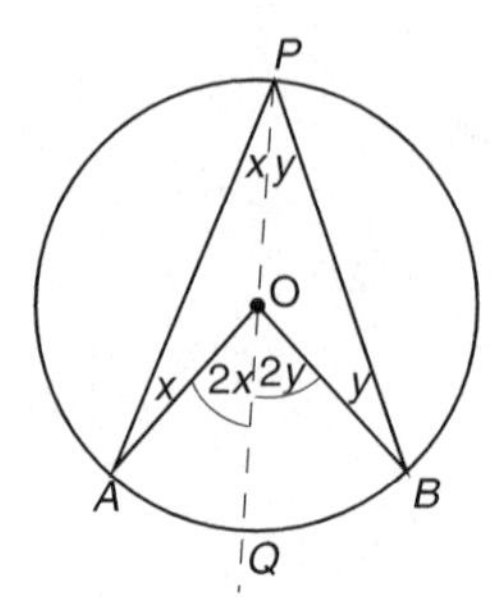

Figure 14.5

Hence, the angle subtended at the centre by AB is *twice* the angle subtended at P by AB. This is always true, if O and P are in the same segment.

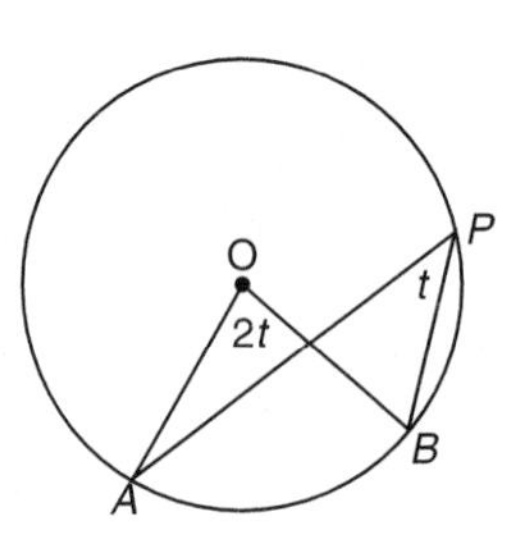

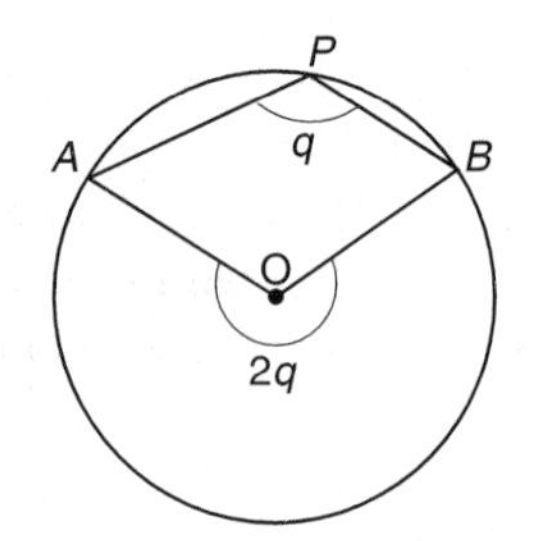

Figure 14.6

In Figure 14.6, you can see two different configurations of the same theorem, which are often forgotten.

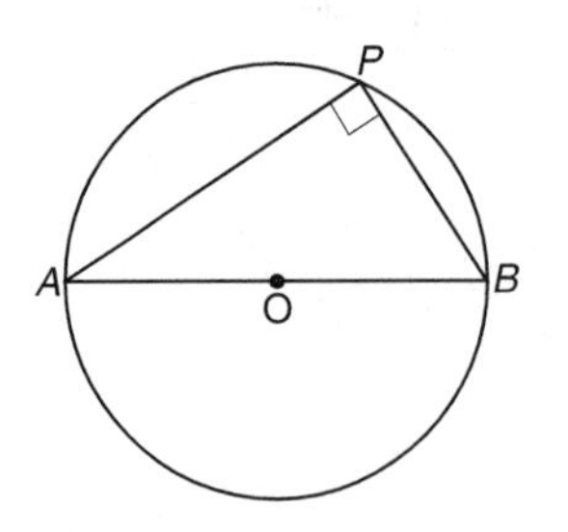

Figure 14.7

If $\angle AOB$ is a diameter, then angle $AOB = 180^\circ$.

$$\text{Hence } \angle APB = \tfrac{1}{2} \times 180^\circ = 90^\circ.$$

This is often stated as **the angle in a semi-circle is 90°**.

(iii) Segments and cyclic quadrilaterals

If AB is a chord, and P_1 and P_2 are points on the circumference both in the same segment, then $\angle AP_1B = \angle AP_2B$. **Angles in the same segment are equal** (see Figure 14.8). This is true, because in each case the angle will be half of the angle at the centre.

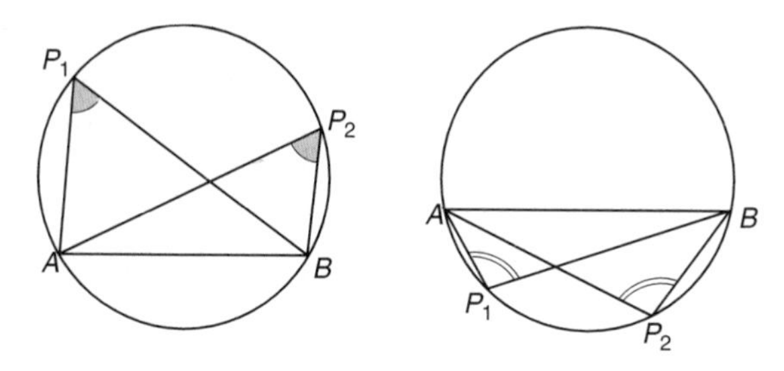

Figure 14.8

If P_1 and P_2 are in different (opposite) segments, then the points AP_1BP_2 form a quadrilateral, called a cyclic quadrilateral (see Figure 14.9). At the centre, $2x + 2y = 360°$, i.e. $x + y = 180°$.

Hence **opposite angles of a cyclic quadrilateral add up to 180°**.

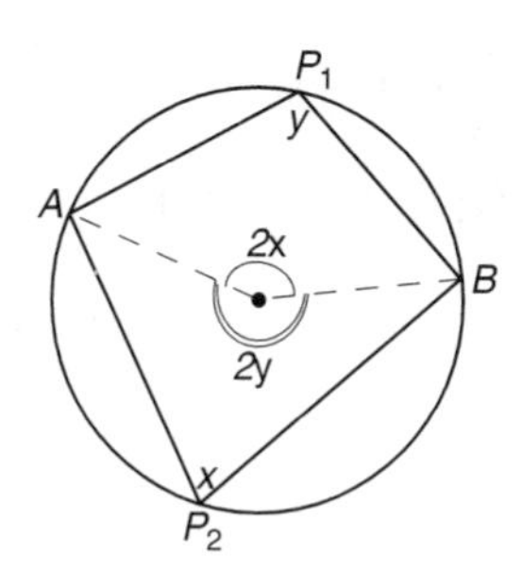

Figure 14.9

(iv) Angles between tangent and chord

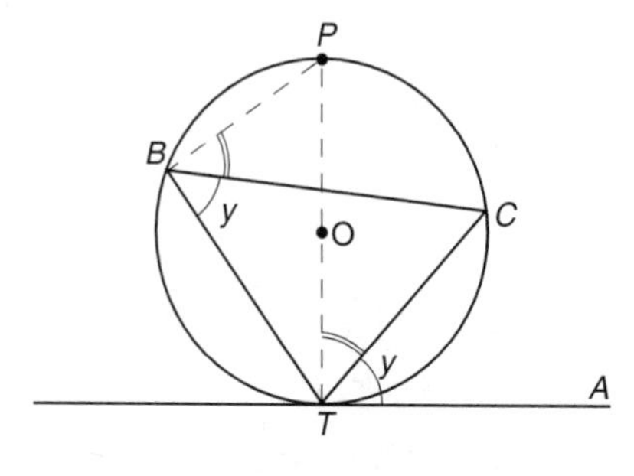

Figure 14.10

If a tangent touches a circle at T, and a chord is drawn through T, then the angle between the tangent and chord CTA is equal to any angle drawn on the chord in the alternate segment, i.e. $\angle TBC$ (see Figure 14.10). This can be proved by joining TO to meet the circle at a point P.

$\angle PBC = \angle PTC$ (angles in the same segment), and $\angle PBT = 90°$ (angle in a semi-circle) $= \angle OTA$.

Hence $\angle TBC = \angle ATC$.

This last theorem can be quite useful, but is often forgotten.

The following examples show how these theorems can be used. These are not always easy, and you will need to persevere if you are to succeed.

Example 14.1

In Figure 14.11, AOB is a diameter, and $\angle CAO = 40°$. Find x.

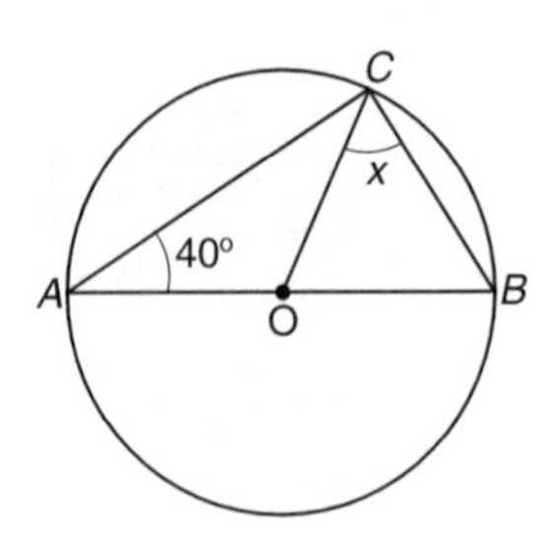

Figure 14.11

Solution

$OC = OA$ (radii of a circle)

Hence $\triangle OAC$ is isosceles, so $\angle ACO = 40°$.

But $\angle ACB = 90°$ (angle in a semi-circle)

So $x = 90° - 40° = 50°$.

Example 14.2

In Figure 14.12, $ABCD$ are four points on a circle, and ABE is a straight line. If $\angle ADC = 100°$, find $\angle CBE$.

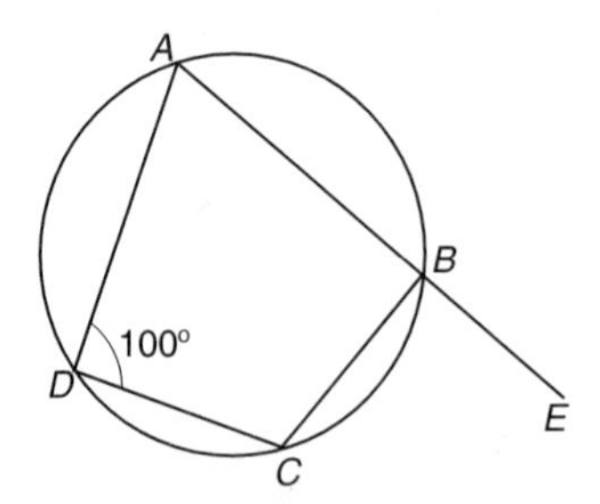

Figure 14.12

Solution

$$\begin{aligned} \angle ABC &= 180° - 100° \text{ (opposite angles of a cyclic quadrilateral)} \\ &= 80° \\ \angle ABC + \angle CBE &= 180° \text{ (angles on a straight line)} \\ \text{Hence} \quad \angle CBE &= 100°. \end{aligned}$$

Example 14.3

Referring to Figure 14.13, O is the centre of the circle and DE is a tangent to the circle, $\angle EAO = 20^\circ$ and $\angle BCA = 40^\circ$.

Find: (a) $\angle ADE$ (b) $\angle OEB$.

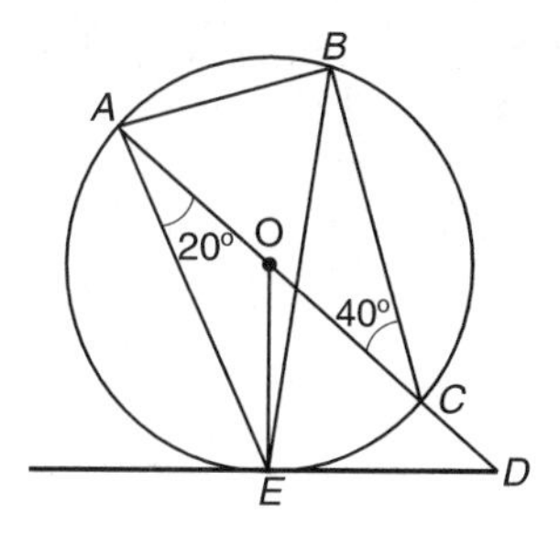

Figure 14.13

Solution

You need to spot (i) $AO = EO$ (radii of the circle) hence triangle AOE is isosceles; (ii) $\angle OED = 90^\circ$ (angle between tangent and radius).

(a) $\angle AOE = 180^\circ - 20^\circ - 20^\circ = 140^\circ$

Hence $\angle EOD = 180^\circ - 40^\circ = 40^\circ$

$\angle ODE = 180^\circ - 40^\circ - 90^\circ = 50^\circ = \angle ADE.$

(b) $\angle AEB = \angle ACB$ (angles in the same segment)

Hence $\angle AEB = 40^\circ$

$\angle OEB = \angle AEB - \angle AEO = 40^\circ - 20^\circ = 20^\circ.$

Example 14.4

A cyclic quadrilateral $PQRS$ has $PQ = PS$. The diagonal $QS = QR$.

If $\angle QSR = 52^\circ$, find $\angle PQR$.

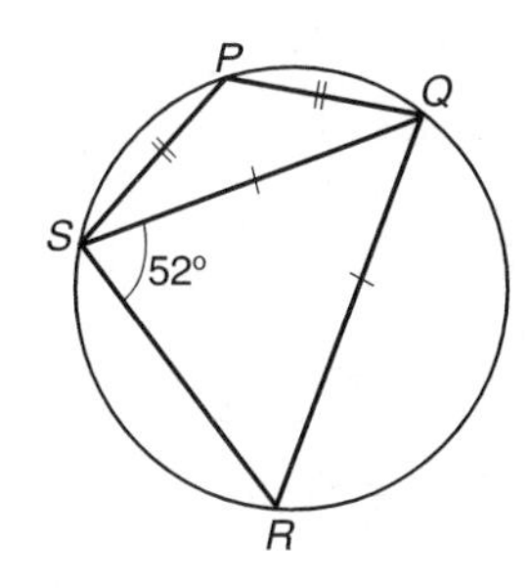

Figure 14.14

Solution

Since $\triangle SRQ$ is isosceles, $\angle SQR = 180^\circ - 52^\circ - 52^\circ = 76^\circ$.

$\angle SPQ$ and $\angle SRQ$ are in opposite segments, hence they are supplementary.

So $\angle SPQ = 180^\circ - 52^\circ = 128^\circ$

$\triangle SPQ$ is isosceles, so $\angle PQS = \frac{1}{2}(180^\circ - 128^\circ) = 26^\circ$

So $\angle PQR = 76^\circ + 26^\circ = 102^\circ$.

Example 14.5

Referring to Figure 14.15, EA and ED are tangents to the circle, $\angle TAB = 46°$ and $\angle AED = 110°$.

Find p and q, given that AB and DC are parallel.

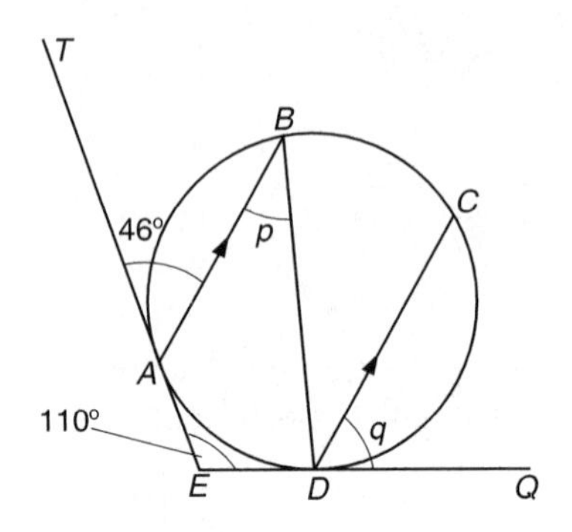

Figure 14.15

Solution

First attempts at this question might give difficulty in using the recognised circle theorems. However, tangents drawn to a circle from a given point have the same length (Figure 14.2), hence if AD is joined, $\triangle ADE$ is isosceles.

$\angle EAD = \angle EDA = 35°$

$\angle TAE = 180°$, hence

$\angle BAD = 180° - 46° - 35° = 99°$

Now $\angle BDA = \angle BAT = 46°$ by the tangent-chord theorem (see Figure 14.16).

Hence in $\triangle ABD$,

$p = 180° - 99° - 46° = 35°$

But $\angle BDC = \angle ABD = p$

So $35° + 46° + p + q = 180°$

So $q = 64°$

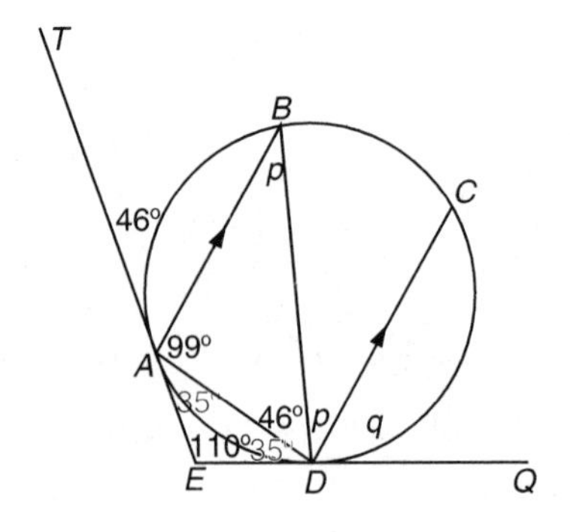

Figure 14.16

(v) Chords and secants

If two chords of a circle cross at X, as shown in Figure 14.17, then it can be shown that

$$XA \times XB = XC \times XD$$

(This follows because $\triangle ACX$ and $\triangle DBX$ are similar and so $\dfrac{XB}{XC} = \dfrac{XD}{XA}$.)

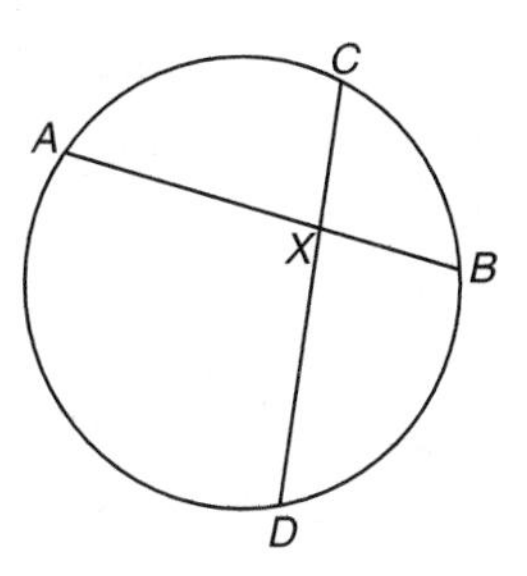

Figure 14.17

Any line that passes through a circle is called a **secant**. Referring to Figure 14.18 , PAB and PCD are secants. It can be shown that

$$PA \times PB = PC \times PD = PT^2$$

(where PT is a tangent).

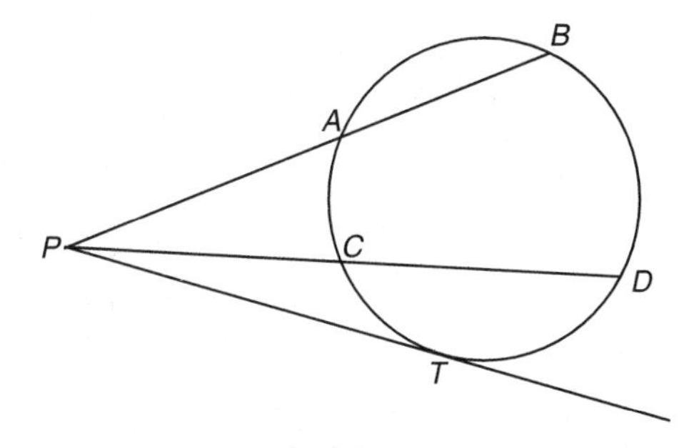

Figure 14.18

Example 14.6

In Figure 14.19, PAB and PCD are straight lines. It is known that PA is 0.6 cm shorter than PC. Find PA and PC.

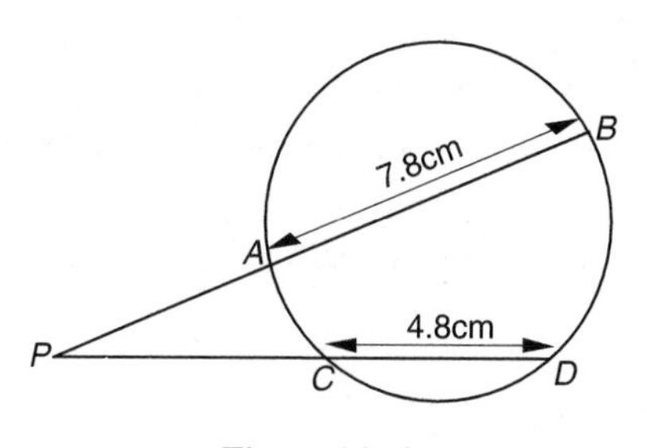

Figure 14.19

Solution

Let $PA = x$ cm, then $PC = x + 0.6$

Using the secant theorem,

$$x \times (x + 7.8) = (x + 0.6)(x + 0.6 + 4.8)$$
$$= (x + 0.6)(x + 5.4)$$

Hence $\quad x^2 + 7.8x = x^2 + 6x + 3.24$

So $\quad 1.8x = 3.24$

$$x = 1.8$$

Hence $PA = 1.8$ cm, $PC = 2.4$ cm.

Before trying Exercise 14(a), the following investigation gives an interesting different look at circles.

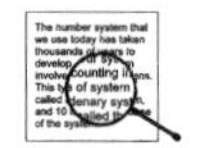

Investigation XXVI: Investigation with circles

A is the centre of circle 1

B is the centre of circle 2

C is the centre of circle 3

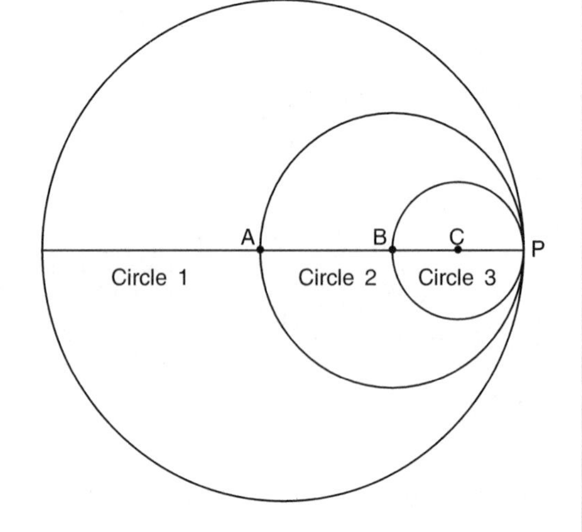

(i) The radius of circle 3 is 4 cm.
  (a) Write down the length of the radius of circle 2.
  (b) Write down the length of the radius of circle 1.

(ii) The circumference of circle 2 is twice the circumference of circle 3.
  Copy and complete each of the following:
  (a) The circumference of circle 1 is . . . . the circumference of circle 2.
  (b) The circumference of circle 1 is . . . . the circumference of circle 3.

(iii) If circle 4 is drawn with centre at the mid-point of CP, write down the length of the radius of circle 4.

(iv) If this pattern continues, what is the length of the radius of circle 7?

(v) Can you construct similar patterns with shapes other than circles?

Exercise 14(a)

1 Find x and y in the following questions. In each case, O is the centre of the circle.

(i)

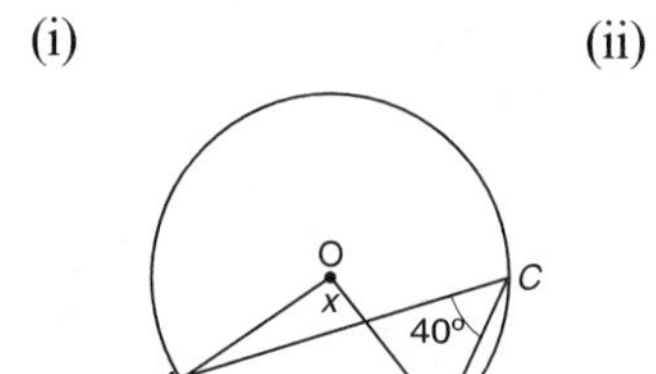

(ii)

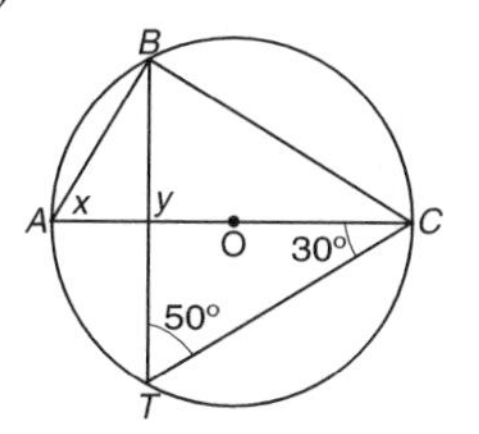

(iii)

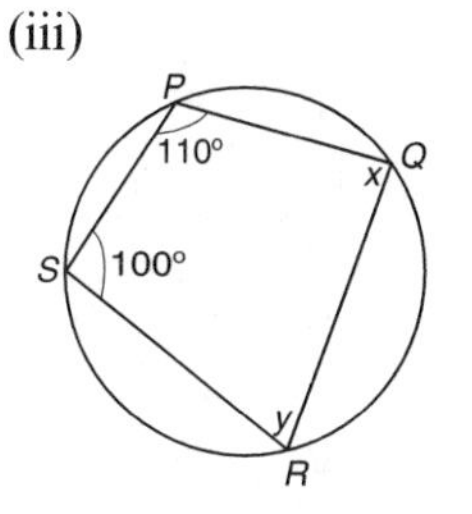

(iv)

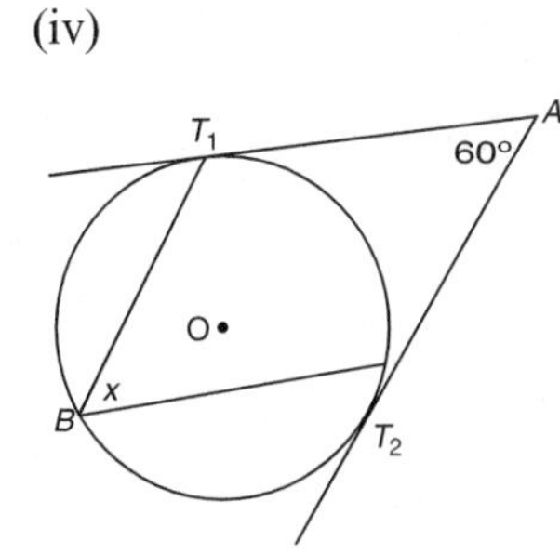

(v)

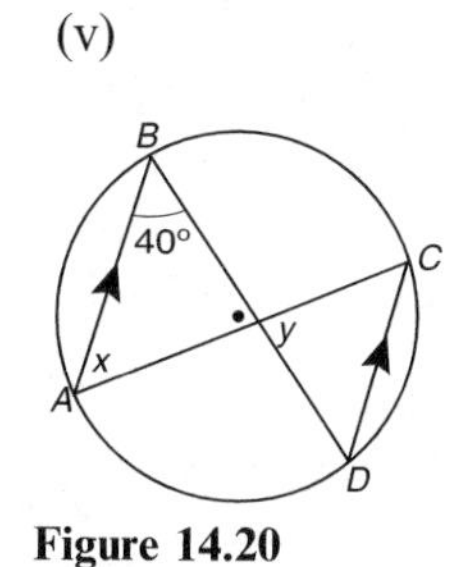

(vi)

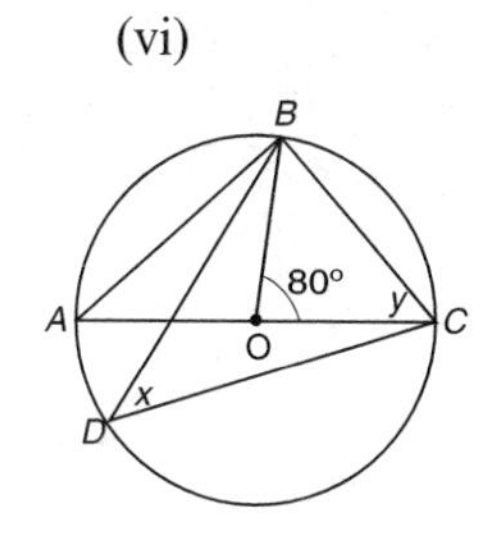

Figure 14.20

2 In Figure 14.21, O is the centre of the circle, $ABOC$ is a straight line and AT is the tangent to the circle at T. Angle $BAT = 28°$.

Calculate:

(i) $\angle ATO$
(ii) $\angle TOA$
(iii) $\angle BCT$
(iv) $\angle OTC$
(v) $\angle CDB$.

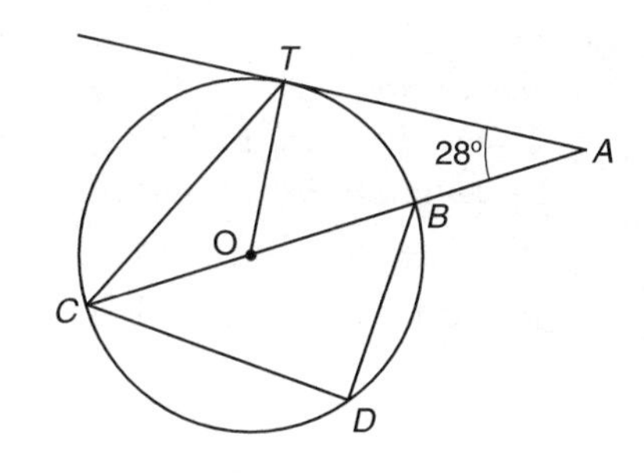

Figure 14.21

3 A, B, C and D are points on a circle with BOC a diameter, $AD = DC$ and $\angle DBC = 40°$.

Find the values of the angles marked a, b, c and d in Figure 14.22.

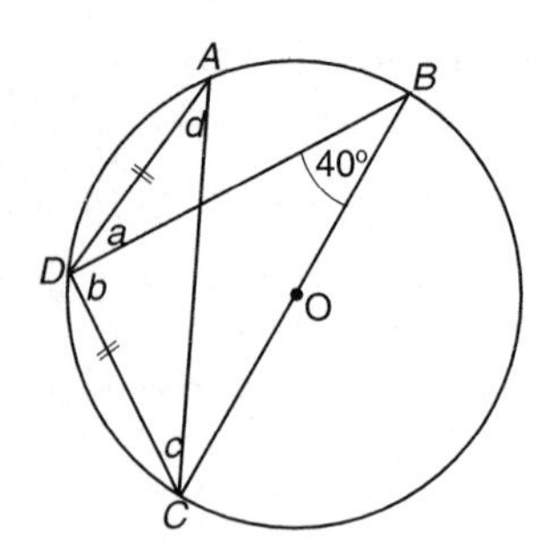

Figure 14.22

4 P, Q, R, S are four points on a circle, centre O. $\angle QPS = 65°$ (see Figure 14.23).

Calculate:

(i) $\angle QOS$

(ii) $\angle OQS$

(iii) $\angle QRS$. State clearly your reasons.

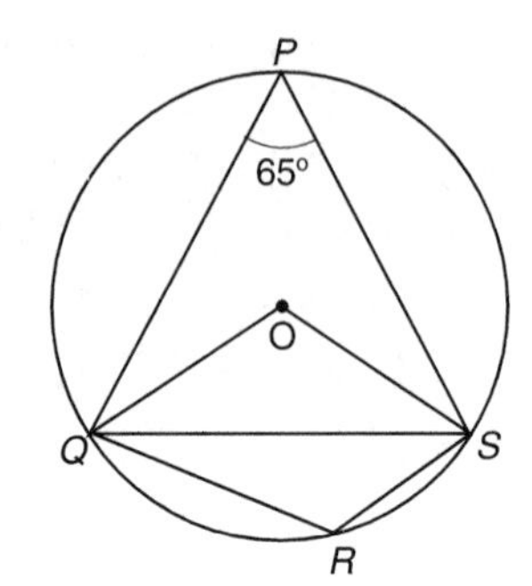

Figure 14.23

5 P, Q and R are points on a circle with centre O. TQ and TR are tangents to the circle from the point T (see Figure 14.24). Calculate: p, q and t, giving full reasons for your answers.

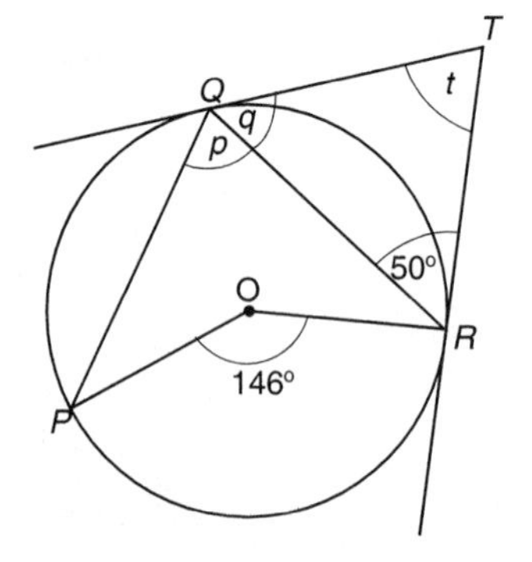

Figure 14.24

6 In Figure 14.25, A and B are points on a circle, centre O and diameter 26 cm. The perpendicular from O on to AB meets AB at C. If $AB = 10$ cm:

(i) find the length of CB

(ii) calculate the length of OC.

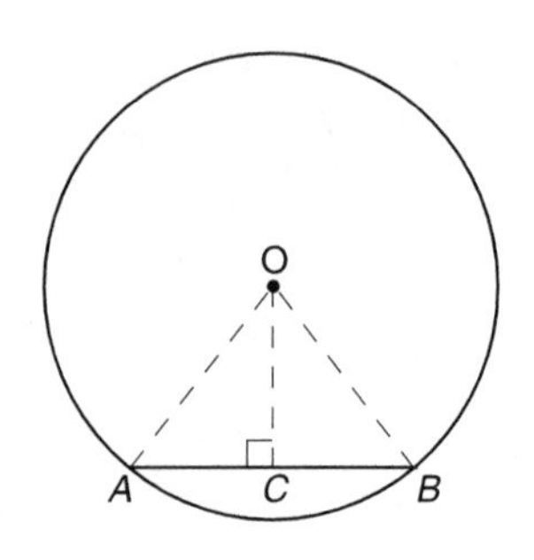

Figure 14.25

7 Referring to Figure 14.26, find the length of PT, where PT is a tangent to the circle, PAB is a straight line, $AB = 8$ cm, and $PA = 6$ cm.

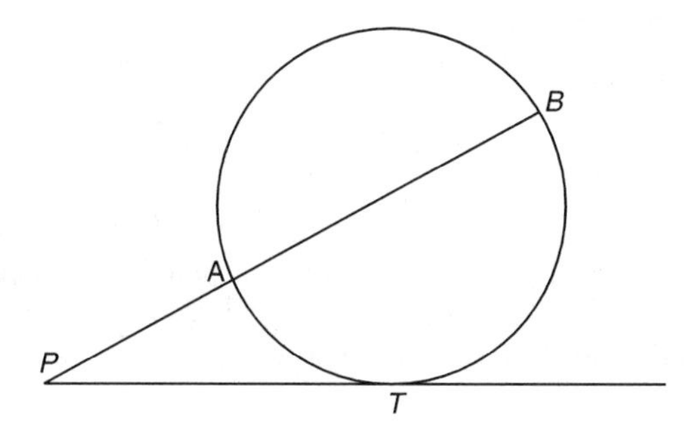

Figure 14.26

8 The chords of a circle XY and PQ cross at R. If $XR = 3$ cm, $RY = 5$ cm, and $RQ = 4$ cm, find RP.

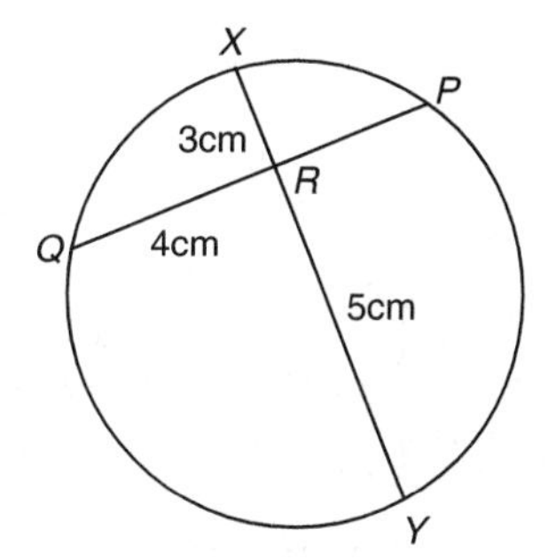

Figure 14.27

9 (i)

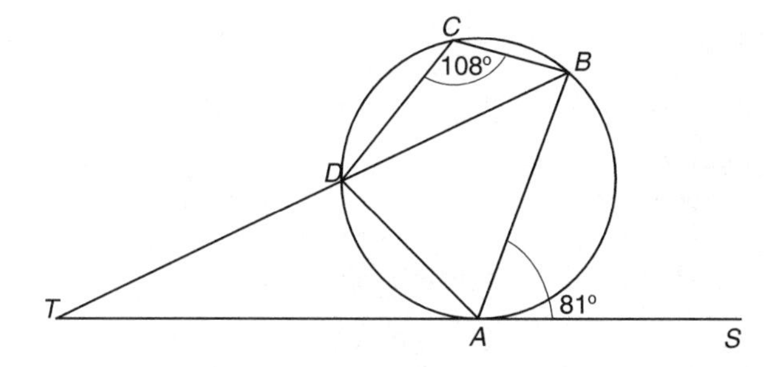

Figure 14.28

In Figure 14.28, which is not drawn to scale, $ABCD$ is a cyclic quadrilateral. SA is the tangent to the circle at A, BD produced meets SA produced at T. Given that $\angle BAS = 81°$ and $\angle DCB = 108°$, calculate the size of:

(a) $\angle BDA$

(b) $\angle DAB$

(c) $\angle DTA$.

(ii) P, Q and R are three points on a circle centre O. The tangent to the circle at R meets PQ produced at T. Given that $RT = 6$ cm, $OT = 6.5$ cm and $QT = 5$ cm, calculate the length of:
(a) OR (b) PT.

(iii) In triangle ABC, K is a point on AC such that BK bisects ABC. The point H is on BA such that HK is parallel to BC. Given that $AH = 2.0$ cm, $AK = 2.4$ cm and $KC = 3.6$ cm, calculate the length of: (a) HB (b) BC.

10 In Figure 14.29 AB and CD are two perpendicular chords crossing at E, with $AE = EB = 2$ cm and $CE = 1$ cm. If R is the radius of the circle, show that R satisfies the equation $2R - 1 = 4$. Hence find R.

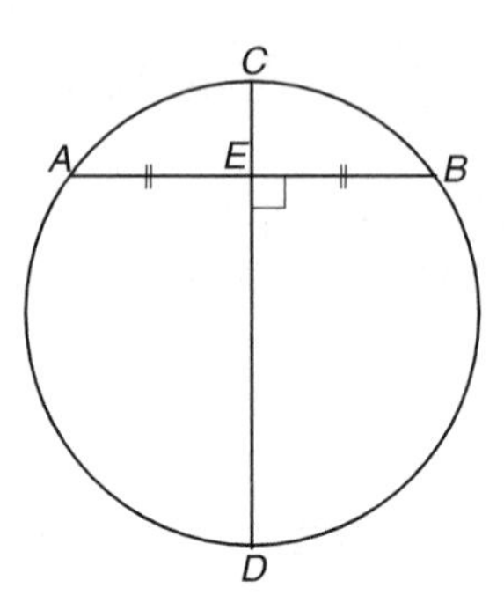

Figure 14.29

14.2 The sine rule

We have looked at how to solve right angled triangles by means of trigonometric ratios, or Pythagoras' theorem in Chapter 7. In this section and the next, we will look at what happens if the triangles are not right angled.

In triangle ABC (Figure 14.30) it can be shown that

Rule

$$\frac{a}{\sin A} = \frac{b}{\sin B} = \frac{c}{\sin C} = 2R$$

where R is the radius of the circle that passes through A, B and C (the circumcircle).

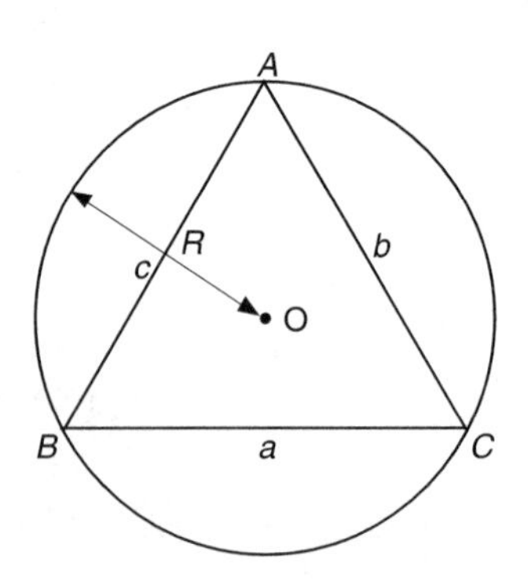

Figure 14.30

Example 14.7

Referring to Figure 14.31, in triangle PQR where $PQ = 6$ cm, $\angle PQR = 74°$ and $\angle PRQ = 69°$, find PR.

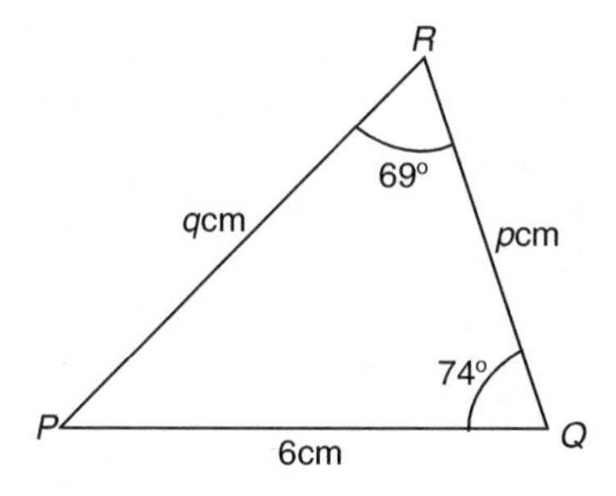

Figure 14.31

Solution

The sine rule states:

$$\frac{6}{\sin 69°} = \frac{q}{\sin 74°}$$

So

$$\frac{6}{0.9336} = \frac{q}{0.9613}$$

Hence

$$\frac{6 \times 0.9613}{0.933} = q \qquad (1)$$

So $q = 6.18$, and the length of $PR = 6.18$ cm.

If one of the angles of a triangle is greater than 90°, the method is the same. More explanation about obtuse angles can be found in Section 14.4.

If the calculator is used throughout, then (1) above becomes

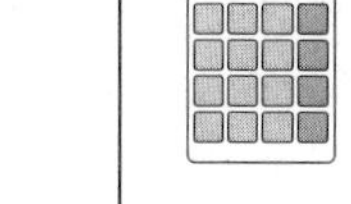

$$q = \frac{6 \times \sin 74°}{\sin 69°}.$$

The key sequence is

Display

74 sin    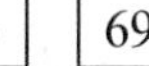 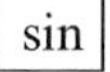 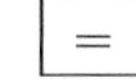 6.1779039

74	sin	×	6	÷	69	sin	=	6.1779039

Example 14.8

In triangle DEF (Figure 14.32) $\angle DEF = 130°$, $\angle EFG = 18°$, and $DF = 12.9$ cm. Find DE.

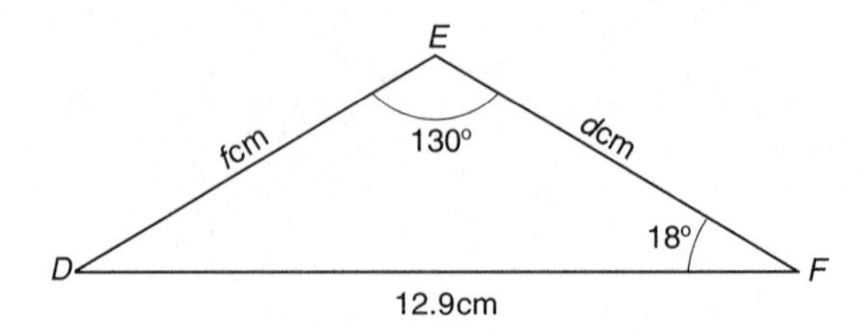

Figure 14.32

Solution

$$\frac{f}{\sin 18°} = \frac{12.9}{\sin 130°}$$

$$\frac{f}{0.309} = \frac{12.9}{0.766}$$

Hence $f = \dfrac{12.9 \times 0.309}{0.766} = 5.20$, and the length of $DE = 5.20$ cm.

When the sine rule is used to find angles, you have to be careful of the **ambiguous** case, demonstrated in Example 14.10.

Example 14.9

In triangle PQR (Figure 14.33), $PQ = 6$ cm, $PR = 8$ cm, and $\angle PQR = 48°$. Find $\angle PRQ$.

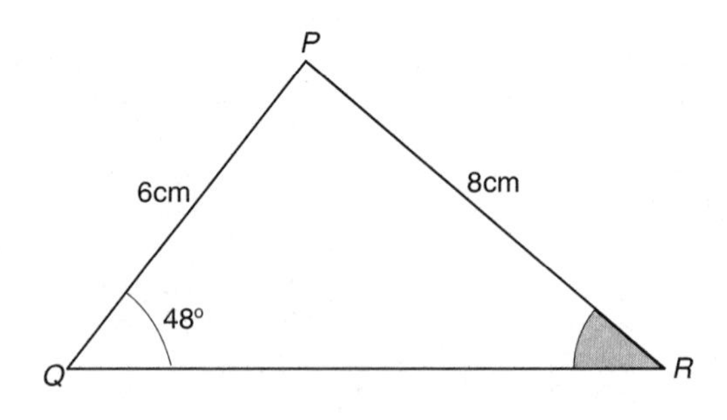

Figure 14.33

Solution

$$\frac{8}{\sin 48^\circ} = \frac{6}{\sin R}, \text{ so } \frac{8}{0.7431} = \frac{6}{\sin R}$$

Hence $\sin R = \dfrac{6 \times 0.7431}{8}$ (careful with this rearrangement) (1)

$= 0.5573$

So $R = 33.9^\circ$.

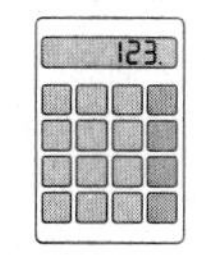

Using the calculator, since (1) can be rewritten

$$\sin R = \frac{6 \times \sin 48^\circ}{8},$$

the key sequence is

48	sin	×	6	÷	8	=	$\sin^{-1}$	**Display** 33.873324

Example 14.10

In triangle RST, $\angle R = 20^\circ$, $RS = 100$ m and $ST = 80$ m. Find T.

Solution

Using the sine rule,

$$\frac{80}{\sin 20^\circ} = \frac{100}{\sin T}$$

So $\sin T = \dfrac{100 \times \sin 20^\circ}{80} = 0.4275$

Hence $T = 25.3^\circ$.

If you try and draw the triangle with the given information, then two possible triangles exist, namely $\triangle RST_1$ and $\triangle RST_2$. These are shown in Figure 14.34.

Since triangle T_1ST_2 is isosceles, then the other angle T is found by subtracting 25.3 from 180°. Hence

$$\angle RT_1S = 154.7^\circ.$$

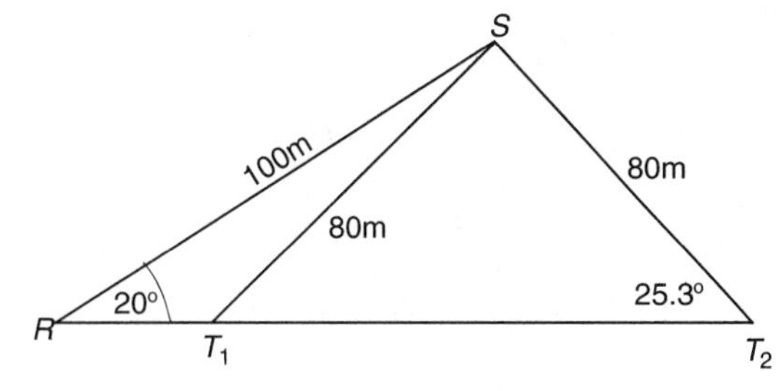

Figure 14.34

(i) Area of a triangle

At this point it is worth stating another formula for finding the area of a triangle.

Referring to Figure 14.35, we have:

$$\text{Area of the } \triangle = \tfrac{1}{2}ah$$
$$= \tfrac{1}{2}a \times b \sin C.$$

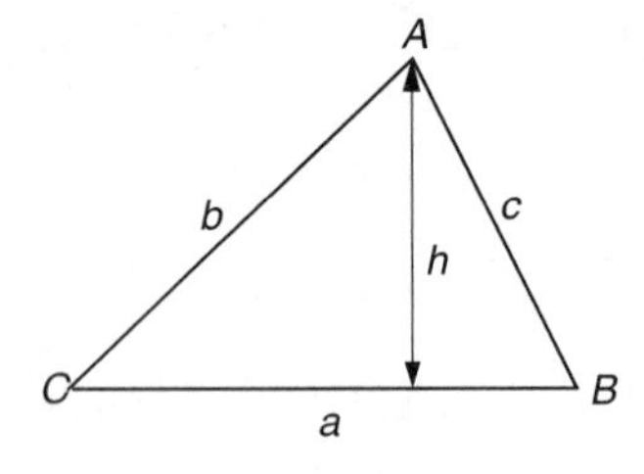

Figure 14.35

Rule

The area of a triangle $= \frac{1}{2}ab\sin C$

This formula is used if you know two sides and the angle between those sides (i.e. the **included** angle).

Example 14.11

Farmer Giles has a triangular field XYZ where $XZ = 100$ m, $XY = 80$ m, and angle $YXZ = 80°$.

Find the area of the field in hectares.

[*Note*: 1 hectare = 10000 m^2.]

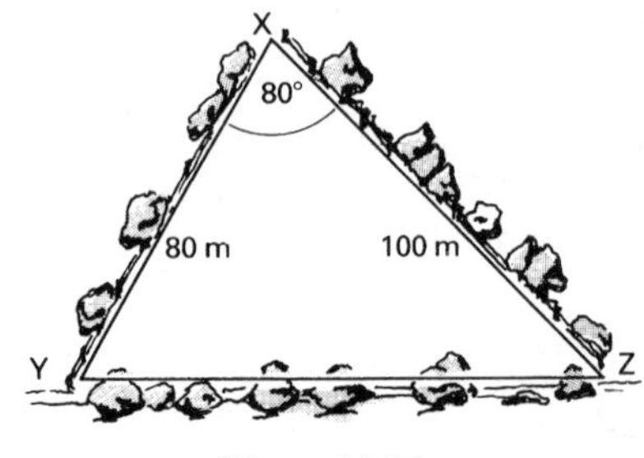

Figure 14.36

Solution

Using area $= \frac{1}{2}yz\sin X$,

$$= \tfrac{1}{2} \times 80 \times 100 \sin 80° \text{ m}^2$$

$$= 3939 \text{ m}^2$$

Hence in hectares we divide by 10 000, to give:

Area = 0.4 hectare.

14.3 The cosine rule

The sine rule which was used in the last section does not cope with all the situations that arise with non-right angled triangles. Using the notation of Figure 14.30, it can be shown that:

Rule

$$a^2 = b^2 + c^2 - 2bc\cos A$$

and $$b^2 = c^2 + a^2 - 2ca\cos B$$

and $$c^2 = a^2 + b^2 - 2ab\cos C$$

or $$\cos A = \frac{b^2 + c^2 - a^2}{2bc}$$

and $$\cos B = \frac{a^2 + c^2 - b^2}{2ac}$$

and $$\cos C = \frac{a^2 + b^2 - c^2}{2ab}$$

This is known as the cosine rule.

Example 14.12

In Figure 14.37, find:

(a) BC (b) $\angle ABC$

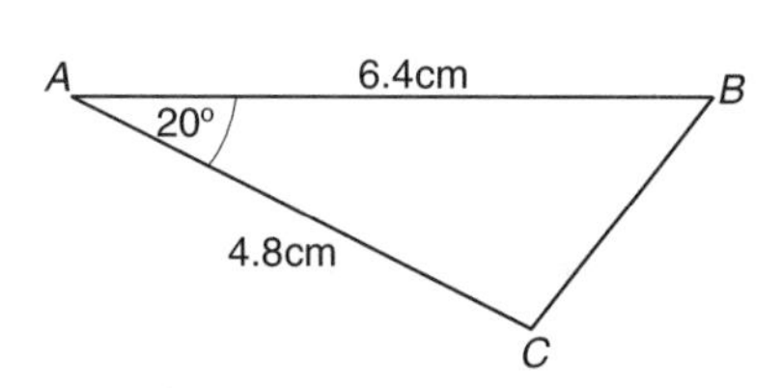

Figure 14.37

Solution

(a) Using $a^2 = b^2 + c^2 - 2bc\cos A$

$$a^2 = 6.4^2 + 4.8^2 - 2 \times 6.4 \times 4.8 \times \cos 20^\circ$$

$$= 40.96 + 23.04 - 57.73 = 6.27$$

Hence $a = \sqrt{6.27} = 2.5$, so $BC = 2.5$ cm (1 d.p.).

This long calculation can be done on the calculator as follows:

Display

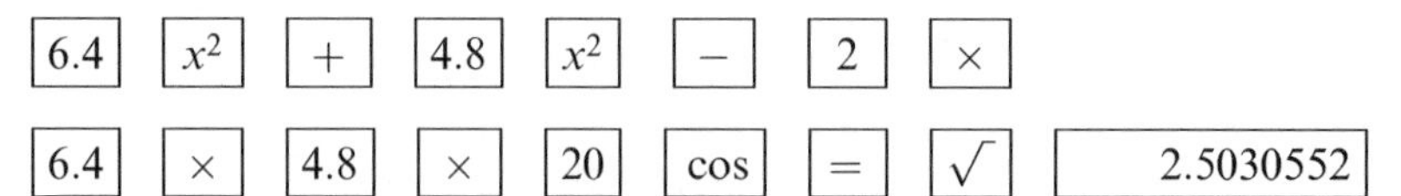

(b) This part is carried out in the same way as Example 14.9.

We can now use the sine rule

$$\text{So} \quad \frac{2.5}{\sin 20^\circ} = \frac{4.8}{\sin B}$$

$$\sin B = \frac{4.8 \times \sin 20^\circ}{2.5} = 0.6567$$

Hence $ABC = 41^\circ$.

When dealing with obtuse-angled triangles, the method stays the same.

Example 14.13

In triangle PQR (Figure 14.38), $PQ = 12$ m, $QR = 5$ m, and $\angle PQR = 140^\circ$. Find PR.

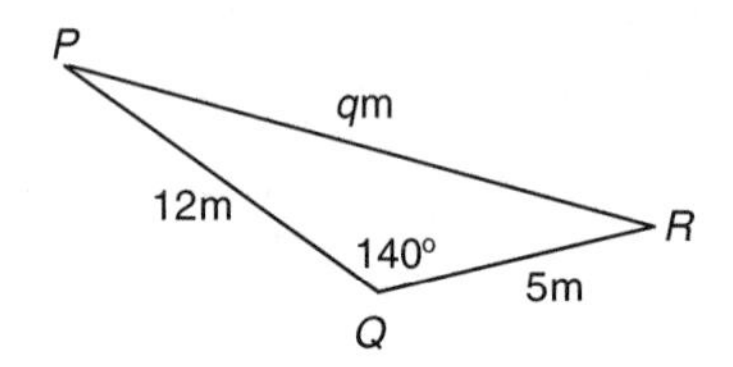

Figure 14.38

Solution

$$\begin{aligned} \text{Using} \quad q^2 &= p^2 + r^2 - 2pr\cos Q \\ &= 12^2 + 5^2 - 2 \times 12 \times 5 \times \cos 140^\circ \\ &= 144 + 25 + 91.93 \end{aligned}$$

[The cosine has become *negative*, changing this sign]

$$q^2 = 260.93$$

Hence $q = 16.2$, so $PR = 16.2$ m.

Example 14.4

Referring to triangle DEF (Figure 14.39), find $\angle DEF$.

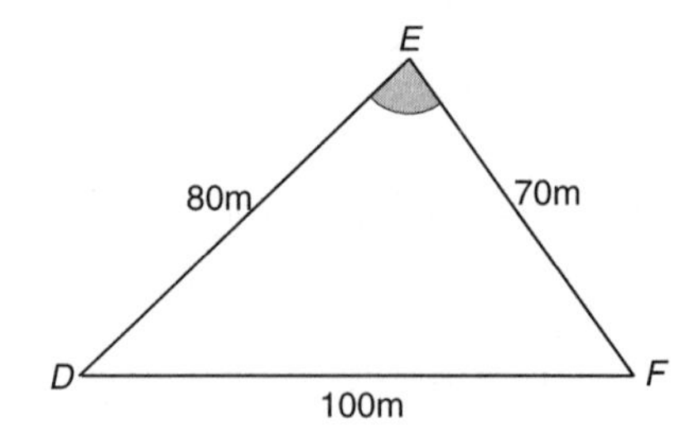

Figure 14.39

Solution

$$\text{Using } \cos E = \frac{d^2 + f^2 - e^2}{2df}$$

$$\cos E = \frac{70^2 + 80^2 - 100^2}{2 \times 70 \times 80}$$

$$= \frac{4900 + 6400 - 10\,000}{11\,200}$$

$$= \frac{1300}{11\,200} = 0.1161$$

Hence $\angle DEF = 83.3^\circ$**.**

Deciding which rule to use is not very difficult and it can be summarised as follows:

Rule

Given three sides, or two sides and the included angle, use the cosine rule. Otherwise, use the sine rule.

Exercise 14(b)

1 Find all the sides marked with a letter and all the shaded angles in the following diagrams.

(i)

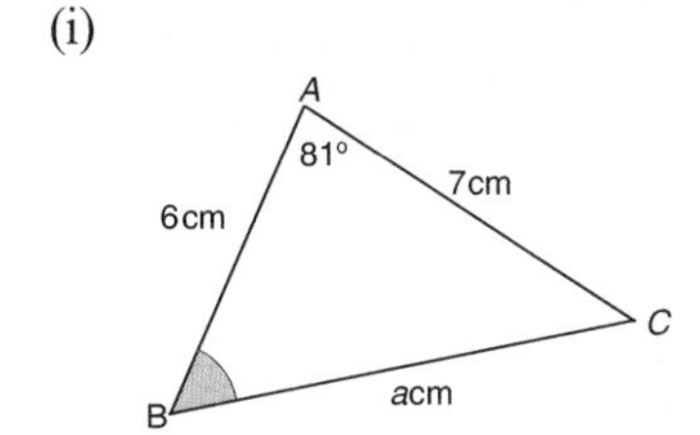

(ii)

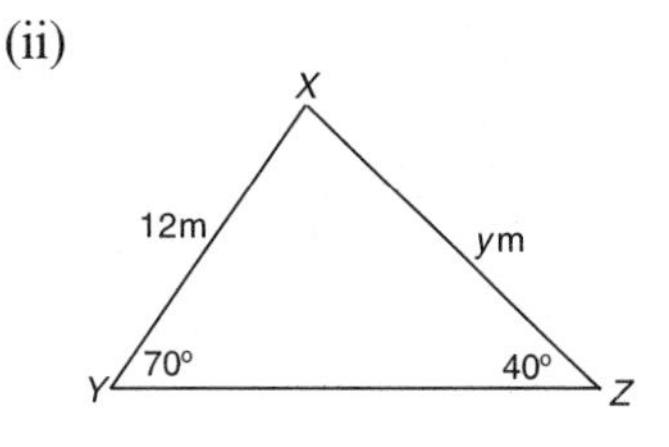

(iii)

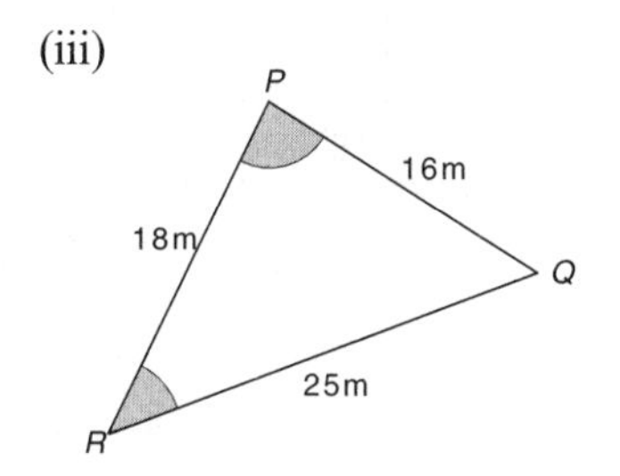

(iv)

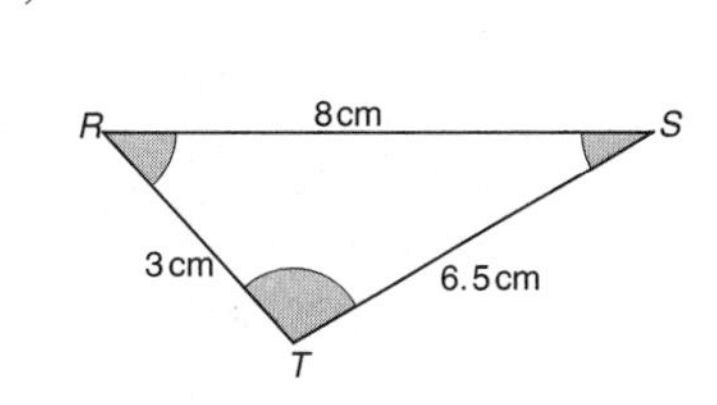

Figure 14.40

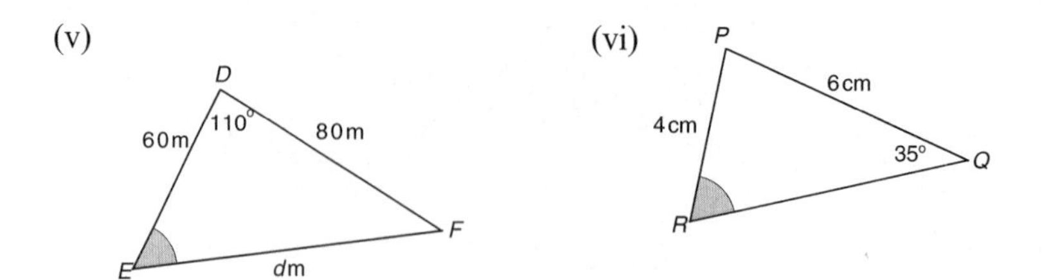

Figure 14.40

2 In a triangle ABC, $AB = 5$ cm, $\angle CAB = 68°$ and $\angle CBA = 30°$. D is the point on the opposite side of BC to A such that $\angle CBD = 65°$ and $BC = CD$. Calculate the lengths of BC and BD, correct to two decimal places.

3 In triangle PQR, $PQ = 5$ cm, $QR = 4$ cm, and $\angle QPR = 40°$. Calculate the two possible values of $\angle QRP$ and the corresponding lengths of PR. Illustrate your answer with a diagram. If only one triangle was possible, what would be the value of QR?

4 A, B and C are three points on a map, $AB = 6.8$ cm, $AC = 5.3$ cm and $\angle BAC = 44.6°$.

(i) Calculate the length of BC.

(ii) Given that the actual distance represented by AB is 13.6 km, calculate the scale of the map in the form $1 : n$, and the actual distance represented by BC in kilometres.

(iii) Given that B is due east of A and that C is north of the line AB, calculate the bearing of C from A.

5 In Figure 14.41, ABC is a straight line, $CD = 6.4$ cm and $AB = 7.3$ cm. Calculate:

(i) BD

(ii) AD

(iii) the area of $\triangle ABD$

(iv) the perpendicular distance of B from AD.

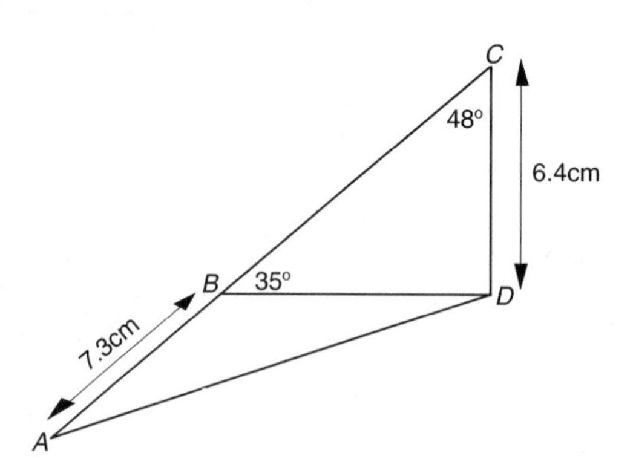

Figure 14.41

6 (Do not solve this equation by scale drawing.)

$ABCD$ is a horizontal field (Figure 14.42).

$AB = 310$ m, $DA = 125$ m, angle $A = 140°$, angle $C = 65°$ and angle $DBC = 60°$.

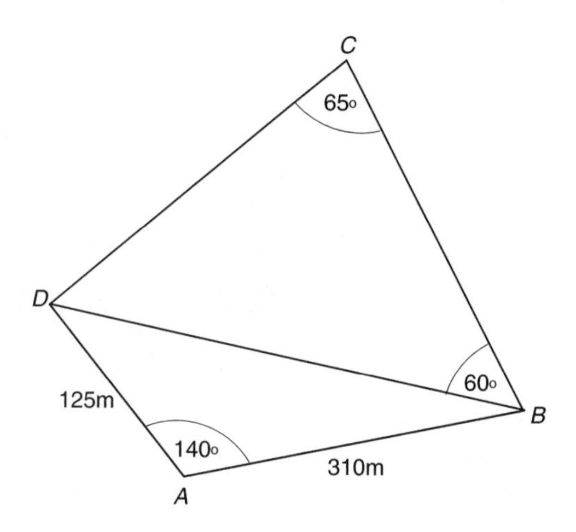

Figure 14.42

Calculate the area, in hectares, of the field, giving your answer correct to 2 significant figures. ($10000 \text{ m}^2 = 1$ hectare).

[MEG]

14.4 Angles greater than 90°

The sine of an angle in a right angled triangle is defined as $\frac{\text{opposite}}{\text{adjacent}}$. If an angle becomes greater than 90°, then we cannot use the triangle definition. If we consider a line of length 1 unit rotating anticlockwise about one end O, and the other end to have coordinates (x, y), then if θ is the angle of rotation,

define $\sin\theta = y$

and $\cos\theta = x$

(see Figure 14.43).

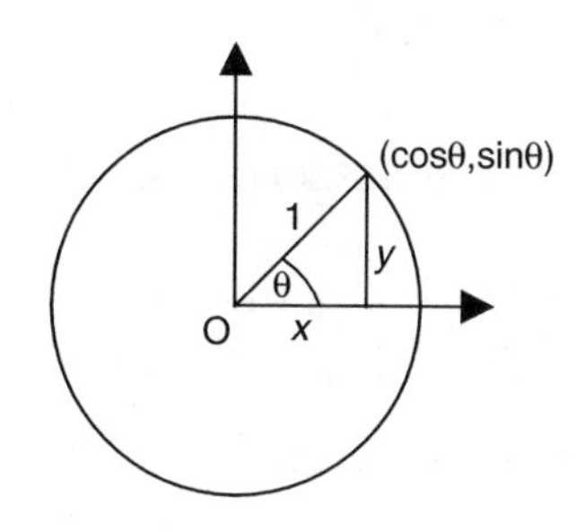

Figure 14.43

As θ increases from 0 to 180° in an anticlockwise direction, y increases from 0 to 1 and back to 0. Between 180° and 360°, y decreases to −1 and back to 0. In Figure 14.44 we can see the resulting curve known as a **sine wave** and a similar curve for cosine, namely the **cosine wave**.

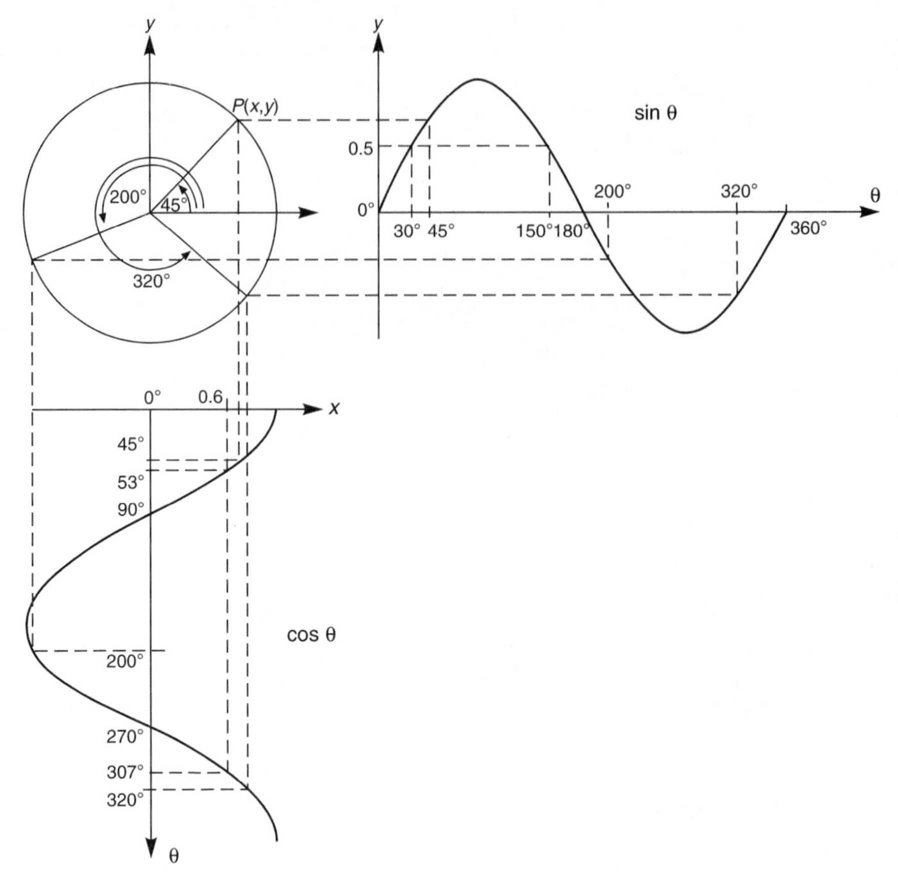

Figure 14.44

These waves occur in many areas in everyday life, including tides, electricity, and magnetism. Your calculator will take care of any angle, and the curve in fact extends beyond 360°, or below 0°. A negative angle refers to notation in a clockwise direction.

14.5 Trigonometric equations

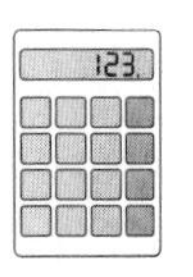

The equation $\sin x = 0.5$ is called a **trigonometric equation**. If we press 0.5 INV sin on the calculator, we get an angle of 30°. But because of the nature of the sine curve shown in Figure 14.44, then x can also equal 150°.

Hence the solution of $\sin x = 0.5$ between 0° and 360° is 30°, 150°.

A second example now:

$$\cos 2x = 0.6$$

From the calculator, using 0.6 INV cos,

$$2x = 53^\circ \text{ (nearest whole number)}$$

However, looking at the cosine curve, in Figure 14.44

$$2x = 53^\circ, 307^\circ, \underbrace{413^\circ, 667^\circ}$$

by extending the curve to the next cycle.

Hence $x = 26.5^\circ, 153.5^\circ, 206.5^\circ, 333.5^\circ$.

Care needs to be taken with negative values.

If $\sin(2x - 15^\circ) = -0.2$

Using 0.2  +/– INV sin, we get -12° (nearest whole number)

Hence $2x - 15^\circ = -12^\circ, 192^\circ, 348^\circ, 552^\circ, \ldots$

Hence $2x = 3^\circ, 207^\circ, 363^\circ, 567^\circ, \ldots$

So $x = 1.5^\circ, 103.5^\circ, 181.5^\circ, 283.5^\circ$.

Exercise 14(c)

1 Solve the following trigonometric equations, giving all answers between 0° and 360° (to the nearest 0.5°).

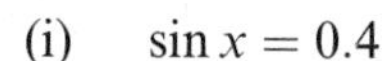

(i) $\sin x = 0.4$ (ii) $\cos x = -0.5$

(iii) $\cos x = 0.8$ (iv) $\sin x = -0.6$

(v) $\cos 2x = 0.4$ (vi) $\sin 2x = 0.1$

(vii) $\cos(2x - 10^\circ) = 0.4$ (viii) $\sin(2x + 40^\circ) = 0.3$

(ix) $\sin(30^\circ - x) = -0.4$ (x) $\cos(2x - 45^\circ) = -1$

2 (i) Given that $y = 2\cos x^\circ$, copy and complete the following table:

x	0	30	60	90	120	150	180	210	240	270
y	2.0	1.73		0	–1.0		–2.0			0

(ii) Using a scale of 2 cm to represent 30 units on the x-axis and 5 cm to represent 1 unit on the y-axis, draw the graph of $y = 2\cos x^\circ$ for values of x from 0 to 270 inclusive.

(iii) Using the same axes and the same scales, draw the graph of the straight line

$$y = \frac{9}{5} - \frac{x}{60}.$$

(iv) Use your graphs to estimate the solutions of the equation

$$2\cos x = \frac{9}{5} - \frac{x}{60}$$

in the range $0° \leqslant x \leqslant 270°$.

(v) By drawing another straight line on the graph, deduce the smallest value of C for which the equation

$$2\cos x = C - \frac{x}{60}$$

has a solution in the range $0° \leqslant x \leqslant 270°$ and write down this value of C.

3 The table shows the depth of water, in metres, at the mouth of Sheringport harbour at various times of a day.

Time (no. of hours after midnight)	0	1	2	3	4	5	6
Depth of water (metres)	12.0	11.7	11.0	10.0	9.0	8.3	8

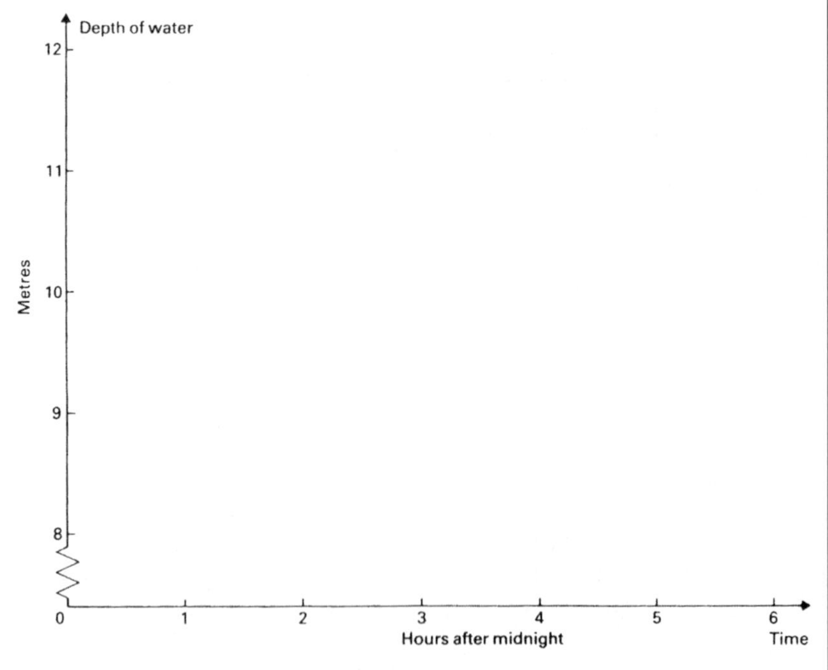

Figure 14.45

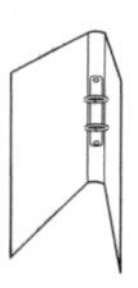

(i) (a) Copy Figure 14.45 and plot the points representing the information in the table.

(b) Join these points to form a smooth curve.

(ii) From your graph, estimate the depth of water at 4.15 am.

(iii) A ship needs at least 9.2 metres of water to be able to leave the harbour.

What is the latest time before 6am that the ship could leave?

(iv) The formula for finding the depth of water is

$$D = 10 + k\cos(30t)^\circ$$

where D is the depth of water in metres, t is the number of hours after midnight and k is a constant. Using the values $t = 0$ and $D = 12.0$, find the value of k.

[LEAG]

15 Matrices

15.1 Basic definitions

A **matrix** is a convenient mathematical device for storing **data**. It consists of information that can be described using rows and columns. The matrix is used extensively in computing.

	Physics	*Chemistry*	*Biology*
Janice	48	52	80
Joseph	63	61	49
Depak	85	48	31
Alice	41	20	63

Table 15.1

The student test information shown in Table 15.1 consists of 4 **rows** and 3 **columns**. We say that it is a 4 by 3 matrix, or 4×3 matrix. The **order** of the matrix is 4×3. If the headings are left out, the matrix can be written:

$$\begin{pmatrix} 48 & 52 & 80 \\ 63 & 61 & 49 \\ 85 & 48 & 31 \\ 41 & 20 & 63 \end{pmatrix}$$

Suppose that a week later, the students took another test, and their results are summarised in the matrix:

$$\begin{pmatrix} 44 & 58 & 70 \\ 50 & 63 & 53 \\ 68 & 54 & 39 \\ 45 & 48 & 65 \end{pmatrix}$$

then we could find the *total* marks obtained in the two tests for each subject, by carrying out a simple addition.

$$\begin{pmatrix} 48 & 52 & 80 \\ 63 & 61 & \boxed{49} \\ 85 & 48 & 31 \\ 41 & 20 & 63 \end{pmatrix} + \begin{pmatrix} 44 & 58 & 70 \\ 50 & 63 & \boxed{53} \\ 68 & 54 & 39 \\ 45 & 48 & 65 \end{pmatrix} = \begin{pmatrix} 92 & 110 & 150 \\ 113 & 124 & \boxed{102} \\ 153 & 102 & 70 \\ 86 & 68 & 128 \end{pmatrix}.$$

Each **element** (or **cell**) in the first matrix is added to the corresponding element in the second matrix. We can see for example that Joseph scored a total of $49 + 53 = 102$ in the two Biology tests. Subtraction would be carried out in a similar way.

Rule

Two matrices can only be added or subtracted if they have the same order.

In a local indoor cricket league, the points system was 5 points for a win, 3 for a draw, and 1 for a match lost. The Rover's Return team played 24 matches. They won 10, drew 6 and lost 8. How many points did they get? We can represent the two sets of information by matrices

$$(5 \quad 3 \quad 1) \text{ and } (10 \quad 6 \quad 8)$$

Clearly, the number of points obtained can be evaluated by multiplication and addition as follows:

$$5 \times 10 + 3 \times 6 + 1 \times 8 = 76 \text{ points}$$

This process is known as **matrix multiplication** and *must be* written in the form:

$$(5 \quad 3 \quad 1)\begin{pmatrix} 10 \\ 6 \\ 8 \end{pmatrix} = (5 \times 10 + 3 \times 6 + 1 \times 8) = (76)$$

In the same season, The Queen Vic won 8, drew 9 and lost 7. The number of points obtained by The Queen Vic can be evaluated in a similar manner.

$$(5 \quad 3 \quad 1)\begin{pmatrix} 8 \\ 9 \\ 7 \end{pmatrix} = (5 \times 8 + 3 \times 9 + 1 \times 7) = (74)$$

The two sets of results can be combined in the following manner:

$$(5 \quad 3 \quad 1)\begin{pmatrix} 10 & 8 \\ 6 & 9 \\ 8 & 7 \end{pmatrix} = (76 \quad 74)$$

The following year, the points system was altered so that there were only 4 points for a win, and by a coincidence both teams again produced the same results. The matrix representation is therefore:

$$(4 \quad 3 \quad 1)\begin{pmatrix} 10 & 8 \\ 6 & 9 \\ 8 & 7 \end{pmatrix} = (66 \quad 66)$$

We can combine these results as follows.

$$\begin{array}{c} (5 \quad 3 \quad 1) \\ \uparrow \\ (4 \quad 3 \quad 1) \end{array}\begin{pmatrix} 10 & 8 \\ 6 & 9 \\ 8 & 7 \end{pmatrix} = \begin{array}{c} (76 \quad 74) \\ \uparrow \\ (66 \quad 66) \end{array}$$

becomes

$$\begin{pmatrix} 5 & 3 & 1 \\ 4 & 3 & 1 \end{pmatrix}\begin{pmatrix} 10 & 8 \\ 6 & 9 \\ 8 & 7 \end{pmatrix} = \begin{pmatrix} 76 & 74 \\ 66 & 66 \end{pmatrix}$$

This is a matrix multiplication.

To obtain the elements of the resultant matrix in a multiplication problem, we must combine rows of the first matrix with columns of the second matrix.

Rule

Two matrices can only be multiplied if the number of columns in the first equals the number of rows in the second

Now try the following question.

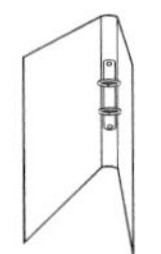

Exercise 15(a)

1 The table below shows the purchases made from a School Tuck Shop during one week.

	Choco bars	*Crisps*	*Fizzo*	*Apples*	*Fruito*	*Oat crunch*
Monday	20	75	59	36	19	13
Tuesday	19	80	49	24	21	17
Wednesday	16	91	55	24	15	19
Thursday	10	46	25	10	7	10
Friday	21	83	70	0	30	21

(i) Express the purchases made during this week as a 5×6 matrix. Call this matrix S.

(ii) The cost of the items during that week was

Choco bars	20p each
Crisps	15p packet
Fizzo	25p can
Apples	12p each
Fruito	8p each
Oat Crunch	17p each

Express the cost of the items as a column matrix. Call this matrix C.

(iii) (a) Evaluate the matrix product SC.

(b) Write down the order of the matrix product SC.

(c) What information is given by the matrix product SC?

(iv) The cost of Choco bars was increased by 5% and the cost of apples went down by 2p each. Calculate the new cost of

(a) 1 Choco bar

(b) 1 apple.

(v) Following the price changes the purchases on the next Monday were:

Choco bars	*Crisps*	*Fizzo*	*Apples*	*Fruito*	*Oat crunch*
17	80	50	41	20	12

Calculate the difference between the takings on the two Mondays. [LEAG]

15.2 Route matrices

A very interesting application of a matrix is in the study of networks, or routes. Figure 15.1 shows a network of roads consisting of four **nodes** (A, B, C, D) connected by seven **arcs**. You can travel in either direction along the roads, indicated by the arrows. The arcs divide the flat plane into 5 regions (x, y, z, t, v; including the *outside*).

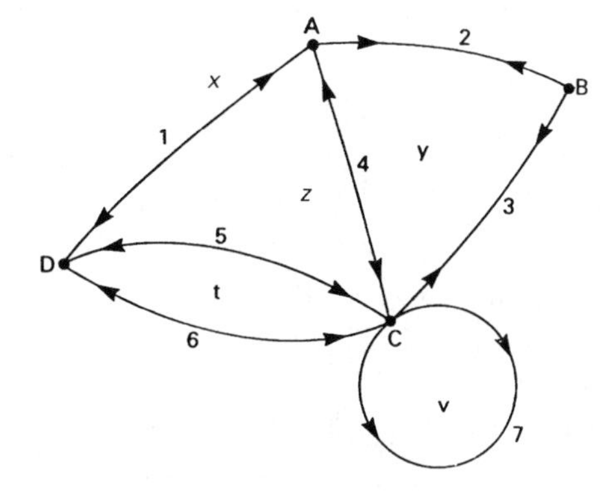

Figure 15.1

A **one-stage** route matrix indicates how many routes there are between any two nodes, where a route is not allowed to pass through any other node.

The one-stage matrix R for Figure 15.1 is as follows:

$$R = \text{From} \begin{array}{c} \\ A \\ B \\ C \\ D \end{array} \begin{array}{c} \text{To} \\ \begin{array}{cccc} A & B & C & D \end{array} \\ \begin{pmatrix} 0 & 1 & 1 & 1 \\ 1 & 0 & 1 & 0 \\ 1 & 1 & \boxed{2} & 2 \\ 1 & 0 & 2 & 0 \end{pmatrix} \end{array}$$

leading diagonal

There are 2 routes from C to C, because you can travel in either direction. [*Note*: The matrix is symmetrical about the leading diagonal.]

A **two-stage** route between two nodes passes through one other node on the way. The two-stage routes between A and C are shown in Figure 15.2. It can be seen that there are five.

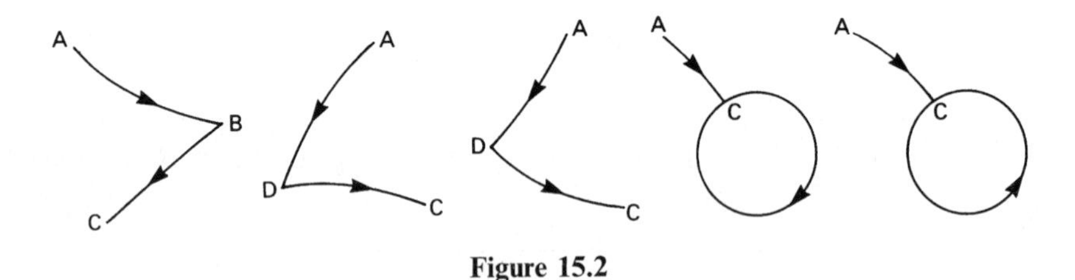

Figure 15.2

If you now work out $R \times R$, written R^2, you will find that:

$$R^2 = \begin{array}{c} \\ A \\ B \\ C \\ D \end{array} \begin{array}{c} \begin{array}{cccc} A & B & C & D \end{array} \\ \begin{pmatrix} 3 & 1 & \textcircled{5} & 2 \\ 1 & 2 & 3 & 3 \\ 5 & 3 & 10 & 5 \\ 2 & 3 & 5 & 5 \end{pmatrix} \end{array}$$

five two-stage routes between A and C

This matrix gives the two-stage routes. You may like to try and find the ten routes from C to C. Similarly, R^3 gives the three-stage routes for the network.

Let us look again at Figure 15.2. Node A lies on (is **incident** on) arcs 1, 2, 4. Node C is incident on 3, 4, 5, 6 and 7 (twice). The matrix N representing the incidence of nodes on arcs is given below:

$$N = \begin{array}{c} \\ A \\ B \\ C \\ D \end{array} \begin{array}{c} \begin{array}{ccccccc} 1 & 2 & 3 & 4 & 5 & 6 & 7 \end{array} \\ \begin{pmatrix} 1 & 1 & 0 & 1 & 0 & 0 & 0 \\ 0 & 1 & 1 & 0 & 0 & 0 & 0 \\ 0 & 0 & 1 & 1 & 1 & 1 & 2 \\ 1 & 0 & 0 & 0 & 1 & 1 & 0 \end{pmatrix} \end{array}$$

If we form the **transpose** N^T of the matrix (interchange rows and columns), we get:

$$N^T = \begin{array}{c} \\ 1 \\ 2 \\ 3 \\ 4 \\ 5 \\ 6 \\ 7 \end{array} \begin{array}{c} \begin{array}{cccc} A & B & C & D \end{array} \\ \begin{pmatrix} 1 & 0 & 0 & 1 \\ 1 & 1 & 0 & 0 \\ 0 & 1 & 1 & 0 \\ 1 & 0 & 1 & 0 \\ 0 & 0 & 1 & 1 \\ 0 & 0 & 1 & 1 \\ 0 & 0 & 2 & 0 \end{pmatrix} \end{array}$$

This gives the incidence of the arcs on the nodes. The reader should look at NN^T, and compare it with the one-stage route matrix R.

Rule

Euler's theorem states:

For any network, if R is the number of regions, N the number of nodes, and A the number of arcs, then

$$R + N = A + 2$$

RULE

If a network can be drawn without taking the pen off the paper and without following same arc twice, it is said to be **traversable**. The order of a node is the number of arcs leaving the node.

Rule

A network is traversable if:

(i) all its junctions are even nodes

or

(ii) only two of its junctions are odd.

RULE

Referring to Figure 15.3, the order of the nodes is $A(4)$, $B(4)$, $C(4)$, $D(4)$ and $E(4)$. Since these are all even, the network is traversable.

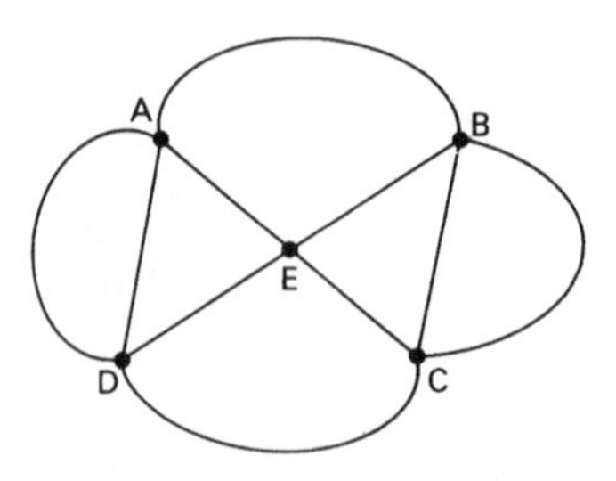

Figure 15.3

Exercise 15(b)

1 A factory sells ice-cream in four carton sizes, E, F, G and H. Delivery is made to three shops:

Asdah Stores receives 60 of E, 80 of F, 40 of G and 40 of H.

Bucket Discount receives 40 of F, 20 of G and 60 of H.

Cashcart receives 80 of E, 80 of F and 50 of G.

(i) The cost to the shops of cartons E, F, G and H is 60p, 90p, £1.20 and £2.50 respectively.

Write down two matrices only such that the elements of their product under matrix multiplication give the cost of the ice-cream delivered to each shop. Find this product.

(ii) During a particular year, Asdah Stores has 30 such deliveries, Bucket Discount 20 deliveries, and Cashcart 40 deliveries.

Write down two matrices only such that the elements of their product give the total number of cartons of each size leaving the factory. Obtain this product and hence find the total number of cartons supplied by the factory to these shops.

2 Figure 15.4 shows the network of roads linking the towns A, B and C. Information concerning this network is to be fed into a computer in the form of a 3×3 route matrix M.

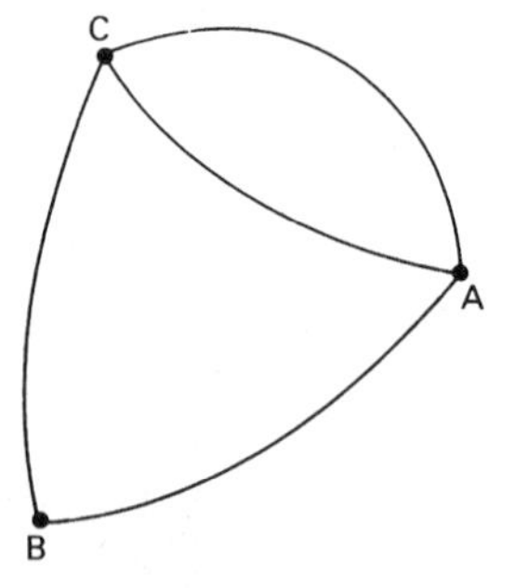

Figure 15.4

(i) Complete the matrix

$$M = \text{From} \begin{matrix} A \\ B \\ C \end{matrix} \overset{\displaystyle \text{To} \atop \displaystyle A \quad B \quad C}{\begin{pmatrix} 0 & & 2 \\ & & \\ & 1 & \end{pmatrix}}$$

(ii) The computer prints out the matrix M^2, where

$$M^2 = \begin{pmatrix} 5 & 2 & 1 \\ 2 & 2 & 2 \\ 1 & 2 & 5 \end{pmatrix}$$

Explain, in terms of routes between the towns, the entry in the bottom right-hand corner of the matrix M^2.

3 Which of the networks in Figure 15.5 are traversable?

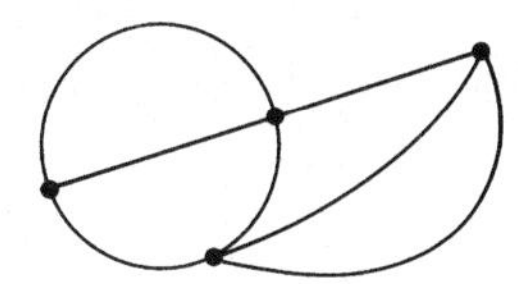
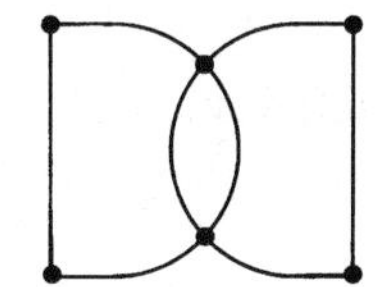
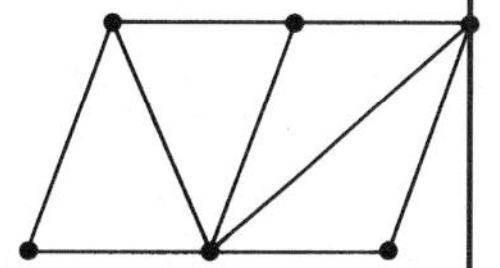

Figure 15.5

4 Figure 15.6 shows routes from two places Nottingham (N) and Morpeth (M) to three airports, Leeds (L), Edinburgh (E) and Stansted (S).

It also shows aeroplane routes from L, E and S to three European cities, Paris (P), Dusseldorf (D) and Copenhagen (C).

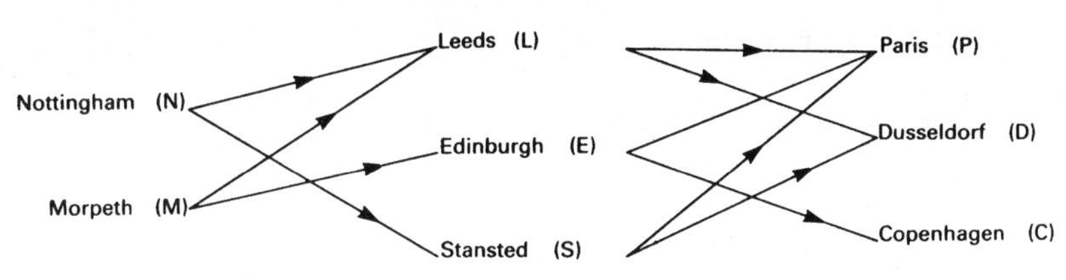

Figure 15.6

In this question route matrices show the number of routes available and 0 indicates no route.

(i) Copy and complete the following route matrix for journeys from N or M to L, E or S.

$$M = \text{From} \begin{matrix} N \\ M \end{matrix} \overset{\begin{matrix} & \text{To} & \\ L & E & S \end{matrix}}{\begin{pmatrix} 1 & 0 & 1 \\ & & \end{pmatrix}}$$

(ii) Copy and complete the following route matrix for flights from L, E or S to P, D or C.

$$M = \text{From} \begin{matrix} L \\ E \\ S \end{matrix} \overset{\begin{matrix} & \text{To} & \\ P & D & C \end{matrix}}{\begin{pmatrix} & 1 & \\ & 0 & \\ & 1 & \end{pmatrix}}$$

(iii) (a) List the ways in which it is possible to get from N to P via L, E or S.

(b) Copy and complete the following route matrix for journeys from N or M to P, D or C via one of the airports.

$$M = \text{From} \begin{matrix} N \\ M \end{matrix} \overset{\text{To}}{\overset{\begin{matrix} P & D & C \end{matrix}}{\begin{pmatrix} 2 & & 0 \\ & 1 & \end{pmatrix}}}$$

(c) How many different routes are there from Morpeth to either Paris, Dusseldorf or Copenhagen?

[MEG]

15.3 Algebra of matrices

(i) Addition and subtraction

We have already seen, in Section 15.1, that if

$$A = \begin{pmatrix} 2 & 1 \\ 3 & 4 \end{pmatrix} \quad \text{and} \quad B = \begin{pmatrix} 4 & 1 \\ 2 & 2 \end{pmatrix}$$

then

add corresponding cells

$$A + B = \begin{pmatrix} 2+4 & 1+1 \\ 3+2 & 4+2 \end{pmatrix} = \begin{pmatrix} 6 & 2 \\ 5 & 6 \end{pmatrix} = B + A.$$

We say addition of matrices is **commutative**.

Similarly, for three matrices A, B, C $(A + B) + C = A + (B + C)$, and so addition of matrices is said to be **associative**.

$$3A = A + A + A$$

$$\begin{pmatrix} 6 & 3 \\ 9 & 12 \end{pmatrix} = \begin{pmatrix} 3 \times 2 & 3 \times 1 \\ 3 \times 3 & 3 \times 4 \end{pmatrix}$$

Hence each element of the matrix is multiplied by 3

Also,

$$A - B = \begin{pmatrix} 2-4 & 1-1 \\ 3-2 & 4-2 \end{pmatrix} = \begin{pmatrix} -2 & 0 \\ 1 & 2 \end{pmatrix}.$$

(ii) Multiplication

To multiply two matrices is a more complicated exercise. The numbers in the following example have been carefully chosen to show the method clearly:

$$\begin{pmatrix} 2 & 5 \\ 6 & 7 \\ 9 & 8 \end{pmatrix} \begin{pmatrix} 10 & 0 & 4 \\ 3 & 16 & 1 \end{pmatrix} = \begin{pmatrix} 2 \times 10 + 5 \times 3 & 2 \times 0 + 5 \times 16 & 2 \times 4 + 5 \times 1 \\ 6 \times 10 + 7 \times 3 & 6 \times 0 + 7 \times 16 & 6 \times 4 + 7 \times 1 \\ 9 \times 10 + 8 \times 3 & 9 \times 0 + 8 \times 16 & 9 \times 4 + 8 \times 1 \end{pmatrix}$$

add

$$= \begin{pmatrix} 35 & 80 & 13 \\ 81 & 112 & 31 \\ 114 & 128 & 44 \end{pmatrix}.$$

Two matrices can only be multiplied if the number of colums in the first matrix equals the number of rows in the second.

Changing the order of multiplication:

$$\begin{pmatrix} 10 & 0 & 4 \\ 3 & 16 & 1 \end{pmatrix} \begin{pmatrix} 2 & 5 \\ 6 & 7 \\ 9 & 8 \end{pmatrix} = \begin{pmatrix} 56 & 82 \\ 111 & 135 \end{pmatrix}.$$

Clearly, multiplication of matrices is *not* commutative.

Rule

If an $m \times n$ matrix is multiplied by an $n \times p$ matrix, the result is an $m \times p$ matrix.

(iii) Identity

A matrix with 1s on the leading diagonal, and zeros elsewhere, is called an **identity matrix** or **unit matrix**. So

$$\begin{pmatrix} 1 & 0 \\ 0 & 1 \end{pmatrix}$$

is the 2×2 identity matrix

$$\begin{pmatrix} 1 & 0 & 0 \\ 0 & 1 & 0 \\ 0 & 0 & 1 \end{pmatrix}$$

is the 3×3 identity matrix and so on, denoted by I.

Rule

If you multiply a matrix by an identity matrix, it remains unchanged

So

$$\begin{pmatrix} 1 & 0 \\ 0 & 1 \end{pmatrix}\begin{pmatrix} 2 & 4 & 3 \\ 1 & -1 & 2 \end{pmatrix} = \begin{pmatrix} 2 & 4 & 3 \\ 1 & -1 & 2 \end{pmatrix}.$$

(iv) Inverse

The **inverse** of a matrix is a very useful quantity. We will only consider 2×2 matrices as the theory gets quite complicated for matrices of higher order. If

$$M = \begin{pmatrix} a & b \\ c & d \end{pmatrix}$$

then we first find the value of an expression denoted by $D = ad - bc$. This is called the **determinant** of the matrix.

Rule

$M^{-1} = \begin{pmatrix} \dfrac{d}{D} & -\dfrac{b}{D} \\ -\dfrac{c}{D} & \dfrac{a}{D} \end{pmatrix}$ is the inverse of M where $MM^{-1} = I$

Its uses are demonstrated in some of the following sections. The following example clarifies how an inverse is calculated.

Example 15.1

Consider the matrix

$$A = \begin{pmatrix} 2 & 3 \\ 4 & 5 \end{pmatrix}$$

Find the inverse and show that AA^{-1} gives the identity matrix.

Solution

To find the determinant

$$D = 2 \times 5 - 3 \times 4$$
$$= -2$$

2 3
4 5

Notice the pattern

The inverse is

$$A^{-1} = \begin{pmatrix} \frac{5}{-2} & \frac{-3}{-2} \\ \frac{-4}{-2} & \frac{2}{-2} \end{pmatrix} = \begin{pmatrix} -2\frac{1}{2} & 1\frac{1}{2} \\ 2 & -1 \end{pmatrix}$$

If you multiply A by A^{-1}, you get

$$AA^{-1} = \begin{pmatrix} 2 & 3 \\ 4 & 5 \end{pmatrix}\begin{pmatrix} -2\frac{1}{2} & 1\frac{1}{2} \\ 2 & -1 \end{pmatrix} = \begin{pmatrix} 1 & 0 \\ 0 & 1 \end{pmatrix}$$

Also

$$A^{-1}A = \begin{pmatrix} 1 & 0 \\ 0 & 1 \end{pmatrix}$$

(You should check this.)

Rule

If the determinant $D = 0$, the matrix does not have an inverse

For example,

$$B = \begin{pmatrix} 3 & 2 \\ 6 & 4 \end{pmatrix}$$

then

$$\begin{aligned} D &= 3 \times 4 - 2 \times 6 \\ &= 0 \end{aligned}$$

Hence if we try to find B^{-1}, we get

$$B^{-1} = \begin{pmatrix} \frac{4}{0} & \frac{-2}{0} \\ \frac{-6}{0} & \frac{3}{0} \end{pmatrix}$$

which is impossible to work out (you cannot divide by zero).

Because usually $AB \neq BA$ for matrices, then we say that multiplication of matrices is *not* **commutative**, and so this has to be taken into account in manipulating matrix equations.

For example,

if $AC = B$, where A, B and C are matrices,

multiply both sides on the *left* by A^{-1}:

So $A^{-1}AC = A^{-1}B$

$IC = C$

But $A^{-1}A = I$, hence $IC = A^{-1}B$

i.e. $C = A^{-1}B$. (Provided $A^{-1}B$ is possible.)

We can now make use of this to develop another method of solving simultaneous linear equations.

(v) Simultaneous equations

Example 15.2

Solve the equations $\begin{aligned} 2x + 4y &= 12 \\ 3x - 7y &= 5 \end{aligned}$ using matrices.

Solution

These equations can be represented using matrices in the following way.

$$\begin{pmatrix} 2 & 4 \\ 3 & -7 \end{pmatrix}\begin{pmatrix} x \\ y \end{pmatrix} = \begin{pmatrix} 12 \\ 5 \end{pmatrix}$$

If

$$A = \begin{pmatrix} 2 & 4 \\ 3 & -7 \end{pmatrix}, \quad B = \begin{pmatrix} 12 \\ 5 \end{pmatrix} \quad \text{and} \quad C = \begin{pmatrix} x \\ y \end{pmatrix}$$

then we have $AC = B$

$$\text{i.e.} \quad \begin{pmatrix} 2 & 4 \\ 3 & -7 \end{pmatrix}\begin{pmatrix} x \\ y \end{pmatrix} = \begin{pmatrix} 12 \\ 5 \end{pmatrix}$$

$$\text{Now} \quad A^{-1} = \begin{pmatrix} \dfrac{-7}{-26} & \dfrac{-4}{-26} \\ \dfrac{-3}{-26} & \dfrac{2}{-26} \end{pmatrix} = \begin{pmatrix} \dfrac{7}{26} & \dfrac{2}{13} \\ \dfrac{3}{26} & \dfrac{-1}{13} \end{pmatrix}.$$

Now multiply $AC = B$ each side on the left by A^{-1}

$$\begin{pmatrix} \frac{7}{26} & \frac{2}{13} \\ \frac{3}{26} & \frac{-1}{13} \end{pmatrix} \begin{pmatrix} 2 & 4 \\ 3 & -7 \end{pmatrix} \begin{pmatrix} x \\ y \end{pmatrix} = \begin{pmatrix} \frac{7}{26} & \frac{2}{13} \\ \frac{3}{26} & \frac{-1}{13} \end{pmatrix} \begin{pmatrix} 12 \\ 5 \end{pmatrix} = \begin{pmatrix} 4 \\ 1 \end{pmatrix}$$

i.e. $$\begin{pmatrix} 1 & 0 \\ 0 & 1 \end{pmatrix} \begin{pmatrix} x \\ y \end{pmatrix} = \begin{pmatrix} 4 \\ 1 \end{pmatrix}$$

So $$\begin{pmatrix} x \\ y \end{pmatrix} = \begin{pmatrix} 4 \\ 1 \end{pmatrix}$$

The solution is $x = 4$, $y = 1$.

Exercise 15(c)

1 $A = \begin{pmatrix} 3 & 0 \\ -1 & 2 \end{pmatrix}$, $B = \begin{pmatrix} 2 & 2 \\ -1 & 0 \end{pmatrix}$, $C = \begin{pmatrix} 3 & -1 & -1 \\ 2 & 2 & 0 \end{pmatrix}$ and $D = \begin{pmatrix} 1 \\ 2 \end{pmatrix}$.

Evaluate if possible:

(i) $A + 2B$ (ii) AB (iii) AC
(iv) $B + C$ (v) CD (vi) DA.

2 If $2M - \begin{pmatrix} 1 & 3 \\ 1 & 4 \end{pmatrix} = \begin{pmatrix} -1 & 2 \\ 1 & 0 \end{pmatrix}$, find M.

3 Find the inverse of the following matrices:

(i) $\begin{pmatrix} 2 & 4 \\ 3 & 5 \end{pmatrix}$ (ii) $\begin{pmatrix} 2 & -1 \\ -1 & -1 \end{pmatrix}$ (iii) $\begin{pmatrix} 5 & -6 \\ 0 & 2 \end{pmatrix}$.

4 Find a and b, if $\begin{pmatrix} a & 2 \\ -1 & 5 \end{pmatrix} \begin{pmatrix} 3 \\ b \end{pmatrix} = \begin{pmatrix} 2 \\ 24 \end{pmatrix}$.

5 $\begin{pmatrix} 2 & 3 \\ 4 & 6 \end{pmatrix}^2 = \begin{pmatrix} 2 & 3 \\ 4 & 6 \end{pmatrix} \begin{pmatrix} 2 & 3 \\ 4 & 6 \end{pmatrix} = \begin{pmatrix} 16 & 24 \\ 32 & 48 \end{pmatrix} = 8 \begin{pmatrix} 2 & 3 \\ 4 & 6 \end{pmatrix}$.

(i) Using the above as an example, evaluate:

(a) k when $\begin{pmatrix} 2 & 3 \\ 2 & 3 \end{pmatrix}^2 = k \begin{pmatrix} 2 & 3 \\ 2 & 3 \end{pmatrix}$

(b) λ when $\begin{pmatrix} 2 & 3 \\ 6 & 9 \end{pmatrix}^2 = \lambda \begin{pmatrix} 2 & 3 \\ 6 & 9 \end{pmatrix}$.

(ii) Evaluate $\begin{pmatrix} 2 & 3 \\ 8 & 12 \end{pmatrix}^2$ similarly.

(iii) Hence (a) find p in terms of x if

$$\begin{pmatrix} 2 & 3 \\ 2x & 3x \end{pmatrix}^2 = p\begin{pmatrix} 2 & 3 \\ 2x & 2x \end{pmatrix}$$

and (b) find q in terms of a, b and x if

$$\begin{pmatrix} ax & bx \\ a & b \end{pmatrix}^2 = q\begin{pmatrix} ax & bx \\ a & b \end{pmatrix}.$$

6 Use a matrix method to solve the simultaneous equations.

(i) $2x + 3y = 1$
$x + 2y = 4$

(ii) $2x - 3y = 5$
$3x - 4y = 6$

(iii) $5x + 7y = 3$
$3x + 5y = 4$

7 Using any three matrices A, B, C of your choice, prove that $(AB)C = A(BC)$.

8 If $\begin{pmatrix} 3 & 1 \\ 1 & 0 \end{pmatrix}\begin{pmatrix} 4 & a \\ a & 2 \end{pmatrix} = \begin{pmatrix} 15 & b \\ c & 3 \end{pmatrix}$ find the values of a, b and c.

9 If $\begin{pmatrix} x & x \\ y & y \end{pmatrix}\begin{pmatrix} x & y \\ y & x \end{pmatrix} = \begin{pmatrix} 2 & t \\ t & 2 \end{pmatrix}$ find x, y and t.

16 Vectors and transformations

16.1 Column vectors

(i) Definitions

Referring to Figure 16.1, the line joining $A(0, 1)$ and $B(3, 2)$ is called a **vector**. It has **length** and **direction** (or **magnitude**).

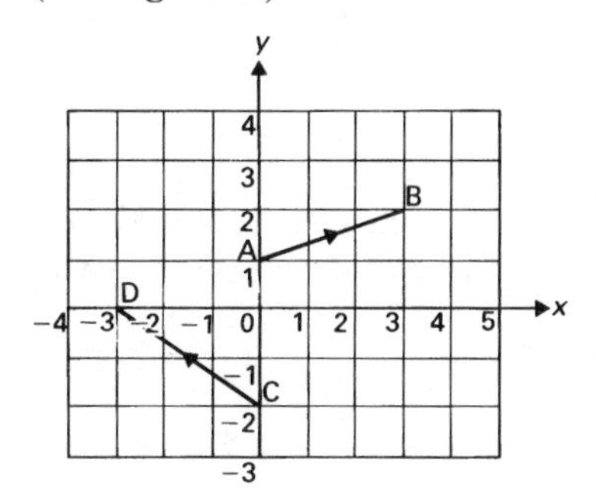

Figure 16.1

There are various ways of writing down a vector. For example $\overrightarrow{AB}$, **AB**, or quite simply, using a column of numbers (a matrix).

$$\text{So}\quad \overrightarrow{AB} = \begin{pmatrix} 3 \\ 1 \end{pmatrix} \begin{matrix} \leftarrow \text{3 along} \\ \leftarrow \text{1 up} \end{matrix}$$

In the same diagram (Figure 16.1), we have the line joining C to D, which would be written

$$\overrightarrow{CD} = \begin{pmatrix} -3 \\ 2 \end{pmatrix} \begin{matrix} \leftarrow \text{3 back} \\ \leftarrow \text{2 up} \end{matrix}$$

The arrow is an important part of a vector, because it is used to describe a direction of motion.

Hence $\overrightarrow{CD}$ and $\overrightarrow{DC}$ would be different vectors.

$$\text{In fact}\quad \overrightarrow{DC} = \begin{pmatrix} 3 \\ -2 \end{pmatrix} \begin{matrix} \leftarrow \text{3 along} \\ \leftarrow \text{2 down} \end{matrix}$$

Hence $\overrightarrow{CD} = -\overrightarrow{DC}$, because each number in the column vector is multiplied by -1.

The idea of moving in a straight line is called a **translation**.

(ii) Addition

In Figure 16.2

$$\overrightarrow{AB} = \begin{pmatrix} 2 \\ 1 \end{pmatrix}, \quad \overrightarrow{BC} = \begin{pmatrix} 3 \\ -2 \end{pmatrix}, \text{ and } \overrightarrow{AC} = \begin{pmatrix} 5 \\ -1 \end{pmatrix}$$

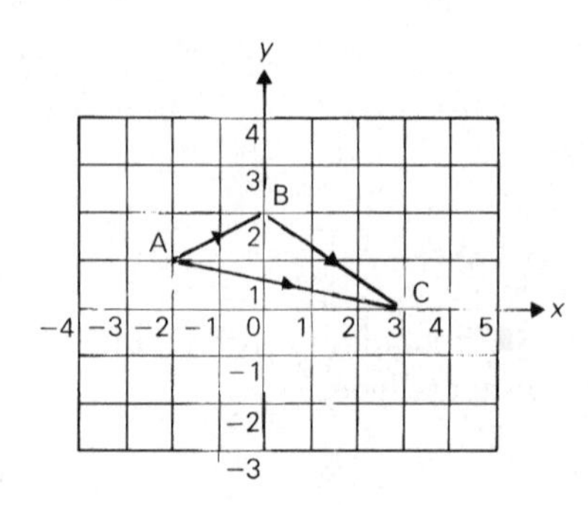

Figure 16.2

If we regard the vectors as describing a *route*, then since the route A to B to C ends up in the same place as the direct route A to C, then we say that

$$\overrightarrow{AC} = \overrightarrow{AB} + \overrightarrow{BC}$$

$$\text{i.e.} \quad \begin{pmatrix} 5 \\ -1 \end{pmatrix} = \begin{pmatrix} 2 \\ 1 \end{pmatrix} + \begin{pmatrix} 3 \\ -2 \end{pmatrix} = \begin{pmatrix} 2+\ \ 3 \\ 1+-2 \end{pmatrix}$$

You can see that to add two vectors you add the top numbers, and add the bottom numbers.

(iii) Subtraction

$\overrightarrow{AB} - \overrightarrow{BC}$ is the same as $\overrightarrow{AB} + \overrightarrow{CB}$, since $\overrightarrow{CB} = -(\overrightarrow{BC})$

$$\text{So} \quad \overrightarrow{AB} - \overrightarrow{BC} = \begin{pmatrix} 2 \\ 1 \end{pmatrix} - \begin{pmatrix} 3 \\ -2 \end{pmatrix} = \begin{pmatrix} 2 \\ 1 \end{pmatrix} + \begin{pmatrix} -3 \\ 2 \end{pmatrix}$$

$$= \begin{pmatrix} -1 \\ 3 \end{pmatrix}$$

In fact, you can see that to subtract two vectors you simply subtract the top number and subtract the bottom numbers.

When a point X is moved to another point Y by means of a vector T, we say that X is **mapped onto** Y by the translation T.

Example 16.1

If A is the point $(1, 3)$ and B is the point $(4, 2)$, find the point D that the vector $3\overrightarrow{AB}$ will map the point $C(-2, 3)$ onto. Describe the shape $ABDC$.

Solution

The vector $\overrightarrow{AB} = \begin{pmatrix} 3 \\ -1 \end{pmatrix}$ (see Figure 16.3).

$$\text{Hence} \quad 3\,\overrightarrow{AB} = \begin{pmatrix} 3 \times \ \ 3 \\ 3 \times -1 \end{pmatrix}$$

$$= \begin{pmatrix} 9 \\ -3 \end{pmatrix}.$$

If we start at $C(-2, 3)$ and go 9 along and 3 down, we end up at the point

$$(-2 + 9, 3 - 3) = (7, 0).$$

Since $AB \| CD$, then $ABDC$ is a trapezium.

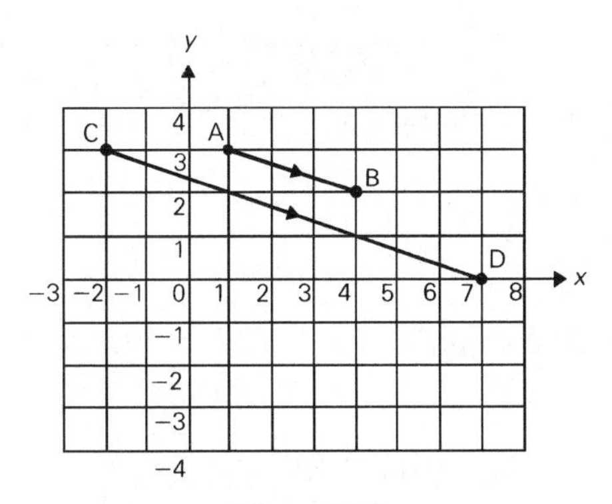

Figure 16.3

(iv) Unit vectors i, j

An alternative way of writing a column vector is by using **unit vectors i** and **j**. These are vectors each of length 1, where **i** is in the direction of the x-axis and **j** is in the direction of the y-axis.

Hence 3**j** means move 3 up, and −2**i** would mean move 2 back.

Hence $\begin{pmatrix} -2 \\ 3 \end{pmatrix}$ can be written $-2\mathbf{i} + 3\mathbf{j}$.

(v) Length or modulus

The **length** or **modulus** of a vector is written $|\overrightarrow{AB}|$ to mean the length of the vector AB. It is found using Pythagoras' theorem.

For example, if $\overrightarrow{AB} = \begin{pmatrix} 3 \\ -4 \end{pmatrix}$

$$|\overrightarrow{AB}| = \sqrt{3^2 + (-4)^2} = 5$$

(see Figure 16.4(i)).

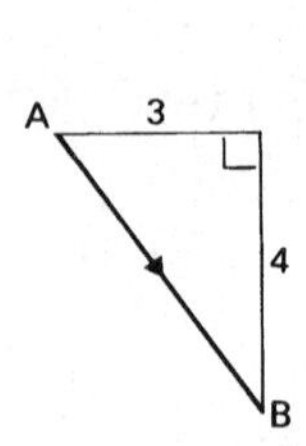

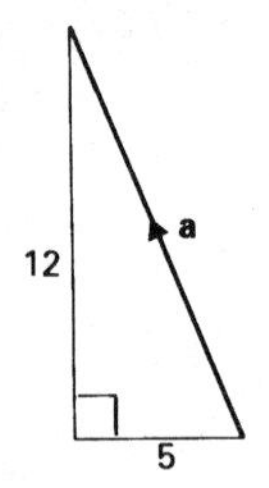

Figure 16.4

If $\mathbf{a} = -5\mathbf{i} = 12\mathbf{j}$

$$\text{then } |\mathbf{a}| = \sqrt{(-5)^2 + 12^2} = 13$$

(see Figure 16.4(ii)).

Example 16.2

$\mathbf{p}$ and $\mathbf{q}$ are two vectors such that $\mathbf{p}$ is perpendicular to $\mathbf{p} + \mathbf{q}$. If $|\mathbf{p}| = 5$ and $|\mathbf{q}| = 13$, find $\mathbf{p} + \mathbf{q}$.

Solution

Since $\mathbf{p} \perp \mathbf{p} + \mathbf{q}$, then draw $\mathbf{p} + \mathbf{q}$ first, call it DB (Figure 16.5). Then DA is perpendicular to DB, where DA represents $\mathbf{p}$.

Now complete the parallelogram $DABC$.

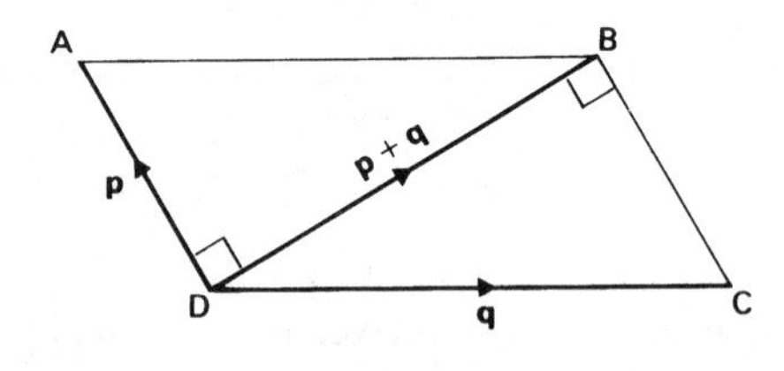

Figure 16.5

In $\triangle DBC$, $\angle DBC = 90°$, $BC = |\mathbf{p}| = 5$ and $DC = |\mathbf{q}| = 13$

$$\text{Hence } |\mathbf{p} + \mathbf{q}| = DB = \sqrt{13^2 - 5^2} = 12.$$

Exercise 16(a)

1 If $\mathbf{a} = \begin{pmatrix} 1 \\ 2 \end{pmatrix}$, $\mathbf{b} = \begin{pmatrix} 2 \\ -1 \end{pmatrix}$ and $\mathbf{c} = \begin{pmatrix} 0 \\ -2 \end{pmatrix}$, find

(i) $\mathbf{a} + \mathbf{b}$ (ii) $\mathbf{b} - \mathbf{c}$ (iii) $\mathbf{a} + \mathbf{b} + \mathbf{c}$

(iv) $\mathbf{a} - 2\mathbf{b}$ (v) $3\mathbf{a} - \mathbf{b} - 2\mathbf{c}$ (vi) $\mathbf{a} - 2\mathbf{b} - 3\mathbf{c}$

(vii) $2\mathbf{a} + 3\mathbf{b} + 5\mathbf{c}$.

2 $A(2, 3)$, $B(4, -1)$ and $C(2, 8)$ are the vertices of a triangle. X is the midpoint of AB, and Y is the midpoint of BC. If XY is extended to T, so that $XY = YT$, find the coordinates of T. What shape is $XBTC$?

3 $O(0, 0)$, $A(3, 1)$ and $B(5, 1)$ are three vertices of a quadrilateral $OABC$. If

$$\overrightarrow{BC} = \begin{pmatrix} 3 \\ -2 \end{pmatrix},$$

what are the coordinates of C? Find $\overrightarrow{AC}$.

4 If $2\begin{pmatrix} x \\ y \end{pmatrix} + 3\begin{pmatrix} 2x \\ y \end{pmatrix} = \begin{pmatrix} 8 \\ -4 \end{pmatrix}$, find x and y.

5 $ABCD$ is a quadrilateral. A, B, D have coordinates $(0, 2)$, $(2, 5)$, $(8, 0)$ respectively. If $\overrightarrow{AD} = 2\overrightarrow{BC}$, find the coordinates of C.

6 Find, by any method, the acute angle between the vectors

$$\begin{pmatrix} 2 \\ 4 \end{pmatrix} \text{ and } \begin{pmatrix} -1 \\ 3 \end{pmatrix}.$$

7 If $\mathbf{x} = 2\mathbf{i} + 3\mathbf{j}$ and $\mathbf{y} = \mathbf{i} - 3\mathbf{j}$, find

(i) $\mathbf{x} + \mathbf{y}$ (ii) $\mathbf{x} - \mathbf{y}$ (iii) $2\mathbf{x} - 3\mathbf{y}$

(iv) $|\mathbf{x}|$ (v) $|\mathbf{x} + \mathbf{y}|$

8 (i) A triangle has vertices $A_1(2, 3)$, $B_1(-1, 0)$, $C_1(0, -2)$. Draw axes showing values of x from -3 to 8 and values of y from -10 to 8, and draw and label $A_1B_1C_1$.

(ii) The image of $A_1B_1C_1$ under an enlargement E is $A_2B_2C_2$. E has centre $(0, 0)$ and scale factor 2. Calculate the coordinates A_2, B_2, C_2, and draw this triangle on the same diagram.

(iii) The image of $A_2B_2C_2$ under a translation with vector

$$\begin{pmatrix} 1 \\ -5 \end{pmatrix}$$

is $A_3B_3C_3$. Draw this on your diagram.

(iv) Describe fully a single transformation mapping $A_1B_1C_1$ onto $A_3B_3C_3$.

9 A is the point $(1, 2)$, B the point $(2, 5)$, C the point $(4, 2)$

(i) Draw axes Ox and Oy from 0 to 6. Plot and label points A, B, C.

(ii) Calculate the area of triangle ABC.

(iii) A translation with vector

$$\begin{pmatrix} -4 \\ 3 \end{pmatrix}$$

maps the points A, B, C on to the points L, M, N respectively.

(a) Write down the area of triangle LMN.

(b) Find the coordinates of the points L, M, N.

(c) Calculate the distance between A and L.

16.2 Geometrical proofs

Mathematicians have found vectors very useful in proving ideas in geometry. The methods used in this section will depend on those in the last section, but you will not find it is necessary to use column vectors, or the **i**, **j** notation.

Example 16.3

In triangle ABC (Figure 16.6), D lies on AB and $DB = 2AD$. E lies on AC and $EC = 2AE$. Let $\overrightarrow{AD} = \mathbf{x}$ and $\overrightarrow{AE} = \mathbf{y}$.

Write down in terms of $\mathbf{x}$ and $\mathbf{y}$

(a) $\overrightarrow{AB}$ (b) $\overrightarrow{AC}$ (c) $\overrightarrow{DE}$ (d) $\overrightarrow{BC}$

What do your results tell you about DE and BC?

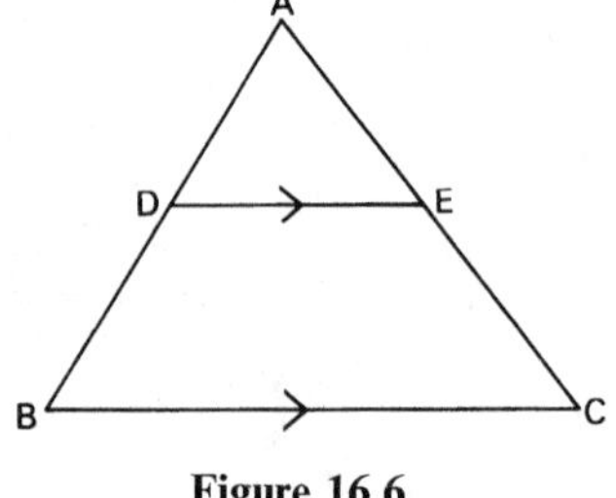

Figure 16.6

Solution

(a) AB is $3 \times AD$ in length, and in the same direction, hence $\overrightarrow{AB} = 3\mathbf{x}$

(b) Similarly $\overrightarrow{AC} = 3\mathbf{y}$

(c) $\overrightarrow{DE} = \overrightarrow{DA} + \overrightarrow{AE} = -\mathbf{x} + \mathbf{y}$

(d) $\overrightarrow{BC} = \overrightarrow{BA} + \overrightarrow{AC} = -3\mathbf{x} + 3\mathbf{y}$

It follows that $\overrightarrow{BC} = 3\overrightarrow{DE}$.

This is only possible if $DE \| BC$, and BC is 3 times the length of DE.

The next example makes use of another simple theorem.

> **Rule**
>
> If **a** and **b** are **not** parallel vectors, and
>
> $h\mathbf{a} + k\mathbf{b} = 0,$
>
> then h and k must both be zero

RULE

Example 16.4

In the triangle OAB, $\overrightarrow{OA} = \mathbf{a}$ and $\overrightarrow{OB} = \mathbf{b}$. L is a point on the side AB, M is a point on the side OB, and OL and AM meet at S. It is given that $AS = SM$ and $OS/OL = \frac{3}{4}$; also that $OM/OB = h$ and $AL/AB = k$.

(a) express the vectors $\overrightarrow{AM}$ and $\overrightarrow{OS}$ in terms of **a**, **b** and h

(b) express the vectors $\overrightarrow{OL}$ and $\overrightarrow{OS}$ in terms of **a**, **b** and k.

Find h and k, and hence find the values of the ratios OM/MB and AL/LB.

Solution

Since $\dfrac{OS}{OL} = \frac{3}{4}$, then $\overrightarrow{OS} = \frac{3}{4}\overrightarrow{OL}$

$\dfrac{OM}{OB} = h$, so $\overrightarrow{OM} = h\overrightarrow{OB}$

$\dfrac{AL}{AB} = k$, so $\overrightarrow{AL} = k\overrightarrow{AB}$

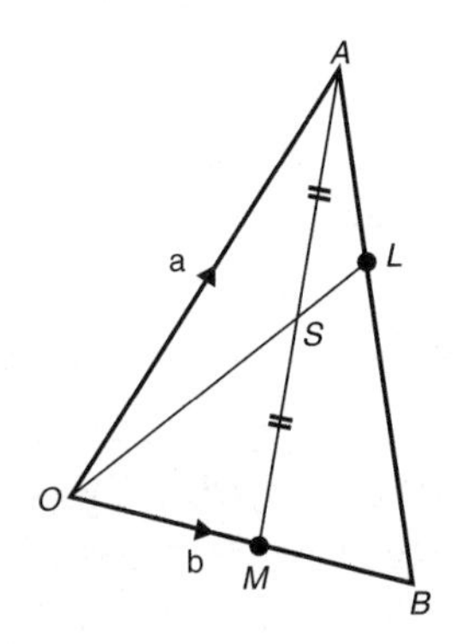

Figure 16.7

Hence $\overrightarrow{OM} = h\mathbf{b}$

Now $\overrightarrow{AB} = -\mathbf{a} + \mathbf{b}$ or $\mathbf{b} - \mathbf{a}$

So $\overrightarrow{AL} = k(\mathbf{b} - \mathbf{a})$

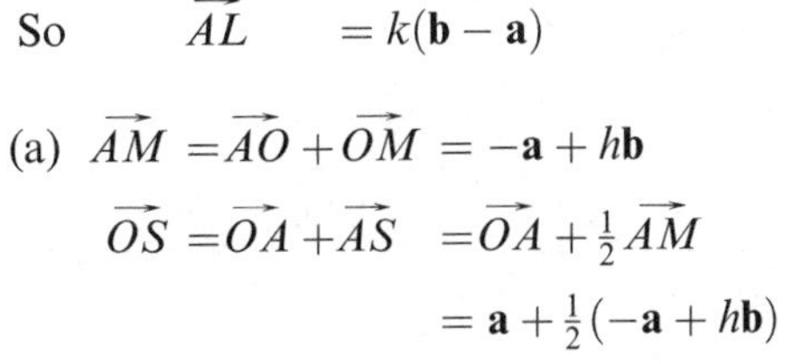

(a) $\overrightarrow{AM} = \overrightarrow{AO} + \overrightarrow{OM} = -\mathbf{a} + h\mathbf{b}$

$$\begin{aligned}\overrightarrow{OS} = \overrightarrow{OA} + \overrightarrow{AS} &= \overrightarrow{OA} + \tfrac{1}{2}\overrightarrow{AM}\\ &= \mathbf{a} + \tfrac{1}{2}(-\mathbf{a} + h\mathbf{b})\\ &= \tfrac{1}{2}\mathbf{a} + \tfrac{1}{2}h\mathbf{b}.\end{aligned}$$

(b) $\overrightarrow{OL} = \overrightarrow{OA} + \overrightarrow{AL} = \mathbf{a} + k(\mathbf{b} - \mathbf{a}) = (1 - k)\mathbf{a} + k\mathbf{b}$

$\overrightarrow{OS} = \frac{3}{4}\overrightarrow{OL} \quad = \frac{3}{4}(1 - k)\mathbf{a} + \frac{3}{4}k\mathbf{b}.$

The expressions in parts (a) and (b) are both for $\overrightarrow{OS}$, hence

$$\tfrac{1}{2}\mathbf{a} + \tfrac{1}{2}h\mathbf{b} = \tfrac{3}{4}(1-k)\mathbf{a} + \tfrac{3}{4}k\mathbf{b}$$

So $(\tfrac{1}{2} - \tfrac{3}{4}(1-k))\mathbf{a} + (\tfrac{1}{2}h - \tfrac{3}{4}k)\mathbf{b} = 0$

Since $\mathbf{a}$ and $\mathbf{b}$ are not parallel,

$$\tfrac{1}{2} - \tfrac{3}{4}(1-k) = 0, \quad \text{so} \quad k = \tfrac{1}{3}$$

and $\tfrac{1}{2}h - \tfrac{3}{4}k = 0, \quad \text{hence } h = \tfrac{1}{2}$

i.e. $\dfrac{OM}{OB} = \dfrac{1}{2}$ so $\dfrac{OM}{MB} = 1$

$\dfrac{AL}{AB} = \dfrac{1}{3}$ so $\dfrac{AL}{LB} = \dfrac{1}{2}$.

Exercise 16(b)

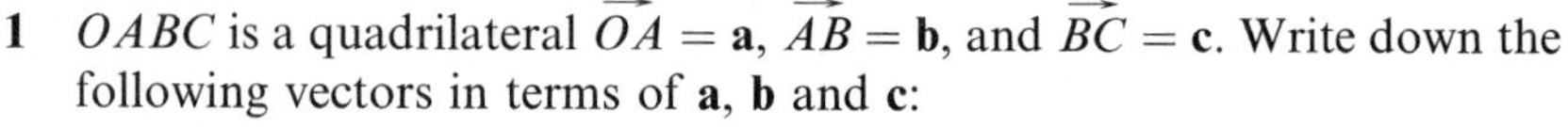

1 $OABC$ is a quadrilateral $\overrightarrow{OA} = \mathbf{a}$, $\overrightarrow{AB} = \mathbf{b}$, and $\overrightarrow{BC} = \mathbf{c}$. Write down the following vectors in terms of $\mathbf{a}$, $\mathbf{b}$ and $\mathbf{c}$:

(i) $\overrightarrow{AC}$ (ii) $\overrightarrow{OC}$ (iii) $\overrightarrow{OB}$

2 $RSTU$ is a rectangle. The diagonals of the rectangle meet in M. If $\overrightarrow{MR} = \mathbf{a}$ and $\overrightarrow{MS} = \mathbf{b}$, write down in terms of $\mathbf{a}$ and $\mathbf{b}$.

(i) $\overrightarrow{RS}$ (ii) $\overrightarrow{ST}$ (iii) $\overrightarrow{US}$

3 ABC is a triangle, and X is the midpoint of BC. If $\overrightarrow{CX} = \mathbf{a}$ and $\overrightarrow{BA} = \mathbf{b}$, write down in terms of $\mathbf{a}$ and $\mathbf{b}$:

(i) $\overrightarrow{XB}$ (ii) $\overrightarrow{XA}$ (iii) $\overrightarrow{AC}$

4 $ABCD$ is a parallelogram. E is the midpoint of BC, and AE is extended to F where $AE = EF$. Show that DCF is a straight line.

5 $OABCDE$ is a regular hexagon. Let $\overrightarrow{OA} = \mathbf{a}$ and $\overrightarrow{OB} = \mathbf{b}$. Write the following vectors in terms of $\mathbf{a}$ and $\mathbf{b}$:

(i) $\overrightarrow{AB}$ (ii) $\overrightarrow{OC}$ (iii) $\overrightarrow{BC}$

(iv) $\overrightarrow{OD}$ (v) $\overrightarrow{OE}$ (vi) $\overrightarrow{AE}$

6 $OABC$ is a parallelogram. M and N are the midpoints of the sides OC and CB respectively. $\overrightarrow{OA} = \mathbf{a}$ and $\overrightarrow{OC} = \mathbf{c}$.

(i) Write down $\overrightarrow{ON}$ and $\overrightarrow{AM}$ in terms of $\mathbf{a}$ and $\mathbf{c}$.

(ii) AM meets ON at P. Denoting $\overrightarrow{OP}$ by $\lambda\overrightarrow{ON}$ and $\overrightarrow{AP}$ by $\mu\overrightarrow{AM}$, use the fact that $\overrightarrow{OA} + \overrightarrow{AP} = \overrightarrow{OP}$ to find the values of λ and μ.

(iii) Hence express the area of triangle OAP as a fraction of the area of the parallelogram $OABC$.

[NEA]

7 In Figure 16.8, $\mathbf{a}$ is the position vector of A and $\mathbf{b}$ is that of B, both with respect to O as origin. M is the midpoint of AB and N is the midpoint of OB.

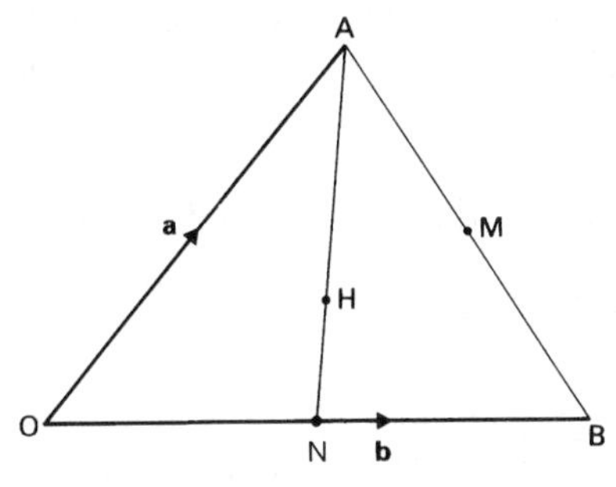

Figure 16.8

(i) Write down in terms of $\mathbf{a}$ and $\mathbf{b}$, the following vectors, simplifying your answers where possible:

(a) $\overrightarrow{AB}$ (b) $\overrightarrow{AM}$ (c) $\overrightarrow{OM}$

(d) $\overrightarrow{ON}$ (e) $\overrightarrow{AN}$

(ii) H is a point on AN and $AH = \frac{2}{3}AN$

(a) Find $\overrightarrow{OH}$ in terms of $\mathbf{a}$ and $\mathbf{b}$.

(b) $\overrightarrow{OH} = k\overrightarrow{OM}$. Find the value of k.

(c) What can you deduce about O, M and H?

(iii) P is the midpoint of OA. Show that BHP is a straight line.

[MEG]

8 In Figure 16.9, $\overrightarrow{OA} = \mathbf{a}$, $\overrightarrow{OB} = \mathbf{b}$, $\overrightarrow{OC} = \mathbf{c}$ and $\overrightarrow{OD} = \mathbf{d}$. X is the midpoint of AB and Y is the midpoint of CD.

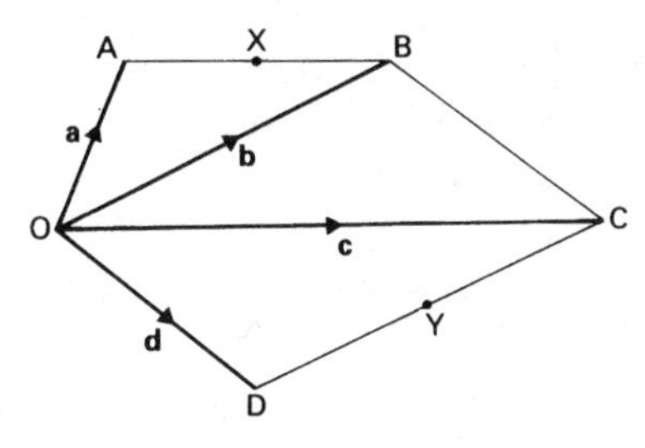

Figure 16.9

(i) Find, in terms of $\mathbf{a}$ and $\mathbf{b}$:

(a) $\overrightarrow{AB}$ (b) $\overrightarrow{AX}$ (c) $\overrightarrow{OX}$.

(iii) If M is the midpoint of XY (not shown in Figure 16.9), write down an expression for $\overrightarrow{OM}$ in terms of $\mathbf{a}$, $\mathbf{b}$, $\mathbf{c}$ and $\mathbf{d}$.

(iv) Describe exactly the position of P if $\overrightarrow{OP} = \frac{1}{2}(\mathbf{a} + \mathbf{d})$.

16.3 Transformations

Matrices can be used very effectively to change one shape into another. This process can be carried out in 2 dimensions using a 2×2 matrix, and in 3 dimensions using a 3×3 matrix. In this book, we shall only be concerned with 2-dimensional problems.

Suppose we want to change the position of the point $(3, -2)$ by using the matrix

$$\begin{pmatrix} 2 & 1 \\ 3 & 2 \end{pmatrix}$$

The coordinates of the points are first written as a column vector

$$\begin{pmatrix} 3 \\ -2 \end{pmatrix}.$$

This is *premultiplied* by the matrix.

Hence

$$\begin{pmatrix} 2 & 1 \\ 3 & 2 \end{pmatrix}\begin{pmatrix} 3 \\ -2 \end{pmatrix} = \begin{pmatrix} 2 \times 3 + 1 \times -2 \\ 3 \times 3 + 2 \times -2 \end{pmatrix} = \begin{pmatrix} 4 \\ 5 \end{pmatrix}$$

Hence the point $(3, -2)$ is *mapped* to the point $(4, 5)$.

(i) Transforming a shape

Any geometrical shape consists of a collection of points, and so to transform a shape means transforming all of the points of the shape. Fortunately, the nature of matrix multiplication means that you only need to consider the vertices of the shape.

Figure 16.10 shows a parallelogram, with vertices at $A(1, 4)$, $B(4, 5)$, $C(6, 2)$ and $D(3, 1)$. The complete shape can be represented by a **single** matrix.

$$\begin{array}{cccc} A & B & C & D \end{array}$$
$$\begin{pmatrix} 1 & 4 & 6 & 3 \\ 4 & 5 & 2 & 1 \end{pmatrix}$$

It saves quite a bit of time to work with this single matrix rather than to transform each point separately.

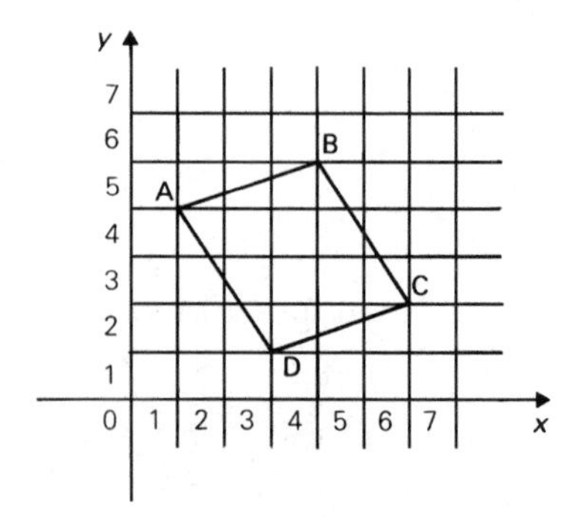

Figure 16.10

Example 16.5

Transform the parallelogram $ABCD$ shown in Figure 16.10 by means of the matrix

$$\begin{pmatrix} 2 & 0 \\ 0 & 2 \end{pmatrix}$$

into the shape $A'B'C'D'$. Draw a diagram to illustrate the outcome, and describe what has happened.

Solution

You still premultiply the shape

$$\begin{matrix} A & B & C & D \end{matrix}$$
$$\begin{pmatrix} 1 & 4 & 6 & 3 \\ 4 & 5 & 2 & 1 \end{pmatrix} \quad \text{by the matrix}$$

Hence

$$\begin{pmatrix} 2 & 0 \\ 0 & 2 \end{pmatrix}\overset{\begin{matrix} A & B & C & D \end{matrix}}{\begin{pmatrix} 1 & 4 & 6 & 3 \\ 4 & 5 & 2 & 1 \end{pmatrix}} = \overset{\begin{matrix} A' & B' & C' & D' \end{matrix}}{\begin{pmatrix} 2 & 8 & 12 & 6 \\ 8 & 10 & 4 & 2 \end{pmatrix}}$$

The new points to plot are $A'(2, 8)$, $B'(8, 10)$, $C'(12, 4)$ and $D'(6, 2)$.

The shape and its image are shown in Figure 16.11.

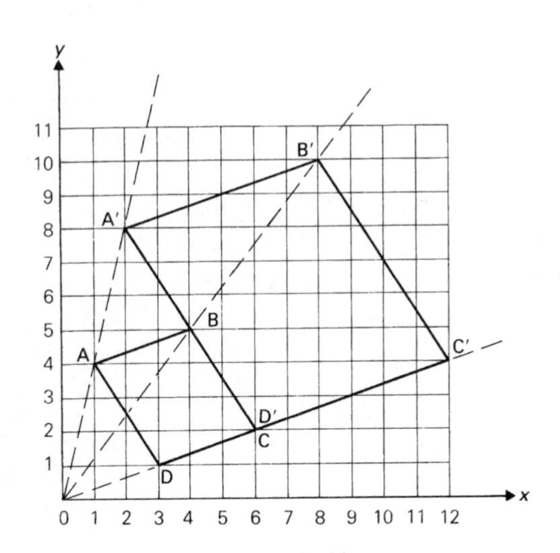

Figure 16.11

You can see that the shape has been *enlarged* by a *scale factor* 2, with O as the *centre of the enlargement*.

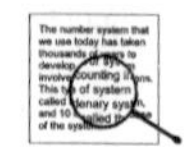

Investigation XXVII: Enlargements

The object of this activity is to help you understand the idea of enlargement. You need a centre of enlargement O, and a scale factor.

(i) Scale factor 2.5

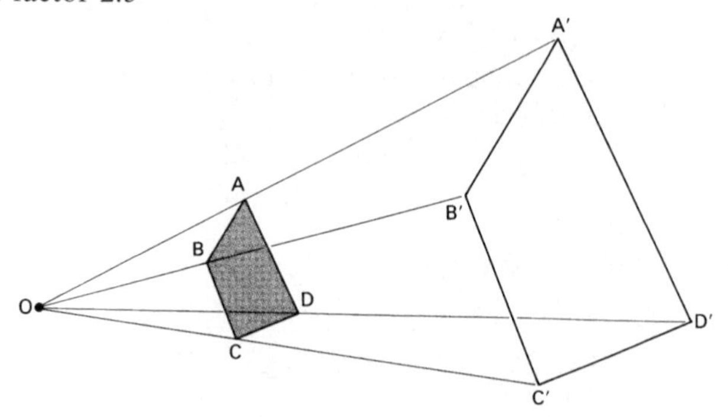

Draw rays from O, through A, B, C, D to the image A', B', C', D' so that $OA' = 2.5 \times OA$, $OB' = 2.5 \times OB$ etc. You will find that all lengths in the image shape are $2.5\times$ corresponding lengths in the object shape.

(ii) Scale factor $\frac{1}{4}$

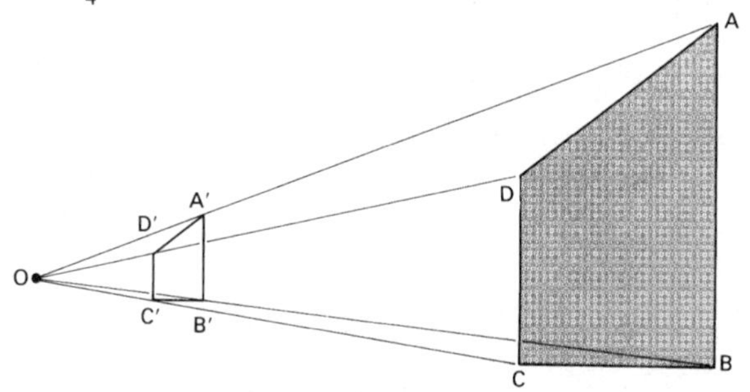

This time, $OA' = \frac{1}{4} \times OA$. The object is reduced in size.

(iii) Scale factor -0.8

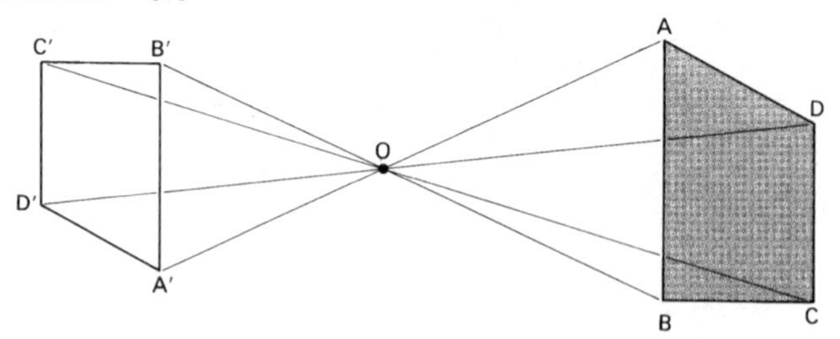

The rays are drawn in the opposite direction for a negative scale factor. This effect occurs in photography. An inverted image is formed.

Try some of your own ideas, it is possible to get some very attractive patterns.

(ii) Classification of matrices

The best way of looking at what a particular matrix will do is to investigate the effect of a matrix on a simple shape. The shape used is the square $O(0,0)$, $A(1,0)$, $B(1,1)$ and $C(0,1)$.

This will be referred to as the *unit square*.

Reflection

1 $\begin{pmatrix} -1 & 0 \\ 0 & 1 \end{pmatrix}$

$$\begin{pmatrix} -1 & 0 \\ 0 & 1 \end{pmatrix} \overset{\begin{matrix} O & A & B & C \end{matrix}}{\begin{pmatrix} 0 & 1 & 1 & 0 \\ 0 & 0 & 1 & 1 \end{pmatrix}} = \overset{\begin{matrix} O & A' & B' & C' \end{matrix}}{\begin{pmatrix} 0 & -1 & -1 & 0 \\ 0 & 0 & 1 & 1 \end{pmatrix}}$$

If you look at the two squares plotted in Figure 16.12, the matrix represents *reflection in the y-axis* (the line $x = 0$).

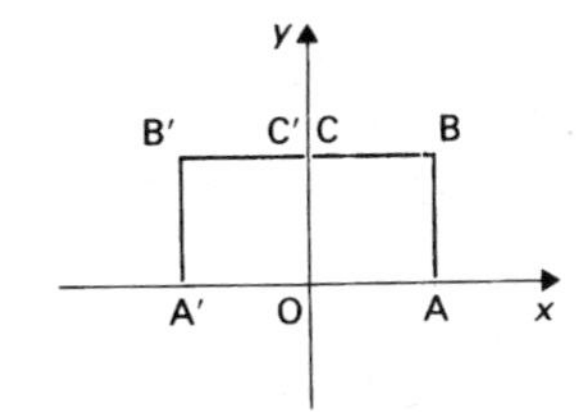

Figure 16.12

2 $\begin{pmatrix} 0 & 1 \\ 1 & 0 \end{pmatrix}$

Here

$$\begin{pmatrix} 0 & 1 \\ 1 & 0 \end{pmatrix} \overset{\begin{matrix} O & A & B & C \end{matrix}}{\begin{pmatrix} 0 & 1 & 1 & 0 \\ 0 & 0 & 1 & 1 \end{pmatrix}} = \overset{\begin{matrix} O & A' & B' & C' \end{matrix}}{\begin{pmatrix} 0 & 0 & 1 & 1 \\ 0 & 1 & 1 & 0 \end{pmatrix}}$$

You can see in Figure 16.3 that the square has been *reflected in the line* $y = x$.

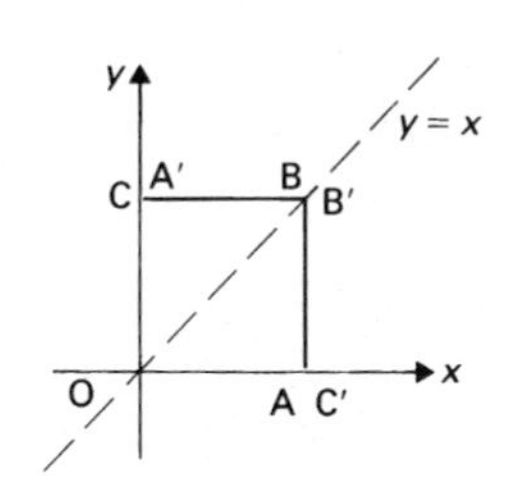

Figure 16.13

3 $\begin{pmatrix} 1 & 0 \\ 0 & -1 \end{pmatrix}$ represents *reflection in the x-axis.*

4 $\begin{pmatrix} 0 & -1 \\ -1 & 0 \end{pmatrix}$ represents *reflection in the line* $y = -x$.

Rotation

1 $\begin{pmatrix} 0 & -1 \\ 1 & 0 \end{pmatrix}$

$$\begin{pmatrix} 0 & -1 \\ 1 & 0 \end{pmatrix} \begin{pmatrix} \overset{O}{0} & \overset{A}{1} & \overset{B}{1} & \overset{C}{0} \\ 0 & 0 & 1 & 1 \end{pmatrix} = \begin{pmatrix} \overset{O}{0} & \overset{A'}{0} & \overset{B'}{-1} & \overset{C'}{-1} \\ 0 & 1 & 1 & 0 \end{pmatrix}$$

Figure 16.14 shows the matrix has been *rotated by 90° in an anticlockwise direction about O*.

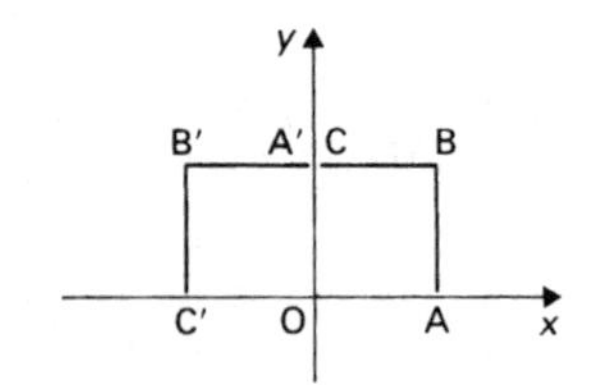

Figure 16.14

2 $\begin{pmatrix} 0 & 1 \\ -1 & 0 \end{pmatrix}$ represents *rotation of 90° clockwise about O*.

3 $\begin{pmatrix} -1 & 0 \\ 0 & -1 \end{pmatrix}$ represents *rotation of 180° about O*.

Shear

1 $\begin{pmatrix} 1 & 2 \\ 0 & 1 \end{pmatrix}$

$$\begin{pmatrix} 1 & 2 \\ 0 & 1 \end{pmatrix} \begin{pmatrix} \overset{O}{0} & \overset{A}{1} & \overset{B}{1} & \overset{C}{0} \\ 0 & 0 & 1 & 1 \end{pmatrix} = \begin{pmatrix} \overset{O}{0} & \overset{A'}{1} & \overset{B'}{3} & \overset{C'}{2} \\ 0 & 0 & 1 & 1 \end{pmatrix}$$

Figure 16.15 shows this transformation as a *shear parallel to the x-axis*. The distance a point moves to the right is proportional to its distance from the x-axis.

Since OA stays in the same place, it is known as the *invariant line*.

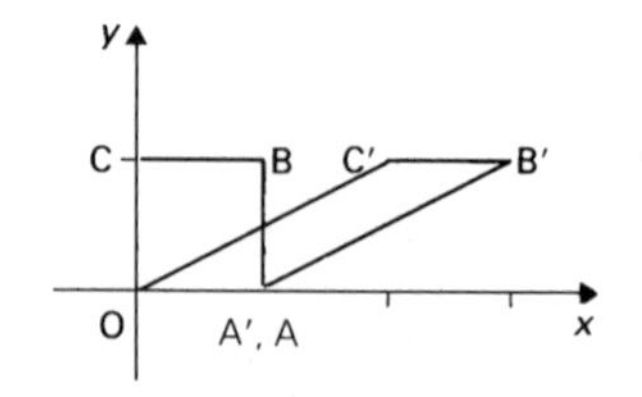

Figure 16.15

2 $\begin{pmatrix} 1 & 0 \\ 3 & 1 \end{pmatrix}$ would be a *shear parallel to the y-axis*.

Enlargement

Any matrix of the form

$$\begin{pmatrix} k & 0 \\ 0 & k \end{pmatrix}$$

will be an *enlargement, centre O, scale factor k.*

Investigation XXVIII: Transformation matrices

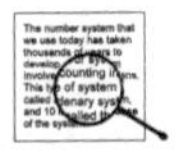

The previous section does not give a complete list of all matrices. If you have understood the work so far, you should try to complete the classification.

Here are some ideas for you to explore.

(i) $\begin{pmatrix} n & 0 \\ 0 & 1 \end{pmatrix}$ and $\begin{pmatrix} 1 & 0 \\ 0 & m \end{pmatrix}$

(ii) $\begin{pmatrix} a & 0 \\ 0 & b \end{pmatrix}$ where a is not equal to b.

(iii) $\begin{pmatrix} \cos\theta & -\sin\theta \\ \sin\theta & \cos\theta \end{pmatrix}$ where θ is any angle between 0° and 360°.

(iv) $\begin{pmatrix} \cos 2\theta & \sin 2\theta \\ \sin 2\theta & -\cos 2\theta \end{pmatrix}$ where θ is any angle between 0° and 180°.

You should also draw the line $y = x\tan\theta$ on the same diagram. So for example, if $\theta = 45°$, the matrix is

$$\begin{pmatrix} \cos 90° & \sin 90° \\ \sin 90° & -\cos 90° \end{pmatrix} = \begin{pmatrix} 0 & 1 \\ 1 & 0 \end{pmatrix},$$

and the line is $y = x\tan 45°$: i.e. $y = x$.

(iii) Combining tranformations

If you want to carry out a transformation T_1, followed by another transformation T_2, then to represent this by a single matrix, you must find T_2T_1. [**Notice the order is reversed.**] This idea is used in the following example.

Example 16.6

R is the transformation with matrix $\begin{pmatrix} 0 & -1 \\ 1 & 0 \end{pmatrix}$

S is the transformation with matrix $\begin{pmatrix} 0 & -1 \\ -1 & 0 \end{pmatrix}$

A is the point $(1, 0)$, B is the point $(3, 0)$ and C is the point $(1, 1)$

(a) Find the coordinates of A_1, B_1, C_1, the images of A, B and C under R.

(b) Using scales of 2 cm for 1 unit on both axes, draw a diagram to show ABC and $A_1B_1C_1$. Label these points.

(c) (i) Find the coordinates of A_2, B_2 and C_2, the images of A_1, B_1 and C_1 under S.

(ii) Add A_2, B_2 and C_2 to your diagram and label the points.

(d) (i) Calculate $\begin{pmatrix} 0 & -1 \\ -1 & 0 \end{pmatrix}\begin{pmatrix} 0 & -1 \\ 1 & 0 \end{pmatrix}$.

(ii) Describe geometrically the transformation T which this represents.

(iii) What is the relationship between R, S and T?

[MEG]

Solution

(a)

$$\overset{}{\begin{pmatrix} 0 & -1 \\ 1 & 0 \end{pmatrix}}\overset{A \;\; B \;\; C}{\begin{pmatrix} 1 & 3 & 1 \\ 0 & 0 & 1 \end{pmatrix}} = \overset{A_1 \;\; B_1 \;\; C_1}{\begin{pmatrix} 0 & 0 & -1 \\ 1 & 3 & 1 \end{pmatrix}}$$

Hence $A_1(0, 1)$, $B_1(0, 3)$, $C_1(-1, 1)$.

(b) See Figure 16.16

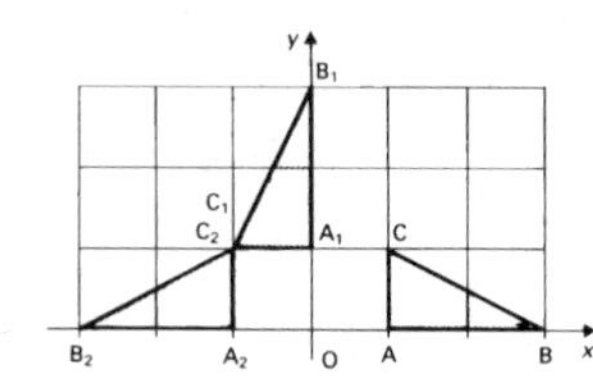

Figure 16.16

(c) (i)

$$\begin{pmatrix} 0 & -1 \\ -1 & 0 \end{pmatrix}\overset{A_1 \;\; B_1 \;\; C_1}{\begin{pmatrix} 0 & 0 & -1 \\ 1 & 3 & 1 \end{pmatrix}} = \overset{A_2 \;\; B_2 \;\; C_2}{\begin{pmatrix} -1 & -3 & -1 \\ 0 & 0 & 1 \end{pmatrix}}$$

(ii) See Figure 16.6.

(d) (i) $\begin{pmatrix} 0 & -1 \\ -1 & 0 \end{pmatrix}\begin{pmatrix} 0 & -1 \\ 1 & 0 \end{pmatrix} = \begin{pmatrix} -1 & 0 \\ 0 & 1 \end{pmatrix}$

(ii) This stands for R followed by S, which means that $\triangle ABC$ has become $\triangle A_2B_2C_2$.

So T is reflection in the y-axis.

(iii) Hence T is R followed by S, or

$$SR = T.$$

(iv) Finding a matrix for a given transformation

If you want a matrix to carry out a particular transformation, proceed as in the following example.

Example 16.7

Find the matrix which will rotate a shape by 180° about O.

Solution

You look at the effect on the points $(1,0)$ and $(0,1)$ **in that order**.

$(1,0) \to (-1,0)$ after rotation of 180°

i.e. $\begin{pmatrix} -1 \\ 0 \end{pmatrix}$

$(0,1) \to (0,-1)$ after rotation of 180°

i.e. $\begin{pmatrix} 0 \\ -1 \end{pmatrix}$

The two answers are written after each other in the same order, to give the required matrix

$$R = \begin{pmatrix} -1 & 0 \\ 0 & -1 \end{pmatrix}.$$

Exercise 16(c)

1 (i) The vertices of a square A are $(0,0)$, $(4,0)$, $(4,4)$ and $(0,4)$. Using x- and y-axes with values from 0 to 8 draw and label A.

(ii) Transformation S is defined by

$$S : \begin{pmatrix} x \\ y \end{pmatrix} \to \begin{pmatrix} 1 & 1 \\ 0 & 1 \end{pmatrix} \begin{pmatrix} x \\ y \end{pmatrix}$$

(a) Draw the image of A under transformation S. Label it B.

(b) Describe the transformation S.

(iii) Transformation T is reflection in the line $y = x$.

(a) Draw the image of B under transformation T. Label it C.

(b) Write down the matrix of transformation T.

(iv) Transformation U is transformation S followed by transformation T. By evaluating a suitable matrix product, or otherwise, find the matrix of transformation U.

(v) Transformation V is defined by

$$V : \begin{pmatrix} x \\ y \end{pmatrix} \to \begin{pmatrix} 1 & 0 \\ 1 & 1 \end{pmatrix} \begin{pmatrix} x \\ y \end{pmatrix}$$

(a) Draw the image of A under transformation V. Label it D.

(b) Explain why your answer to part (iv) is not the same as the matrix for transformation V.

[NEA]

2 The vertices of an isosceles triangle have coordinates $A(5, 0)$, $B(8, 4)$ and $C(8, -1)$

(i) Using a scale of 1 cm to 1 unit, draw x- and y-axes, taking values of x from 0 to 16 and values of y from -2 to 14.

(ii) Transformation X is defined as

$$X : \begin{pmatrix} x \\ y \end{pmatrix} \to \begin{pmatrix} 2 & 0 \\ 0 & 2 \end{pmatrix} \begin{pmatrix} x \\ y \end{pmatrix}$$

(a) Draw and label the isosceles triangle X such that ABC maps onto $A'B'C'$.

(b) Describe fully the single transformation X.

(iii) Transformation K is defined as

$$K : \begin{pmatrix} x \\ y \end{pmatrix} \to \begin{pmatrix} 1.2 & 1.6 \\ 1.6 & -1.2 \end{pmatrix} \begin{pmatrix} x \\ y \end{pmatrix}.$$

(a) Draw and label the isosceles triangle K such that ABC maps onto $A''B''C''$.

(b) Describe fully the single transformation that maps the triangle X onto the triangle K such that $A'B'C' \to A''B''C''$, stating the equation of any invariant line.

3 The vertices of a quadrilateral $OABC$ are $O(0, 0)$, $A(4, 0)$, $B(4, 2)$ and $C(1, 3)$

(i) Taking 1 cm to represent 1 unit on each axis and marking your x-axis from -5 to 9 and your y-axis from -4 to 5, draw and label the quadrilateral $OABC$.

(ii) The quadrilateral $OABC$ is mapped onto another quadrilateral $O_2A_2B_2C_2$ by the transformation represented by the matrix

$$\begin{pmatrix} -1 & 0 \\ 0 & -1 \end{pmatrix}$$

Draw and label quadrilateral $O_1A_1B_1C_1$ on your diagram.

(iii) The original quadrilateral $OABC$ is mapped onto another quadrilateral $O_2A_2B_2C_2$by the transformation given by

$$\begin{pmatrix} x \\ y \end{pmatrix} \to \begin{pmatrix} 8 \\ 4 \end{pmatrix} - \begin{pmatrix} x \\ y \end{pmatrix}.$$

(a) Draw and label quadrilateral $O_2A_2B_2C_2$ on your diagram.

(b) Describe fully, in geometrical terms, the **single** transformation which maps quadrilateral $O_1A_1B_1C_1$ onto quadrilateral $O_2A_2B_2C_2$.

17 The calculus

17.1 Basic ideas of differentiation

The invention of the calculus was one of the most far-reaching events in the history of mathematics. It was developed simultaneously in the seventeenth century by Isaac Newton in England, and Gottfried von Leibnitz in Germany. The notation that we use today is the one used by Leibnitz.

The differential calculus is used to find the rate of change of variable quantities such as the speed and acceleration of a moving body, or the instantaneous speed of an accelerating object. It has a very wide range of uses in graphical work, including the calculation of the gradient of a curve at a point.

The Greek letter δ is used to denote a small change. Hence δx means a small change in x.

Figure 17.1 shows the graph of $y = x^2$. $P(1, 1)$ and $Q(1.01, 1.0201)$ are two points on the curve which are close together. To move from P to Q, you have to increase x by 0.01. This is a small change, and so we can write $\delta x = 0.01$.

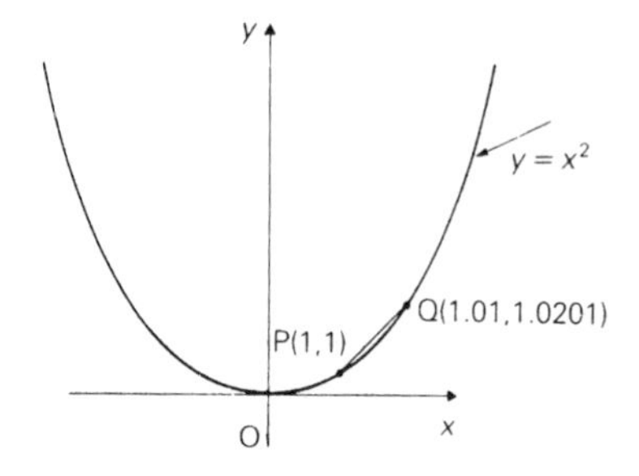

Figure 17.1

To get from P to Q you also have to increase y by 0.0201. This again is a small change, so we could write $y = 0.0201$.

Now look at the magnified picture in Figure 17.2.

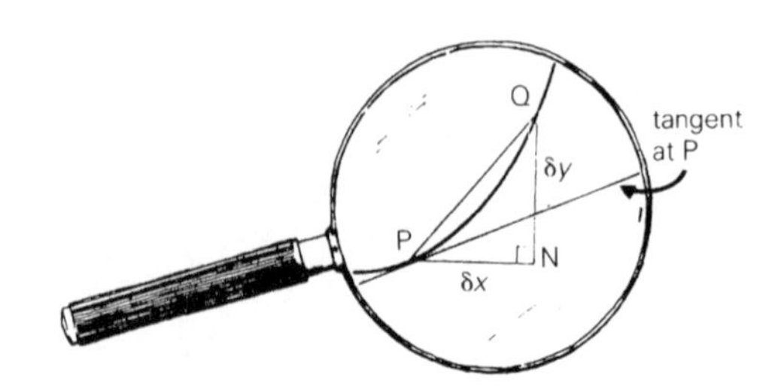

Figure 17.2

The gradient of the line $PQ = \frac{QN}{PN} = \frac{\delta y}{\delta x} = \frac{0.0201}{0.01} = 2.01$

The tangent at P has very nearly the same gradient. In fact, the gradient of the tangent is 2.

Since this is nearly equal to $\frac{\delta y}{\delta x}$ as x gets closer and closer to zero, it is written $\frac{dy}{dx}$ (pronounced d y by d x). We have that $\frac{dy}{dx} = 2$ at the point P. If this process is repeated for different values of x, a table is obtained as follows:

x	1	2	3	−1	0
$\frac{dy}{dx}$	2	4	6	−2	0

The pattern is simple, $\frac{dy}{dx} = 2 \times x$.

Hence if $y = x^2$, the gradient of the tangent is given by $\frac{dy}{dx} = 2x$. We can also refer to $\frac{dy}{dx}$ as the **derivative** of y with respect to x, or the **differential coefficient** of y with respect to x. If the whole exercise is repeated for different powers of x, we obtain the general result:

Rule

If $y = x^n$ then $\frac{dy}{dx} = nx^{n-1}$

RULE

Note: This result is true for all values of n (fractional or integer) positive or negative. If the power of x is multiplied by a number, then so is the derivative.

Rule

If $y = kx^n$ then $\frac{dy}{dx} = knx^{n-1}$

RULE

The process of finding $\frac{dy}{dx}$ is called **differentiation**.

Example 17.1

Differentiate the following functions of x with respect to x.

(a) $2x^6$ (b) $\frac{1}{x^2}$ (c) $3x$ (d) 12.

Solution

(a) $y = 2x^6$, so $n = 6$ and $k = 2$

Hence $\frac{dy}{dx} = 2 \times 6x^5 = 12x^5$.

(b) $y = \frac{1}{x^2} = x^{-2}$, so $n = -2$

Hence $\frac{dy}{dx} = -2 \times x^{-2-1} = -2x^{-3} = \frac{-2}{x^3}$.

(c) $y = 3x$

This is really $y = 3x^1$, so $n = 1$, $k = 3$

Hence $\frac{dy}{dx} = 3 \times x^0 = 3$ (since $x^0 = 1$).

This result helps to confirm the work on the equation of a straight line in the form $y = mx + c$, where the gradient is m.

(d) $y = 12$

This could be written as $y = 12x^0$

So $\frac{dy}{dx} = 12 \times 0 \times x^{-1} = 0$.

Rule

If y consists of more than one power of x added together, then each power is differentiated separately, and the results are added together.

Example 17.2

Differentiate with respect to x:

(a) $3x^2 + 5x + 6$ (b) $\frac{1}{x} + 2$

(c) $(x^2 + 1)^2$ (d) $\frac{(x^2 + 1)}{x^2}$

Solution

(a) $y = 3x^2 + 5x + 6$

$$\frac{dy}{dx} = 3 \times 2x + 5 \times 1x^0 + 0$$
$$= 6x + 5.$$

(b) $y = \frac{1}{x} + 2 = x^{-1} + 2$

$$\frac{dy}{dx} = -1x^{-2} + 0 = \frac{-1}{x^2}.$$

(c) $y = (x^2 + 1)^2$

$$= (x^2 + 1)(x^2 + 1)$$
$$= x^4 + x^2 + x^2 + 1$$
$$= x^4 + 2x^2 + 1$$

← The brackets must be expanded out first

So $\frac{dy}{dx} = 4x^3 + 4x = 4x(x^2 + 1)$.

(d) $y = \frac{x^2 + 1}{x^2}$ means $\frac{x^2}{x^2} + \frac{1}{x^2} = 1 + \frac{1}{x^2} = 1 + x^{-2}$

Hence $\frac{dy}{dx} = -2x^{-3} = \frac{-2}{x^3}$.

Example 17.3

If $p = 4V^2$, find

(a) the value of $\frac{dp}{dV}$ when $V = 4$

(b) the value of p when $\frac{dP}{dV} = 80$.

Solution

Do not be confused by the different letters, we are simply differentiating p with respect to V.

(a) $\frac{dp}{dV} = 4 \times 2V = 8V$

If $V = 4$, $\frac{dp}{dV} = 8 \times 4 = 32$.

(b) $\frac{dp}{dV} = 8V = 80$, hence $V = 10$

If $V = 10$, $p = 4 \times 10^2 = 400$.

Example 17.4

Find the equation of the tangent and normal to the curve $y = x^3 - 4x + 1$ at the point $P(2, 1)$.

Solution

$\frac{dy}{dx} = 3x^2 - 4 = 3 \times 2^2 - 4 = 8$ at the point P.

The equation of the tangent is a straight line with gradient 8.

Using $y = mx + c$,

$y = 8x + c$.

Now $x = 2$ when $y = 1$

Hence $c = -15$

The equation of the tangent is $y = 8x - 15$.

A **normal** is a line which is perpendicular to the tangent.

Rule

If m is the gradient of the tangent, and n the gradient of the normal, then $mn = -1$

So in Example 17.4, the gradient of the normal is found by:

$$n = \frac{-1}{m} = \frac{-1}{8}$$

Hence

$$y = \frac{-1x}{8} + c$$

passes through $(2, 1)$, so

$$1 = \frac{-1}{8} \times 2 + c$$

Hence $c = \frac{5}{4}$

The equation of the normal is $y = \frac{-1x}{8} + \frac{5}{4}$.

Exercise 17(a)

1 Differentiate with respect to the letter used for the variable:

(i) $x^3 + 2x$ (ii) $x^4 - \frac{2}{x^2}$

(iii) $(x^2 + 4)^2$ (iv) $\frac{v^2 + 1}{v^2}$

(v) $t^2 + 2t + \frac{1}{t}$ (vi) $(x^3 - 1)(x^2 + 1)$

2 If $x = 2t^2 + 1$, find $\frac{dx}{dt}$ if $t = 2$.

3 If $p = 2V^3$, find:

(i) $\frac{dp}{dV}$ when $V = 2$

(ii) p when $\frac{dp}{dV} = 1$.

4 Find the equation of the tangent to the curve $y = x^3 + 1$ at the point $(1, 2)$.

5 Find the equation of the tangent and normal to the curve $y = x + \frac{1}{x}$ at the point $(2, 2.5)$.

17.2 Maxima and minima

The **turning** points on a graph are called **maximum** points or **minimum** points (see Figure 17.3). At these points, the curve is parallel to the x-axis, hence $\frac{dy}{dx} = 0$.

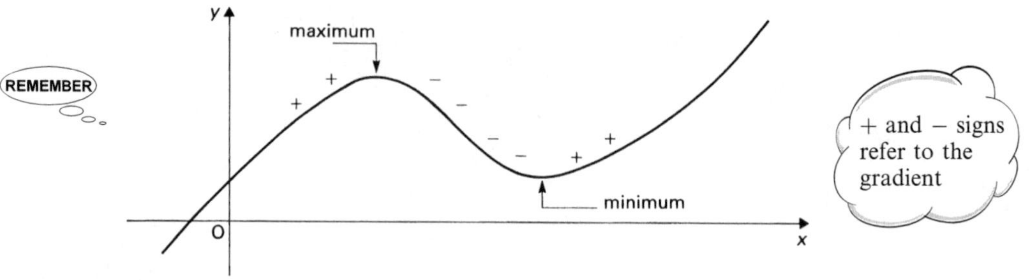

Figure 17.3

Look at the gradient on each side of the point to find whether it is a maximum or a minimum.

Example 17.5

Find the coordinates of the turning points on the curve $y = \frac{1}{x} + x$, and distinguish between maximum and minimum.

Solution

If $\quad y = \frac{1}{x} + x$

$$\frac{dy}{dx} = \frac{-1}{x^2} + 1 = 0 \quad \text{(at turning points)}$$

So $\quad 1 = \frac{1}{x^2}$

i.e. $\quad x^2 = 1$ and hence $x = \pm 1$.

Consider $x = 1$.

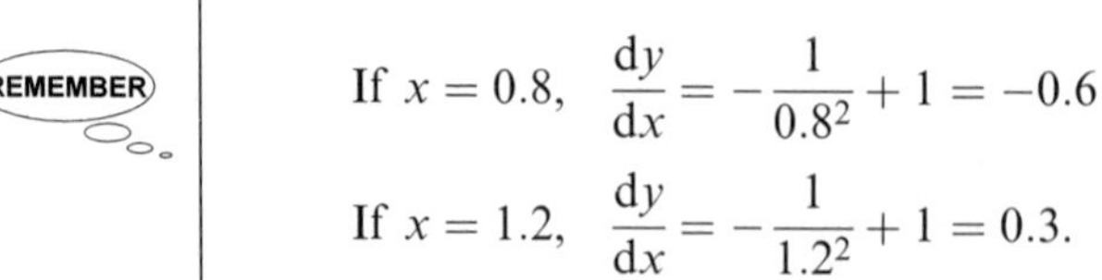

If $x = 0.8$, $\quad \frac{dy}{dx} = -\frac{1}{0.8^2} + 1 = -0.6$

If $x = 1.2$, $\quad \frac{dy}{dx} = -\frac{1}{1.2^2} + 1 = 0.3$.

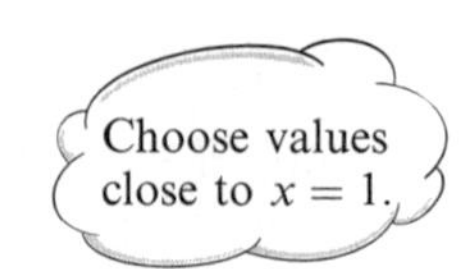

The gradient goes from negative to positive as x increases (see Figure 17.3). Hence $x = 1$ is a minimum. If $x = 1$, $y = \frac{1}{1} + 1 = 2$. The minimum point is $(1, 2)$.

Consider $x = -1$.

If $x = -1.2$, $\dfrac{dy}{dx} = 0.3$

If $x = -0.8$, $\dfrac{dy}{dx} = -0.6$.

Hence the gradient goes from positive to negative (see Figure 17.3) as x increases. It is a maximum, and the coordinates are $(-1, -2)$.

Exercise 17(b)

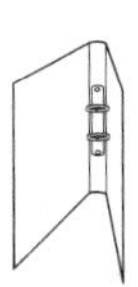

Find the coordinates of the turning points on the following curves distinguishing between maxima and minima:

1 $y = x^2 + 4x + 1$

2 $3 - x^2$

3 $y = x^3 - 3$

4 $y = x + \dfrac{1}{4x}$

5 $y = x^2 - \dfrac{1}{x}$

6 $y = 2x^3 - 15x^2 + 36x + 7$.

17.3 Integration

Integration is the name given to the process which is the reverse of differentiation.

In Section 17.1 we showed that if $y = x^4$, then $\dfrac{dy}{dx} = 4x^3$. The reverse situation would be, given $\dfrac{dy}{dx} = 4x^3$, what did we have to differentiate to get this as the answer? This is called the **integral** of $4x^3$ with respect to x.

REMEMBER

Do not leave out the dx

The symbol used for integration is an elongated s, written $\int$.

We write $\int 4x^3\,dx$ to mean the integral of $4x^3$ with respect to x.

We have then, $\int 4x^3 dx = x^4 + C$, where C stands for an unknown number, since differentiation of C would give zero. C is called the **constant of integration**.

In general, we have:

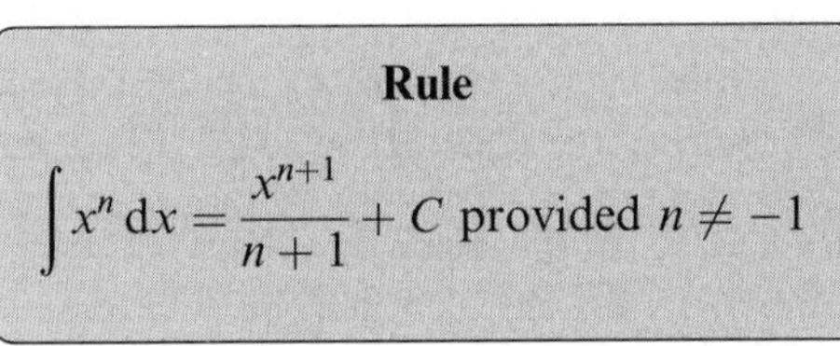

Rule

$$\int x^n\,dx = \frac{x^{n+1}}{n+1} + C \text{ provided } n \neq -1$$

As with differentiation, the integral of a sum of terms is the same as the sum of the integrals, and if a power of x is multiplied by a constant, the integral is multiplied by a constant.

Example 17.6

Integrate the following with respect to x:

(a) $x^3 + 3x^2 + 7x + 4$

(b) $\dfrac{4}{x^2}$

(c) $(x^2 + 2)^2$

(d) $\dfrac{x^4 + 1}{x^2}$

Solution

(a) $\int (x^3 + 3x^2 + 7x + 4)\, dx$

$$= \frac{x^4}{4} + \frac{3x^3}{3} + \frac{7x^2}{2} + \frac{4x^1}{1} + C$$

$$= \frac{x^4}{4} + x^3 + \frac{7x^2}{2} + 4x + C.$$

(b) $\int \frac{4}{x^2}\, dx = \int 4x^{-2}\, dx$

$$= \frac{4x^{-1}}{-1} + C$$

$$= \frac{-4}{x} + C.$$

(c) $\int (x^2 + 2)^2\, dx = \int (x^2 + 2)(x^2 + 2)\, dx$

$$= \int (x^4 + 4x^2 + 4)\, dx$$

$$= \frac{x^5}{5} + \frac{4x^3}{3} + 4x + C.$$

(d) $\int \left(\frac{x^4 + 1}{x^2}\right) dx = \int \left(x^2 + \frac{1}{x^2}\right) dx$

$$= \int (x^2 + x^{-2})\, dx = \frac{x^3}{3} - \frac{1}{x} + C.$$

Integrals which contain $+C$ in the answer are called **indefinite** integrals. The following section deals with the idea of a **definite** integral.

17.4 Area and volume

The integral sign is really an elongated S, standing for *sum*. It can be used for finding any quantity that can be broken down into small parts. Any area under a curve can be divided into rectangles of height y, and thickness δx (see Figure 17.4).

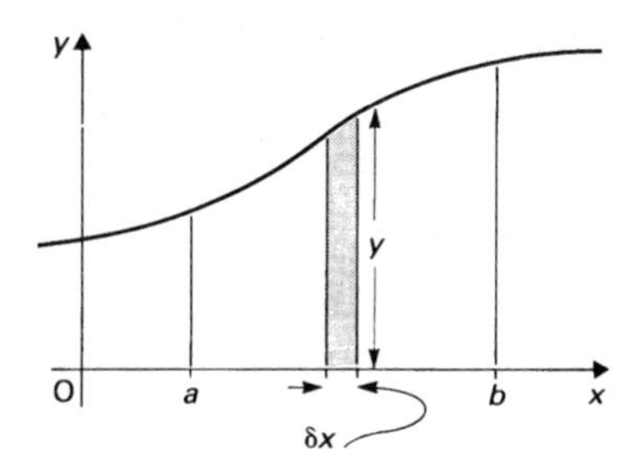

Figure 17.4

The total area = Sum ($y\delta x$) from $x = a$ to $x = b$.

It can be shown that this equals

$$\int_a^b y\,dx$$

where a and b are the **limits** of the integral. To evaluate a definite integral, you proceed as follows:

$$\text{upper limit} \rightarrow \int_2^5 x^2\,dx \leftarrow \text{lower limit} = \left[\frac{x^3}{3}\right]_2^5 = \left[\frac{5^3}{3}\right] - \left[\frac{2^3}{3}\right]$$

(carry out the integral; replace x by upper limit; replace x by lower limit)

$$= \frac{125}{3} - \frac{8}{3}$$

$$= 39$$

Figure 17.5 shows the area shaded that has been found.

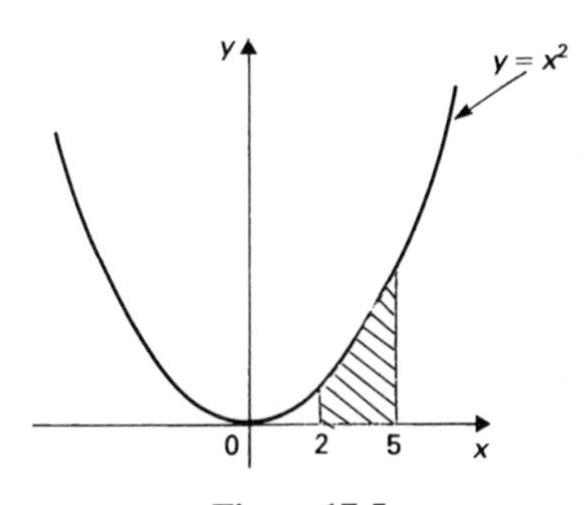

Figure 17.5

If a curve is below the axis, the area will become negative, and you must be very careful.

Example 17.7

Find the area enclosed by the curve $y = x^3 - x$ and the x-axis.

Solution

As part of the curve is below the axis (see Figure 17.6), you must split the integral into two parts.

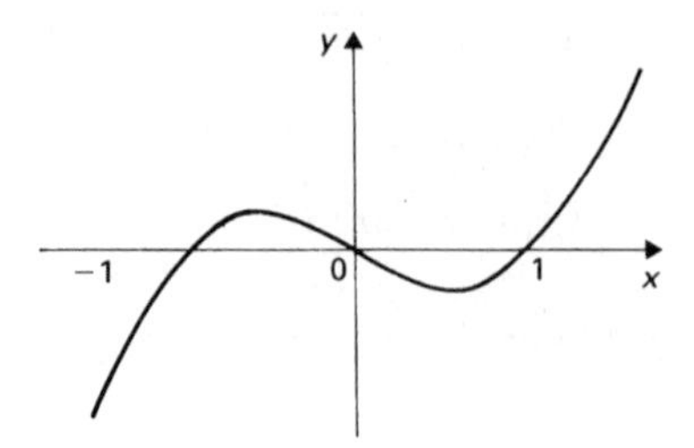

Figure 17.6

$$\int_{-1}^{0} x^3 - x\,dx = \left[\frac{x^4}{4} - \frac{x^2}{2}\right]_{-1}^{0} = \left[0\right] - \left[\frac{1}{4} - \frac{1}{2}\right] = \frac{1}{4}$$

$$\int_{0}^{1} x^3 - x\,dx = \left[\frac{x^4}{4} - \frac{x^2}{2}\right]_{0}^{1} = \left[\frac{1}{4} - \frac{1}{2}\right] - \left[0\right] = -\frac{1}{4}$$

The value $-\frac{1}{4}$ means the area is *below* the axis.

The actual *area* of this second part is $\frac{1}{4}$.

The total area $= \frac{1}{4} + \frac{1}{4} = \frac{1}{2}$ unit^2.

If an area between a and b is rotated by 360° about the x-axis, a **solid** is formed, called a **solid of revolution**.

$$\text{Its volume} = \pi \int_a^b y^2\,dx.$$

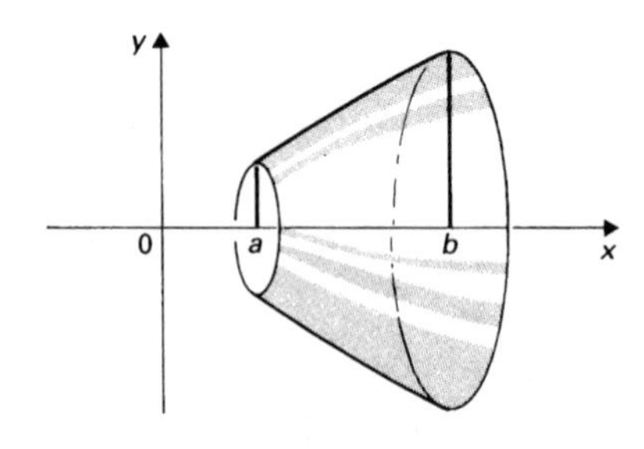

Figure 17.7

Example 17.8

Find the volume of a cone obtained when the line $y = 2x$ from $x = 0$ to $x = 4$ is rotated by 360° about the x-axis.

Solution

$$\text{Volume} = \pi \int_0^4 (2x)^2 \, dx = \pi \int_0^4 4x^2 \, dx$$

$$= \pi \left[\frac{4x^3}{3}\right]_0^4 = \pi \left[\frac{256}{3}\right] - \pi \left[0\right]$$

$$= 268 \text{ cubic units.}$$

Exercise 17(c)

1 Integrate the following functions with respect to x:

(i) $x^2 + 1$ (ii) $3x^2 - 5x + 1$

(iii) $\sqrt{x}$ (iv) $\dfrac{x+1}{x^3}$

(v) $(x-1)^2$ (vi) $(x+2)\left(x + \dfrac{2}{x^3}\right)$

(vii) $x^3 - 3x - \dfrac{1}{2x^2}$ (viii) $(x^3 - 1)(x^3 + 1)$.

2 Find the following definite integrals:

(i) $\displaystyle\int_1^2 x^2 + 3 \, dx$ (ii) $\displaystyle\int_0^1 x^4 - 5 \, dx$

(iii) $\displaystyle\int_{-1}^1 x^2 + 2x + 2 \, dx$ (iv) $\displaystyle\int_{-1}^3 x + \frac{2}{x^2} \, dx$

(v) $\displaystyle\int_1^3 2x(x+3) \, dx$ (vi) $\displaystyle\int_{-2}^0 x(x-2) \, dx$

(vii) $\displaystyle\int_1^4 (x^2+1)^2 \, dx$ (viii) $\displaystyle\int_1^2 \frac{x^2+1}{x^2} \, dx$.

3 Find the area between the curve $y = x^3$, the x-axis and the line $x = 1$ and $x = 4$. Also find the volume when this area is rotated by 360° about the x-axis.

4 Find the area between the two lines $y = 4x$, and $y = x^2$. Illustrate how you have worked this out with a diagram.

5 By rotating the line $y = r$ between a and h through 360° about the x-axis, find the formula for the volume of a certain shape. What is this shape?

6 The area between the curve $y = x^3 - x$, the lines $x = 4$ and $x = 6$ and the x-axis, is rotated by 360° about the x-axis. Find the volume of the solid formed.

Hints and solutions to investigations

I (i)

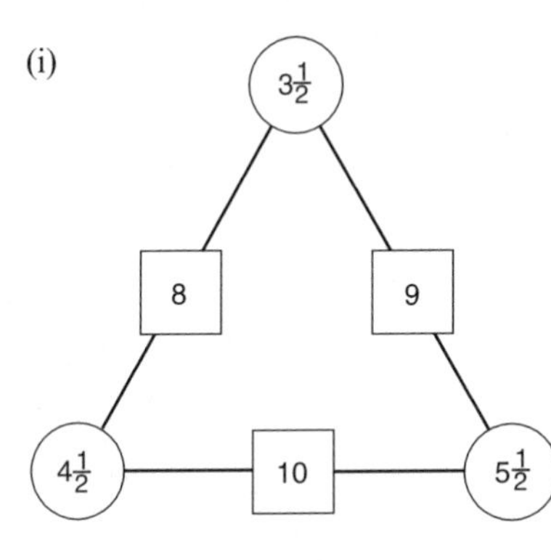

(ii)

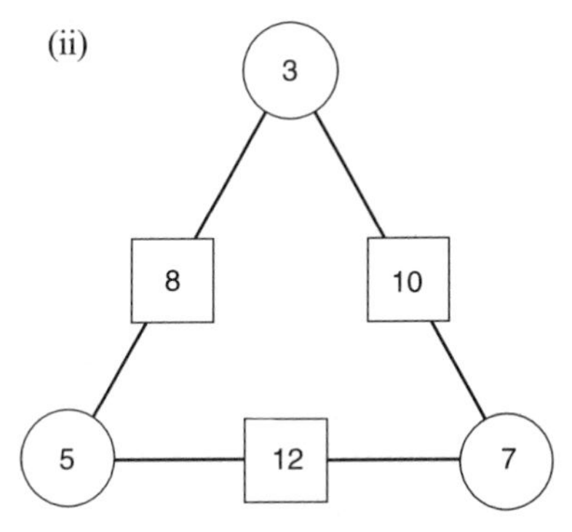

(iii)

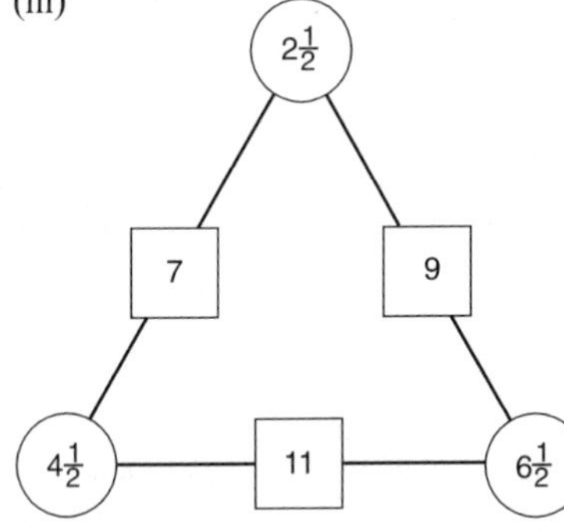

(iv)

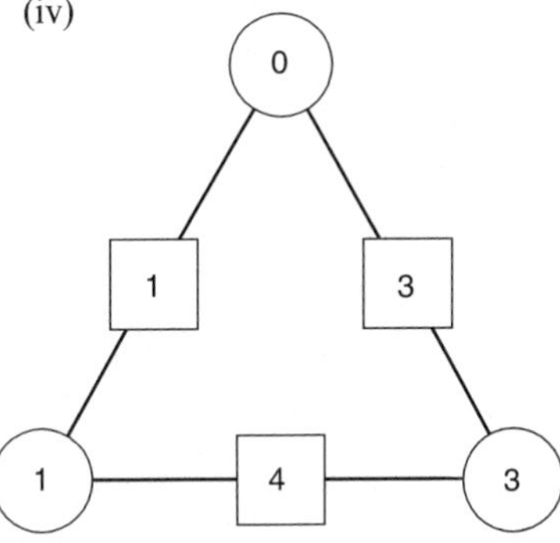

(v) The number in the bottom left hand position is always $(a + c - b) \div 2$. The rest can easily be calculated.

II (i)

$$\begin{array}{r} 124 \\ +\ 24 \\ \hline 148 \\ \hline \end{array} \quad \text{or} \quad \begin{array}{r} 149 \\ +\ 49 \\ \hline 198 \\ \hline \end{array}$$

(ii) $808 \times 88 = 71104$

(iii) $9506 \div 97 = 98$

(iv) u must be zero and $t \times r$ must end in t, so try $2 \times 6 = 12$.
The final solution is $346 \times 12 = 4152$.

III Probably 41, 24, 25.

IV

1 (i) shade 4 squares (ii) shade 12 parallelograms

(iii) shade 4 sectors (iv) shade 25 triangles

2 (i) $\frac{2}{7}$ (ii) $\frac{2}{5}$

(iii) $\frac{3}{8}$ (iv) $\frac{8}{27}$

V

$\frac{1}{9} = 0.11111111 = 0.\dot{1}$ $\quad \frac{2}{9} = 0.22222222 = 0.\dot{2}$ $\quad \frac{3}{9} = 0.33333333 = 0.\dot{3}$

$\frac{1}{11} = 0.09090909 = 0.\dot{0}\dot{9}$ $\quad \frac{2}{11} = 0.18181818 = 0.\dot{1}\dot{8}$ $\quad \frac{3}{11} = 0.27272727 = 0.\dot{2}\dot{7}$

$\frac{1}{7} = 0.\dot{1}4285\dot{7}$ $\quad \frac{2}{7} = 0.\dot{2}8571\dot{4}$ $\quad \frac{3}{7} = 0.\dot{4}2857\dot{1}$

All fractions produce recurring (repeating) decimals, although they are not always easy to spot, since you might need more decimal places than are shown on your calculator.

VI

A	B	C
1	1.5	2.25
1.5	1.4166666	2.0069444
1.4166666	1.4142157	2.0000060
1.4142157	1.4142136	2.0000000

The method finds the square root of 2, the value of B. The last part is very difficult to see, but if you change the box to $B = \frac{A}{2} + \frac{N}{2A}$, it will find the square root of the number N. [You did very well if you were able to work this out.]

VII (i) $3N + 5$, 305 (ii) $5N - 1$, 499

(iii) $4N - 2$, 398 (iv) $11N - 3$, 1097.

VIII

No. of matches	4	7	10	13	16	31
Perimeter	4	6	8	10	12	22

(a) 80

(b) 45

(c) No. of matches $= 3n + 1$

Perimeter $= 2n + 2$

IX

Percentage	*Fraction*	*Decimal*	*Percentage*	*Fraction*	Decimal
5	$\frac{5}{100} = \frac{1}{20}$	0.05	$12\frac{1}{2}$	$\frac{1}{8}$	0.125
10	$\frac{1}{10}$	0.1	$33\frac{1}{3}$	$\frac{1}{3}$	$0.\dot{3}$
15	$\frac{3}{20}$	0.15	$6\frac{1}{4}$	$\frac{1}{16}$	0.0625
20	$\frac{1}{5}$	0.2	$3\frac{1}{8}$	$\frac{1}{32}$	0.03125
25	$\frac{1}{4}$	0.25	$62\frac{1}{2}$	$\frac{5}{8}$	0.625
40	$\frac{2}{5}$	0.4	$31\frac{1}{4}$	$\frac{5}{16}$	0.3125
50	$\frac{1}{2}$	0.5	$68\frac{3}{4}$	$\frac{11}{16}$	0.6875

XI

No. of sides	Name	Exterior angle	Interior angle	Sum of interior angles
3	triangle	120°	60°	180°
4	square	90°	90°	360°
5	pentagon	72°	108°	540°
6	hexagon	60°	120°	720°
7	heptagon	$51\frac{3}{7}°$	$128\frac{4}{7}°$	900°
8	octagon	45°	135°	1080°
9	nonagon	40°	140°	1260°
10	decagon	36°	144°	1440°
12	dodecagon	30°	150°	1800°

XIII There are two essentially different ways of fitting the pentominoes together. If you find more they will be a rotation or reflection of one of these two.

XV (iv) All approximately equal to 0.64.

(v)

x	0	10	20	30	40	50	60	70	80	90
$y = \sin x$	0	0.17	0.34	0.50	0.64	0.77	0.87	0.94	0.98	1.00

XVI There are over 300 different recorded proofs of this theorem.

Pythagorean triples: e.g. 7, 24, 25; 9, 40, 41.

XVIII You should certainly find many differences in the answers to each section, although perhaps not always what you had expected.

XIX (i) 6, 10, 15: they are triangular numbers.

(ii) The pattern is not easy to spot.

For N people, the number of handshakes is $\dfrac{N \times (N-1)}{2}$.

(iii) Solving $\dfrac{N(N-1)}{2} = 11\,175$ leads to the quadratic equation $N^2 - N - 22\,350 = 0$ which has solutions $N = 150$ or -149. Hence there are 150 people.

XX (ii)

No. of cuts	1	2	3	4	5	6
No. of pieces	2	4	7	11	16	22

For N cuts, the number of pieces is $(N^2 + N + 2) \div 2$.

(iii) 15 cuts.

XXI (ii)

Frequency	1	2	3	4	5	6	5	4	3	2	1
Probability	$\frac{1}{36}$	$\frac{1}{18}$	$\frac{1}{12}$	$\frac{1}{9}$	$\frac{5}{36}$	$\frac{1}{6}$	$\frac{5}{36}$	$\frac{1}{9}$	$\frac{1}{12}$	$\frac{1}{18}$	$\frac{1}{36}$

(iii) Results will not agree; the difference between theoretical and practical probability. Relative frequencies would improve if you increased the number of times the dice were rolled.

XXII $P(0) = \frac{5}{16}$, $P(1) = \frac{3}{16}$, $P(2) = \frac{3}{16}$, $P(3) = \frac{1}{8}$, $P(5) = \frac{1}{8}$, $P(10) = \frac{1}{16}$

Expected return $= \frac{5}{16} \times 0 + \frac{3}{16} \times 1 + \frac{3}{16} \times 2 + \frac{1}{8} \times 3 + \frac{1}{8} \times 5 + \frac{1}{16} \times 10$

$= 2\frac{3}{16}$ p.

The game is clearly worth playing, since this is more than the 2p you pay for the go.

XXIII (i) All lines are parallel, gradient $= 1$

(ii) All lines pass through $(0, 3)$.

XXIV (ii) 39.5

(iv) As the width of each trapezium gets smaller, the approximation for the area changes by a smaller amount, and the value is getting nearer and nearer to the exact value.

XXVI (i) (a) 8 cm (b) 16 cm

(ii) (a) twice (b) 4 times

(iii) 2 cm

(iv) 0.25 cm

(v) Use P as a centre of enlargement.

XXVIII (i) Stretch parallel to the x-axis scale factor n, and stretch parallel to the y-axis scale factor m

(ii) Two way stretch

(iii) Rotation $\theta°$ anticlockwise about O

(iv) Reflection in the line $y = x \tan \theta$.

Answers to exercises

Exercise 1(a)

1 (i) 3^3 (ii) 2^5 (iii) 100^2 or 10^4

2 (i) four hundred and eighty three
(ii) four thousand and ninety six
(iii) eight hundred and one thousand four hundred and sixty nine

3 (i) 8908 (ii) 40 600 079

4 (i) 260 (ii) 30 600 (iii) 409 (iv) 180

5 1600

6 122

7 173

8 (i) 1051 (ii) 7189

9 (i) 72 (ii) 100 (iii) 123 (iv) 14
(v) 7

10 $(6+3) \times (15 \div 3) + 2 = 47$

11 (i) $2^2 \times 3^2$ (ii) 3×5^2 (iii) $2^2 \times 5 \times 7$
(iv) $2 \times 5 \times 29$ (v) $5^2 \times 7$

12 (i) 20 (ii) 15 (iii) 9 (iv) 14

13 (i) 36 (ii) 24 (iii) 150 (iv) 726

14 (i) 13 (ii) K (iii) K, Q (iv) Q, J, K, W
(v) 8 (vi) (a) 42 (b) 26 (vii) 10

Exercise 1(b)

1 3 **2** 0 **3** −12 **4** −5
5 −8 **6** 20 **7** 4 **8** 5
9 28 **10** 1 **11** −5 **12** 41
13 −5 **14** −6 **15** −24 **16** 2
17 0 **18** 80 **19** −5 **20** −6
21 3 **22** −16 **23** −2 **24** −24

25 3 **26** −5 **27** −3 **28** 280

29 −2 **30** −4 **31** 99 **32** 96

33 −256 **34** −264

35 (i) −10°C (ii) Thur (iii) 8°C (iv) 1°C (v) 3°C

Exercise 1(c)

1 (i) 5 (ii) 22 (iii) 11 (iv) 73 (v) 23 (vi) 34

2 (a) (i) 1001 (ii) 1111 (iii) 100011 (iv) 1111000
(b) (i) 11 (ii) 17 (iii) 43 (iv) 170
(c) (i) 9 (ii) F (iii) 23 (iv) 78

3 (i) 10010 (ii) 47 (iii) 110110110 (iv) 216

4 (i) 1010_2 (ii) 1100_2 (iii) 62_8 (iv) 10_2 (v) 100_2 (vi) 21_8 (vii) 1032_4 (viii) 1010_2 (ix) 112_3 (x) 354_8

Exercise 2(a)

1 (i) $\frac{3}{8}$ (ii) $\frac{2}{3}$ (iii) $\frac{3}{8}$ (iv) $\frac{1}{4}$ (v) $\frac{3}{5}$ (vi) $\frac{6}{7}$ (vii) $\frac{4}{11}$ (viii) $\frac{6}{7}$

2 (i) $\frac{4}{7}, \frac{3}{5}$ (ii) $\frac{3}{8}, \frac{5}{12}$ (iii) $\frac{2}{9}, \frac{1}{3}, \frac{4}{5}$ (iv) $\frac{1}{3}, \frac{3}{8}, \frac{2}{5}$ (v) $\frac{7}{10}, \frac{3}{4}, \frac{4}{5}$ (vi) $\frac{1}{3}, \frac{5}{9}, \frac{5}{6}$

3 (i) $2\frac{1}{2}$ (ii) $3\frac{1}{3}$ (iii) $2\frac{1}{4}$ (iv) $3\frac{1}{7}$ (v) $5\frac{3}{8}$ (vi) $1\frac{3}{8}$

4 (i) $\frac{9}{4}$ (ii) $\frac{14}{3}$ (iii) $\frac{29}{9}$ (iv) $\frac{47}{7}$ (v) $\frac{77}{13}$ (vi) $\frac{247}{24}$

5 $\frac{5}{12}$

6 (i) $\frac{11}{15}$ (ii) $1\frac{5}{18}$ (iii) $2\frac{3}{10}$ (iv) $5\frac{71}{84}$ (v) $3\frac{5}{6}$ (vi) $3\frac{31}{40}$ (vii) $\frac{5}{8}$ (viii) $2\frac{2}{9}$ (ix) $1\frac{39}{40}$ (x) $2\frac{9}{10}$ (xi) $\frac{7}{8}$ (xii) $\frac{1}{6}$ (xiii) $1\frac{5}{12}$ (xiv) $5\frac{41}{60}$

7 (i) $\frac{1}{2}$ (ii) $1\frac{3}{4}$ (iii) $\frac{1}{12}$ (iv) $3\frac{1}{16}$ (v) $52\frac{1}{2}$ (vi) $\frac{7}{50}$

8 $1\frac{1}{3}$

9 £36

10	(i) $1\frac{1}{2}$	(ii) $\frac{1}{2}$	(iii) $\frac{6}{7}$	(iv) $1\frac{1}{2}$
	(v) $\frac{5}{12}$			
11	(i) $\frac{47}{60}$	(ii) $\frac{13}{60}$	(iii) £180	
12	$\frac{1}{12}$, $16\frac{2}{3}$ cm			

Exercise 2(b)

1	(i) $\frac{4}{5}$	(ii) $\frac{5}{8}$	(iii) $2\frac{1}{8}$	(iv) $4\frac{3}{20}$
2	(i) 0.375	(ii) 0.35	(iii) 0.1875	(iv) 0.475
3	(a) (i) 0.51		(ii) 0.01	
	(iii) 41.13		(iv) 34.83	
	(v) 1985.03		(vi) 285.80	
	(vii) 100.00			
	(b) (i) 0.507		(ii) 0.00858	
	(iii) 41.1		(iv) 34.8	
	(v) 1990		(vi) 286	
	(vii) 100			
4	(i) 0.225	(ii) 1.571	(iii) 0.381	(iv) 0.556
	(v) 0.929			
5	(i) 6×10^{-2}	(ii) 5.8×10^{-3}	(iii) 8.19×10^{1}	(iv) 6.823×10^{3}
	(v) 2.5×10^{5}			
6	(i) 7.01	(ii) 1.664	(iii) 14.71	(iv) −3.5
	(v) 10.97	(vi) −38.34		
7	(i) 9×10^{2}	(ii) 5.05×10^{3}	(iii) 1.2×10^{-1}	(iv) 1.906×10^{-5}
8	(i) 1.1	(ii) 0.0042	(iii) 0.3	(iv) 0.9
	(v) 6.386	(vi) 0.484	(vii) 651	(viii) 0.72
	(ix) 1500	(x) 3		
9	(i) 21.36	(ii) 7.45	(iii) 74.4	
10	9.46×10^{15} m, 2.37×10^{17} m			

Exercise 2(c)

1	200	**2**	3	**3**	60	**4**	£30
5	3000	**6**	20	**7**	300 000	**8**	0.6 m^2
9	20	**10**	400	**11**	16 400	**12**	6340
13	367	**14**	118	**15**	2.25	**16**	15.4
17	0.430	**18**	10.0	**19**	28.0	**20**	−659
21	−3.30	**22**	218	**23**	6 350 000	**24**	0.0283
25	0.0648	**26**	2.72×10^{-51}				

Exercise 2(d)

1 (i) 64 (ii) 86.84 (iii) 600 (iv) 0.84
(v) 650 000 (vi) 40 000 (vii) 6 (viii) 60
(ix) 4 (x) 200 000 (xi) 0.64 (xii) 0.1

2 (i) 1055 (ii) 0700 (iii) 0530

3 (i) 0839 (ii) 52 mph (iii) 1350 (iv) 25%

4 (i) 0715 (ii) (a) 1150 (b) 2100 (c) £24.10

5 (i) (a) 708 km (b) 336 km
(ii) (a) Aberdeen (b) Cambridge
(iii) (a) 1146 km (b) Birmingham to Norwich by 2 km
(iv) (a) 53 km/h (b) 7 hours 20 mins
(c) 15 km

Exercise 3(a)

1 (i) 1 (ii) 4 (iii) 7 (iv) -2
(v) 3 (vi) 2 (vii) -4 (viii) 8
(ix) 5 (x) 4

2 $A - B + C$

3 (i) $1000d$ (ii) $\frac{k}{1000}$ (iii) $100(a+b)$ (iv) $1000d$

4 (i) $3y$ (ii) $y - 8$ (iii) $\frac{y}{6}$ (iv) $1.1y$

5 $\frac{d}{c}$, $wx + cy$

6 (i) $9x$ (ii) $7a + 3b$ (iii) $6x + 3y$ (iv) $3x^2$
(v) $5x + 7$ (vi) $10p + 1$ (vii) t (viii) $h + 7$
(ix) $7x^2$ (x) $32p^3$ (xi) 1 (xii) 7

Exercise 3(b)

1 7 **2** 23 **3** $3\frac{1}{2}$ **4** 4

5 6 **6** $1\frac{1}{2}$ **7** 4 **8** $\frac{5}{6}$

9 -3 **10** 4 **11** $\frac{3}{10}$ **12** $\frac{1}{2}$

13 -5 **14** $2\frac{2}{5}$ **15** 19 **16** -1

17 16 **18** $-5\frac{1}{2}$ **19** $1\frac{9}{13}$ **20** $1\frac{1}{6}$

Exercise 3(c)

1 4

2 £140 and £280

3 $7x = 45.5$, 6.5 cm, 16.25 cm, 22.75 cm

4 $36s + 13.5s + 9s = 140.4$, hence standard rate $s = £2.40$

5 27

6 $x + 40x + 50(18 - 3x) = 464$. Hence $x = 4$, given 4 pennies, 8 twenty-pence pieces, and 6 fifty-pence pieces

Exercise 3(d)

1 (i) $x = \dfrac{y}{H}$ (ii) $x = y - H$ (iii) $x = H - y$

(iv) $x = \dfrac{yH}{t}$ (v) $x = \dfrac{(y - b)}{a}$ (vi) $x = 4t - q$

(vii) $x = \dfrac{100I}{pT}$ (viii) $x = \dfrac{(v - u)}{2a}$ (ix) $x = \dfrac{(v^2 - u^2)}{2a}$

(x) $x = \dfrac{H}{y}$

2 $h = \dfrac{S - 6r^2}{6r}$; $6\frac{1}{3}$

3 (i) 11, 19, $2n - 1$ (ii) 23, 39, $4n - 1$ (iii) 49, 99, $10n - 1$

(iv) 13, 33, $4n - 7$ (v) 25, 13, $43 - 3n$

4 (i) $2x + 78$ (ii) 44

Exercise 4(a)

1 (i) 25% (ii) 37.5% (iii) $233\frac{1}{3}$% (iv) 80%

(v) 28% (vi) 165% (vii) $11\frac{1}{9}$%

2 (i) $\frac{1}{2}$ (ii) $\frac{29}{100}$ (iii) $\frac{9}{50}$ (iv) $\frac{91}{150}$

(v) $1\frac{1}{5}$ (vi) $\frac{16}{25}$

3 (i) £14.73 (ii) 243 mm (iii) 432 (iv) 33 750

(v) £8.10 (vi) 3220 m

4 1200

5 0.50%

6 22.5%

7 $£3360 \div 70 \times 100 = £4800$

8 52.7%

9 £84

10 31.7%

Exercise 4(b)

1 (i) £50.40 (ii) £217.50 (iii) £109.20

2 £800

3 13.5%, 3 years

4 (i) £2395.80 (ii) £16 488.25 (iii) £988.66

5 £3132.04

6 £17 97713

Exercise 4(c)

1 (i) 2 : 15 (ii) 3 : 250 (iii) 11 : 24 (iv) 1 : 4
(v) 21 : 34 (vi) 3 : 35 (vii) 1 : 75 (viii) 6 : 11
(ix) 32 : 41 (x) 2 : 15 (xi) 2 : 3 : 9

2 (a) (i) 1 : 1.69 (ii) 1 : 1.73 (iii) 1 : 4 (iv) 1 : 10
(v) 1 : 0.73 (vi) 1 : 94.6
(b) (i) 0.59 : 1 (ii) 0.58 : 1 (iii) 0.25 : 1 (iv) 0.1 : 1
(v) 1.375 : 1 (vi) 0.01 : 1

3 £4.50

4 6 kg

5 35

6 34.2 m

7 3.2

8 £21.33, £32, £74.67

9 26.25 cm, 35 cm, 43.75 cm

10 £4.20

11 3000 m, 4800 m

12 35 cm

13 1 : 4

14 73.5 g

15 Bill's

16 80% or $\frac{8}{10}$

17 57%

Exercise 4(d)

1 (i) 5.6 m/s (ii) 8000 kg/m^3 (iii) 144 km/h (iv) 0.25 g/cm^3
(v) 1.2 m/min (vi) 146.7 ft/s (vii) 111 111 cm/s

2 (i) 122.5 (ii) 4.27 (iii) 47.14 (iv) 661.96
(v) 1271.33

3 (i) 20 cm (ii) 4.25 km

4 (i) 7.4 km (ii) distance = 13.2km, time = 11 mins 19 secs

Exercise 4(e)

1 (i) £40 000
(ii) (a) £14 400 (b) £27 600 (c) £3800
(iii) (a) £45 800 (b) 14.5%

2 (i) (a) £176 (b) £6.60
(ii) (a) £225.50 (b) 5
(iii) £132

3 (i) £140
(ii) (a) £1.16 (b) £57.81 (c) £24.48
(iii) £14

4 20%, 26p

5 (i) £596.10 (ii) 20.6% (iii) 4285

6 (i) (a) £21 (b) £18.48
(ii) (a) 27p (b) £1.50
(iii) £3.84
(iv) 24%

Exercise 5(a)

1 122° **2** 52° **3** 287° **4** 66°, 137°

5 147° **6** 128°, 128°, 52° **7** 70°, 70°, 110° **8** 105°

9 70°, 110°, 70° **10** 122°, 58° **11** 50°, 130°, 152° **12** 50°

13 70°, 70° **14** 54° **15** 31.5°

Exercise 5(b)

1 (i) 30°, 50°, 100° (ii) 108°, 72° (iii) 30°, 70°, 20° (iv) 40°, 110°

2 $(180 - x) = 3x$, $x = 45$; 8 sides

3 (i) 120°, 30° (ii) 80°

4 26 sides

5 $4x + 470° = 720°$, $x = 62.5°$, angles are 62.5°, 125°, 112.5°, 130°, 140°, 150°

Exercise 5(c)

1 (i) 3.5, 5 (ii) 10, 3.6 (iii) $\frac{6}{3} = \frac{p+3}{4} = \frac{10}{q}$, hence $p = 5$, $q = 5$

2 $AN = 5.3$ cm, $AB = 7.1$ cm

3 (i) side-angle-side (ii) not congruent (iii) not congruent

Exercise 5(d)

1 (i) 135° (ii) 1080° (iii) 45° (iv) 135°

2 (i) 4
(ii) Triangle EBC rotated by 90° anticlockwise becomes triangle HAB, hence $HB \perp EC$, similarly all angles of $JKLM$ are 90°, and since it has rotational symmetry of order 4, it must be a square
(iii) Produces a cross of five equal squares
(iv) $25 \div 5 = 5$ cm^2

3 (i) 135° (ii) $67\frac{1}{2}°$ (iii) $22\frac{1}{2}°$ (QT must be parallel to RS)
(iv) 540°

Exercise 6(a)

1 (i) 9.4 cm – 9.8 cm (ii) $5.4825 cm^2 - 5.9625 cm^2$

2 (i) 60 cm (ii) 45 approximately (iii) £81

3 (i) $6362 cm^2$ (ii) 84.82 m

4 $389.7 cm^2$

5 4630

6 You need to work out $1 ha = 0.01 km^2$; hence $1 \text{ million } ha = 1 \times 10^4 km^2$

7 About 1400 square miles

8 3.77 m; $1.13 m^2$; 1.39 m

9 $60 cm^2$

10 4

11 (i) $3.53 m^2$ (ii) 7.71 m

12 1.155 ha

Exercise 6(b)

1 $194.4 cm^2$ **2** $175.9 cm^2$ **3** $19.5 cm^2$ **4** $6158 cm^2$

5 37.2% **6** 7.96 cm

Exercise 6(c)

1 (i) 10 litres (ii) 40 secs (iii) $523.6 cm^3$ (iv) 0.26 cm

2 (i) $660 cm^2$ (ii) 120 cm (iii) $600 cm^3$

3 (i) $500 m^3$ (ii) 13 m

4 (i) (a) 45° (b) 8.49 cm
(ii) $216 cm^3$
(iii) 72 cm

5 (i) (a) 2.6 cm (b) 2.3 cm
(ii) $4.62 cm^2$
(iii) $462 cm^3$
(iv) 111 kg

Exercise 6(d)

1 (i) 9 : 25 (ii) 27 : 125

2 35.8

3 (i) 4 cm, $302 cm^2$ (ii) $859 cm^3$

4 (i) 560 mm (ii) 64 : 1

5 1 cm = 0.2 km hence $1 cm^2 = 0.04 km^2$
(i) $0.32 km^2$ (ii) $400 cm^2$

6 area $\triangle ABC : \triangle AED = 16 : 49$,
hence area of $\triangle ABC$: area of trapezium $BCED = 16 : 33$

7 4 : 9

8 $20b^2/a^2$

9 Z is the midpoint XY, hence the area $\triangle AZY$: area of trapezium $XYCB = 1 : 6$

10 Scale is 1 : 1250, area = 937.5 m^2

Exercise 7(a)

1 (i) 32° (ii) 61.9° (iii) 53.1° (iv) 49.5°

2 (i) 2.66 cm (ii) 5.47 m (iii) 138 cm (iv) 8.05 cm

Exercise 7(b)

1 (i) 45.6° (ii) 34.8° (iii) 51.3° (iv) 64.3°

2 (i) 3.69 cm (ii) 3.43 m (iii) 6.50 cm (iv) 21.4 m

Exercise 7(c)

1 (i) 31° (ii) 32.4° (iii) 32.2° (iv) 75.8°

2 (i) 9.98 m (ii) 4.69 cm (iii) 8.70 km (iv) 20.5 cm

Exercise 7(d)

1 (i) 150 m

2 138 m

3 (i) 318° (ii) 138°

4 6.76 km/h

5 (i) 22.4 km (ii) 098.4° (iii) 278.4°

Exercise 7(e)

1 4.1 cm 2 14.9 m 3 2.86 km 4 84.7 m

5 1.27 m 6 872 km

Exercise 7(f)

1 (i) 34.9° (ii) 4.46 m (iii) 5.82 m

2 37.9 cm

3 (i) 7.21 cm, 10.8 cm (iii) 56.3° (iv) 55.3°

4 1.09 m

Exercise 7(g)

1 First find the height of the tower = 92.4 m
(i) 173.8 m (ii) 337°

2 (i) 15.3° (ii) 5.96 m

3 (i) 6.93 cm, 4.62 cm (ii) 48.7°
(iii) $\angle TLP = 66.3°$ (iv) 5.26 cm

4 49.2° (i) 6.63 cm (ii) 574 cm^2

Exercise 8(a)

1 (i) $\angle X = 54.3°$, $\angle Y = 34.2°$, $\angle Z = 91.5°$
(ii) $\angle R = 75°$, $PR = 8.8$ cm, $QR = 10.8$ cm
(iii) $AC = 12.9$ cm, $\angle A = 31.4°$, $\angle C = 23.6°$
(iv) $\angle W = 15°$, $VW = 24.5$ cm, $UW = 34.7$ cm

2 39°

3 85 mm

4 9.5 cm, 26.6°, 7.3 cm

5 parallel

Exercise 8(c)

1 For example

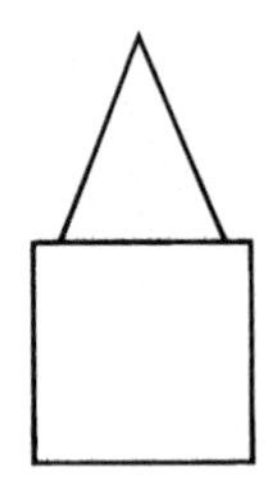

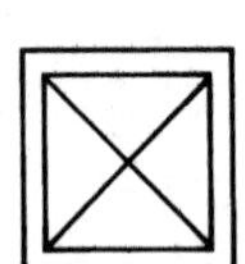

2

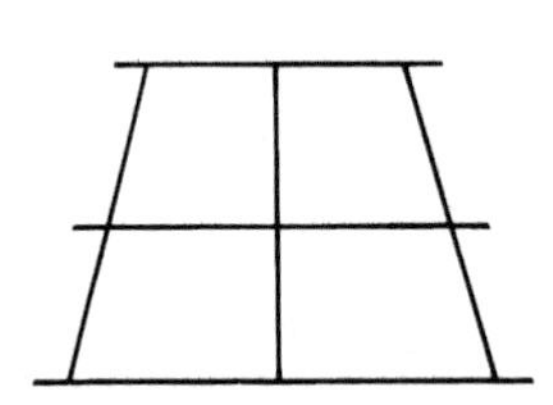

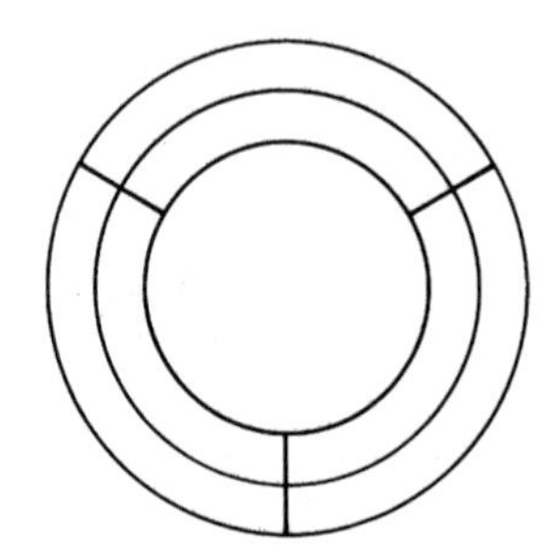

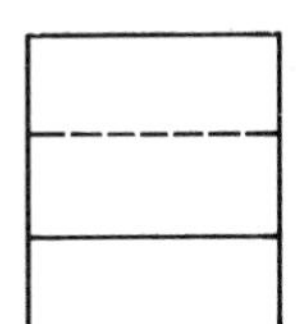

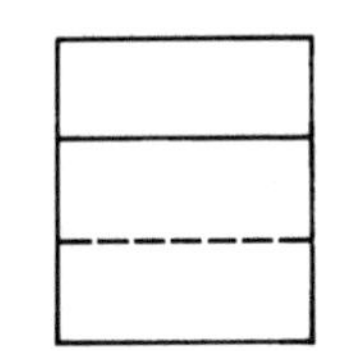

4 Possible answer

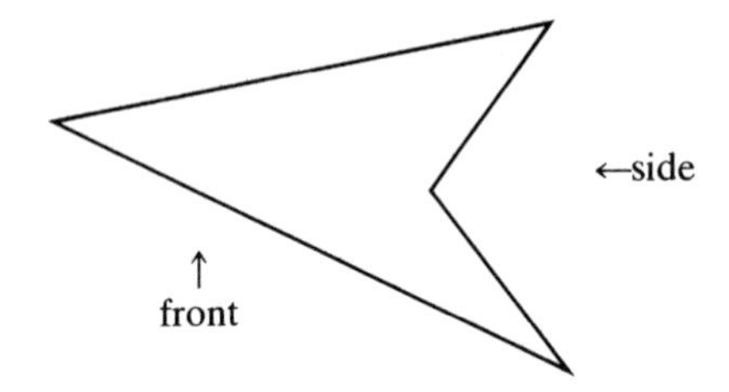

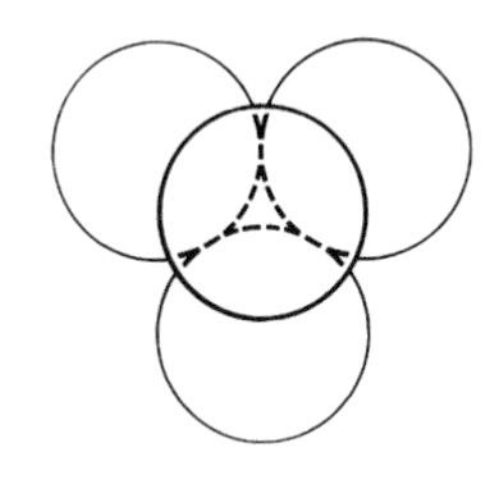

Exercise 8(d)

1 21 600 n. miles

2 6660 n. miles

3 4.5°

4 6 hours

5 (i) 8599 n. miles (ii) 15 930 km

6 (i) 480 n. miles (ii) 889 km

7 1926 km

8 572 km

9 (180°E, 51°S)

10 0°

11 10 h 56 min

12 If O is the centre of the earth, OBA is a right-angled triangle, hence the height = 8703 km

13 (i) 4179 km (ii) 9117 km
(iii) 2460 n. miles (iv) 65°N, 60.5°W

Exercise 8(e)

1 (i) Two straight lines from A at 30° to AB
(ii) A cone with A as the vertex and AB as axis of symmetry

2

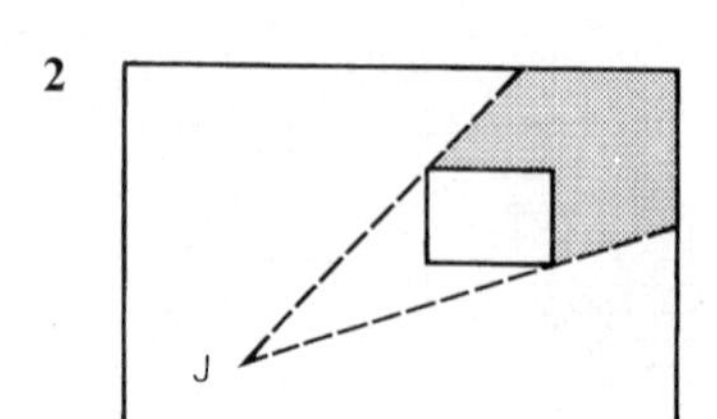

3

A B D C

1.6 cm^2

4 3.3 cm

5 An ellipse (oval)

6 A set of points which are no more than a given distance (the radius) from a given line

Exercise 9(a)

2 (ii) *Mail* 72°, *Guardian* 16°, *The Times* 14°, *Sun* 94°, *Daily Express* 80°, *Daily Mirror* 84°

3 14.6%

4 351

5 *A* 57°, *B* 34°, *C* 116°, *D* 56°, *E* 97°

6 (i) 240 (ii) 25% (ii) 72° (iv) 52

7 (i) Cannot be answered because some of the teams may have played each other
(ii) 10
(iii) 38

Exercise 9(b)

1 5.5, 55.5

2 −2.7, 217.3

3 (i) 81.9 (ii) 52.2 (iii) 5248.6 (iv) −0.90

4 (i) 41 (ii) 64 (iii) 70 (iv) 0.5

5 (a) 6, 5, 4 (b) 15.5, 16.5, 19

6 Lowered by approximately 1 month

8 (i) 3.8 per hour (ii) 2 per hour (iii) 3.5 per hour

9 (ii) 52.5

10 (i) 7 (ii) 10 (iii) 15

11 (i) 30–39 (ii) 36.6

Exercise 9(c)

1 (ii) 2.1

2 3 people

3 median = 51 marks, SIQR = 14 approximately

4 Plot points at < 2.5, < 5.5 etc.

(i) 9.2 cm^2 (ii) 61% (iii) 2.2 cm^2

5 Approximately 175 fail, pass mark = 48, median = 54

6 (iii) (a) 3.1 kg

(b) There are 35 babies, giving a probability of $\frac{35}{100} = 0.35$

Exercise 9(d)

1 (iii) About £25 000

2 (iii) 9 (total of 180 points available) (iv) Probably about 35

3 (ii) About £100 million

4 (ii) Positive

(iii) (a) 7

(b) 13 The rate of producing buds may not be the same for the lower height, hence (b) is likely to be more accurate than (a)

5 (ii) Monday about 500, Saturday about 610; sales unlikely to keep increasing at this rate

6 (ii) 20.4 (iii) 18.5

Exercise 10(a)

1 3 **2** 7 **3** 1 **4** $\frac{1}{2}$

5 25 **6** x^2y **7** 2 **8** 0.1

9 27 **10** $\frac{27}{8}$ **11** $8x^3$ **12** $\frac{1}{x}$

13 32 **14** 8 **15** x^4 **16** $\frac{1}{13}$

17 $\frac{1}{x}$ **18** 243 **19** 0.2 **20** $\frac{243}{3125}$

21 2.83 **22** 2.92 **23** 0.00001 **24** 0.818

25 5.65 **26** 0.405 **27** 1.22 **28** 0.204

29 0.215 **30** 2.43

Exercise 10(b)

1 (i) $ab, 1$ (ii) a^2, a (iii) $4x^2, 2x$
(iv) at^3, t^2 (v) $2xy^2, xy$ (vi) $abcd, bc$

2 (i) $\frac{b+a}{ab}$ (ii) $\frac{2b+3}{ab}$ (iii) $\frac{8t-3}{2t^2}$
(iv) $\frac{13x}{12}$ (v) $\frac{25}{3y}$ (vi) $\frac{28x^2-15y^2}{21xy}$
(vii) $\frac{4x^2+1}{x}$ (viii) $\frac{7a+12b}{8}$ (ix) $\frac{2x-1}{6}$
(x) $\frac{2}{y}$ (xi) b (xii) $\frac{a}{d}$
(xiii) x (xiv) $\frac{4x}{y}$ (xv) $\frac{2}{3x^2}$
(xvi) $\frac{5x}{3}$ (xvii) $\frac{2}{15x}$

Exercise 10(c)

1 x^2+3x+2
2 $x^2+8x+15$
3 $2x^2+5x+2$
4 x^2-5x+6
5 $2x^2-x-3$
6 x^2-81
7 $a^2+3ab+2b^2$
8 $c^2+cd-2d^2$
9 $q^2-qt-2t^2$
10 $12-x-x^2$
11 $4-5x-6x^2$
12 $ac+2ad+bc+2bd$
13 $t-t^3$
14 $2x^2-x-15$
15 $a^2-3a-28$
16 $2p^2-5pq-3q^2$
17 $6t^2+17t+5$
18 q^2-9t^2
19 $p^2+4pq+4q^2$
20 $q^2-6qp+9p^2$

Exercise 10(d)

1 $4(x+2y)$
2 $3(4t+1)$
3 $a(b+c)$
4 $2x(2a+3x)$
5 $p(4q-p)$
6 $8a(3-2a)$
7 $4bt(3-b)$
8 $x^2(1+x)$
9 doesn't factorise
10 $2ab(2c-1)$
11 $xy(y+x)$
12 $100x(1-x)$
13 $(x+y)(q+t)$
14 doesn't factorise
15 $(x-y)(a-b)$
16 $(x+y)(a-b)$
17 $(t-3c)(p+2q)$
18 $(4q-3t)(c+2d)$
19 $(4x+3y)(5p-2q)$
20 $(x^2+y)(x-y^2)$

Exercise 10(e)

1 (i) $(x-8)(x-3)$ (ii) $(x+3)(x-1)$
(iii) $(x+13)(x-2)$ (iv) doesn't factorise
(v) $(2p+3)(p+5)$ (vi) $(2x+1)(2x+3)$
(vii) $(3x-1)(2x-1)$ (viii) $(9t+2)(t-1)$
(ix) $(4y^2+1)(y^2-1)=(4y^2+1)(y-1)(y+1)$
(x) $(2x-5)(2x+5)$
(xi) $(1-x^2)(1+x^2)=(1-x)(1+x)(1+x^2)$
(xii) $\pi(R-r)(R+r)$ (xiii) $(11xy-2)(11xy+2)$
(xiv) $(2x-3y)^2$ (xv) $(x+3)^2$
(xvi) $(2x-5)^2$

2 (i) $-1, 3$ (ii) 2, 8 (iii) $-1, \frac{1}{2}$ (iv) $\frac{1}{3}, 3$
(v) 0, 9 (vi) no solutions
(vii) rewrite as $4y^2-9y-1=0$, hence $y=2.36, -0.11$
(viii) rewrite as $3n^2-6n+1=0$, hence $n=1.82, 0.18$

3 (i) 1.48 (ii) 1.22 or -0.73

4 $(9+2x)(12+2x)-108=72$, hence $x=1.5$

5 $\pi(R+1)^2-\pi R^2=\frac{1}{10}\pi R^2$, leads to the quadratic $R^2-20R-10=0$, hence by formula, $R=20.5$

6 $\frac{400}{x}, \frac{400}{x+40}, \frac{400}{x}=\frac{400}{x+40}+\frac{1}{2}$, simplify to $x^2+40x-32\,000=0$, hence $x=160$

7 (i) $100, n^2$ (ii) $46, (n^2-n+2)\div 2$
(iii) $20, 2n$ (iv) $113, n^2+n+3$

8 (a) 9
(b) (i) 12 (ii) 3 (iii) 15
(d) (i) $P=3L$ (ii) $A=L^2$

9 (ii) 15 (iii) 21 (v) $\frac{n(n-1)}{2}$

Exercise 10(f)

1 (i) $\frac{yt}{2}$ (ii) $\frac{z}{a+b}$ (iii) $\sqrt{\frac{H-b}{a}}$ (iv) $\frac{v^2-u^2}{2a}$
(v) $\frac{uf}{u-f}$ (vi) $\frac{T^2g}{4\pi^2}$ (vii) $\frac{q-y}{2y}$ (viii) $\frac{2q-yt}{2}$
(ix) $\left(\frac{p+2}{k}\right)^2$ (x) $\frac{yb-a}{1-y}$ (xi) $\frac{a+b}{t}$ (xii) $\frac{pbd}{(ad+bc)}$
(xiii) $\frac{y}{\sqrt{1-y}}$ (xiv) $\frac{16a^2y+t^2}{16a^2}$ (xv) $\frac{q^2-2q+17}{16}$

2 (i) (a) 217.5 (b) 1.9 s
(ii) (a) 80 (b) 2010
(iii) The record drops to zero

3 (i) 16.9 (ii) $\dfrac{v^2 - u^2}{14}$ (iii) 35.7 (iv) 13.6

Exercise 10(g)

1 (i) $x = 5, y = 2$ (ii) $t = -2, q = -5$
(iii) $a = 2, b = 1$ (iv) $p = 2, q = 4$
(v) $x = 1.1, y = 0.3$ (vi) $x = 3, y = 2$
(vii) $c = 1.5, d = -3$ (viii) $h = 6, j = -0.5$

2 £1.05

3 (i) $7q + p = 29$, $5q + p = 21$ (ii) $p = 1, q = 4$

Exercise 10(h)

1 (i) 150 (ii) 2 (assuming $x > 0$)

2 12.5

3 (i) 8 (ii) $\frac{8}{9}$

4 64

5 159.3

6 $7\frac{2}{3}$

7 (i) 31.25 (ii) 0.244

8

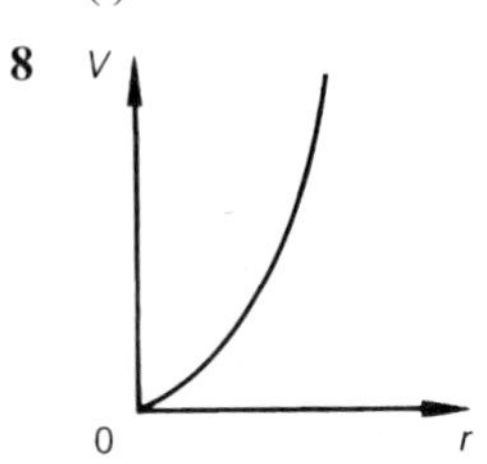

9 (i) 2 (ii) 1 (iii) -1

10 1.5, 6

Exercise 10(i)

1 (i)

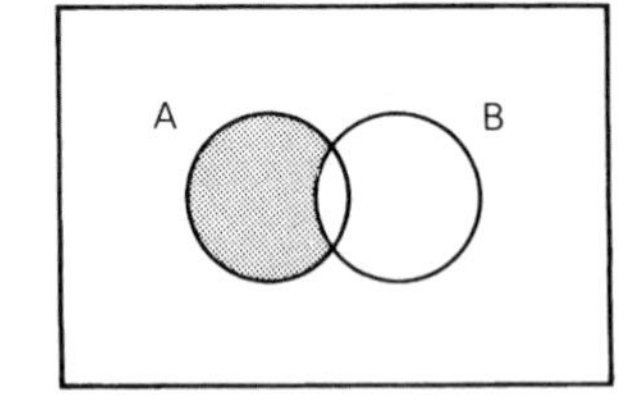

(ii)

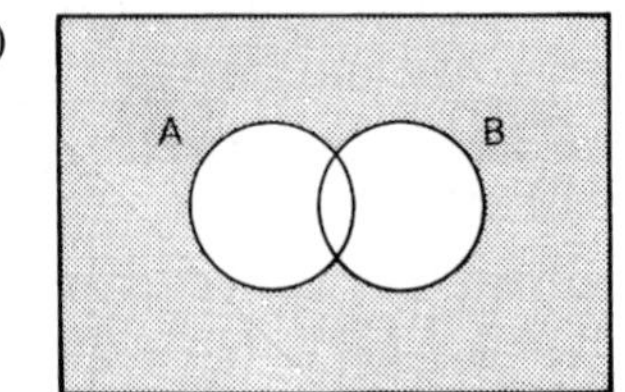

(iii) 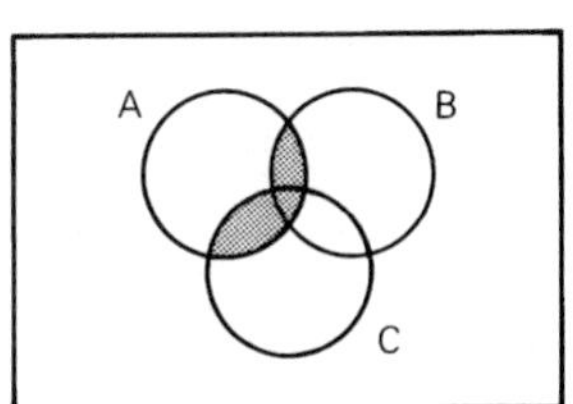

(iv) 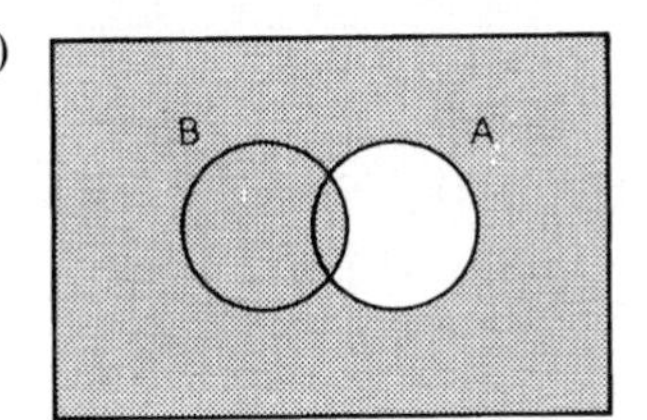

2 8

3 16

4 (i) $\{9\}$ (ii) $\{3, 6, 12, 15, 18, 21, 24\}$
(iii) $\{4, 8, 16, 17, 25\}$

5 (i) 7 (ii) 3

6 (i) 4 (ii) $16 - 51$

7 (i) 34%
(ii) (b) 24 (c) $(6 - x) + (14 - x) + 5 = 19,\ x = 3$

Exercise 11(a)

1 $\frac{5}{6}$

2 (i) $\frac{1}{4}$ (ii) $\frac{1}{13}$ (iii) $\frac{1}{52}$ (iv) $\frac{3}{13}$

3 (i) 3.84
(ii) (a) $\frac{18}{31}$ (b) $\frac{7}{31}$

4 (i) $\frac{1}{4}$ (ii) $\frac{3}{4}; \frac{1}{16}$

5 (i) $\frac{3}{10}$ (ii) $\frac{1}{10}$

6 20

7 7

Exercise 11(b)

1 (i) $\frac{1}{26}$ (ii) $\frac{1}{3}$ (iii) $\frac{4}{11}$ (iv) $\frac{3}{5}$
(v) $\frac{11}{36}$

2 (i) $\frac{1}{3}$ (ii) $\frac{1}{15}$ (iii) $\frac{1}{10}$

3 (i) 37% (ii) 52%

4 (i) $\frac{33}{50}$ (ii) $\frac{26}{50}$ (iii) $\frac{31}{50}$

5 (i) $\frac{7}{90}$ (ii) $\frac{13}{90}$

Exercise 11(c)

1 (i) $\frac{1}{6}$ (ii) $1 - \frac{3}{4} \times \frac{1}{3} = \frac{3}{4}$

2 (i) $\frac{1}{15}$ (ii) $\left(\frac{1}{6} - \frac{1}{15}\right) + \left(\frac{2}{5} - \frac{1}{15}\right) = \frac{13}{30}$

3 $\frac{1}{12}$

4 $P(X \cap Y) = \frac{1}{2} \times \frac{1}{3} = \frac{1}{6}$, hence

(i) $\frac{2}{3}$ (ii) $\frac{1}{3}$ (iii) $\frac{5}{6}$

5 (i) $\frac{1}{4}$ (ii) $\frac{1}{6}$

6 Let the number who take all 3 be x, equation gives $x = 4$

(i) $\frac{1}{25}$ (ii) $\frac{1}{5}$ (iii) $\frac{28}{100} \times \frac{27}{99} + 2 \times \frac{72}{100} \times 2899 = \frac{133}{275}$

Exercise 11(d)

1 (i) (a) 0.4 (b) 0.9
(iii) (a) 0.06 (b) 0.64

2 (i) (a) $\frac{1}{9}$ (b) $N, \frac{1}{3}$
(ii) $\frac{2}{9} \times \frac{1}{8} = \frac{1}{36}$

3 (i) $\frac{1}{2}$ (ii) $\frac{2}{9}$

4 (i) $\frac{8}{27}$ (ii) $\frac{2}{9}$ (iii) $\frac{20}{27}$

5 (i) 0.11 (ii) 0.53

6 Find the area of each region; $\frac{9}{36}, \frac{16}{36}, \frac{11}{36}$; $10 \times \left[\frac{9}{36} \times 100 + \frac{16}{36} \times 50 + \frac{11}{36} \times 20\right] = 533$

7 (i) (a) $\frac{1}{36}$ (b) $\frac{1}{6}$
(ii) (a) $\frac{4}{25}$ (b) $\frac{41}{100}$

8 (i) 0.343 (ii) 0.21 (iii) $3 \times 0.3 \times 0.7^3 = 0.3087$
(iv) 0.6517

9 (i) $6, \frac{1}{2}$
(ii) (a) $\frac{1}{4}$ (b) $\frac{1}{4}$

Exercise 12(a)

1 (iv) (a) 350 miles (b) £35
(vi) MEG by £9
(vii) 500 miles

2 (i) 8 m (ii) 6 m/s (iii) $d = 6t + 8$

3 (iii) 8
(iv) (a) B (b) by 2

4 (i) $4, (0, -7)$ (ii) $-1, (0, 1)$ (iii) $\frac{1}{2}, (0, -2)$ (iv) $-3, (0, 4)$
(v) $-2, \left(0, \frac{1}{2}\right)$ (vi) $\frac{4}{3}, (0, -8)$ (vii) $1, (0, -3)$ (viii) $-\frac{7}{2}, \left(0, \frac{11}{2}\right)$

5 $y = 3x + 4$

6 $y = 2x$, $k = 4$

7 $y = 3x - 5$

8 (i) $x = 0$, $y = 3$ (ii) $p = 3\frac{1}{8}$, $q = -\frac{3}{4}$ (iii) $m = 2.1$, $n = 1.4$

Exercise 12(b)

1 (i) (a) 75 miles (b) 1055 (c) 1 hr 18 min

2 (i) (a) 24 m/s (b) 86.4 km/h
(ii) 140 s
(iii) 0.4 m/s^2
(iv) 0.1 m/s^2

3 6.2 km

4 (i) 45 s (ii) 0.13 ms^{-2} (iii) 2.145 km (iv) 19.3 km/h

Exercise 12(c)

1 (i) 6.2 (ii) 45 approximately

2 380 approximately

3 $k = 0.25$, $t = 5$, $s = 30$

Exercise 12(d)

1 (i) $x < 7$ (ii) $x < 3\frac{2}{3}$ (iii) $x < 1$
(iv) $\{2, 3, 4, 5, 6, 7\}$ (v) $x > 3$

2 $19.31 \leqslant v \leqslant 20.71$

3 $0.77 \leqslant H \leqslant 1.27$

4 (i) 47 (ii) 47

5 $x \geqslant 0$, $y \geqslant 0$, $120x + 75y \leqslant 3600$ which simplifies to $24x + 15y \leqslant 720$,
$15x + 16y \leqslant 450$, $3x + 5y \leqslant 105$; Profit $= 35x + 20y$; $x = 30$, $y = 0$

6 $x + y \leqslant 1000$, $x \geqslant 2y$, $y \geqslant 100$, $x \leqslant 800$; $x = 800$, $y = 200$

Exercise 13(a)

1 (iii) 2 m (iv) 2.30 p.m.
(v) (a) 3.4 m (b) 4.30 p.m.

2 (iii) 163 cm^2, $x = 7.4$ cm

3 (ii) day 23 (iii) 0.06 kg/day

4 (iii) 7.25 (iv) −1.2, 4.2 (vi) −0.7, 2.7

5 (i) 450 l/min (ii) 33 (iii) man; 83.9%

6 (iv) linear graph

7 (iii) (a) 192.5 ft (b) 45 m.p.h.

Exercise 13(b)

1 (i) $\frac{4}{3}$ (ii) 3 (iii) $\frac{3}{4}$

2 (i) 30 (ii) 55

3 $g^{-1} : x \mapsto \dfrac{x-1}{3}$

4 0

5 -1.5

6 (i) 21 (ii) 168

7 (i) 2 (ii) 1, 2 or 3 (iii) 0

8 (i) -3 (ii) 4; 0.5

9 (i) $f^{-1} : x \to 1 - x$ (ii) $g^{-1} : x \to \dfrac{1}{x}$

(iii) $h^{-1} : x \to \sqrt{x}$ (iv) $m^{-1} : x \to \dfrac{1-x}{3}$

(v) $n^{-1} : x \to \dfrac{1-x}{x}$ (vi) $q^{-1} : x \to \dfrac{b-dx}{cx-a}$

10 $a = 1, b = 3$

11 $a = 2, b = 5$

12 Image is $\{0, 1, 7, 5, 3, 9\}$

13 Image is $\{-2, -2, -1, 0, 1, 1, 1, 2, 2, 3\}$

14 $f^{-1} : x \to \dfrac{x-1}{4}$; $g^{-1} : x \to x + 1$; 1.5

15 -4

16 (i) -1.8, 2.8 (ii) -1.4, 3.4 (iii) -2.1, 2.4

Exercise 13(c)

1 (i) 12, 4; 50, 51, 18; 82, 83, 30 (iii) (a) 39 (b) 3

2 1, 1, 2, 3, 1, 3

3 4.47214, $20/x$ becomes N/x

4 20

Exercise 14(a)

1 (i) 80° (ii) 50°, 80° (iii) 80°, 70°
(iv) 60° (v) 40°, 100° (vi) 40°, 50°

2 (i) 90° (ii) 62° (iii) 31° (iv) 31°
(v) 90°

3 10°, 90°, 40°, 40°

4 (i) 130° (ii) 25° (iii) 115°

5 73°, 50°, 80° (Note: $TQ = TR$)

6 (i) 5 cm (ii) 12 cm

7 $PT^2 = 6 \times 14$, so $PT = 9.17$ cm

8 $4 \times RP = 3 \times 5$; $RP = 3.75$ cm

9 (i) (a) 81° (b) 72° (c) 54°
(ii) (a) 2.5 cm (b) 7.2 cm
(iii) (a) 3 cm (b) 7.5 cm (Note: $\triangle HBK$ is isosceles)

10 2.5

Exercise 14(b)

1 (i) $a = 8.48$, $B = 54.6°$ (ii) $y = 17.5$
(iii) $P = 94.5°$, $R = 39.6°$ (iv) $R = 50.2°$, $S = 20.8°$, $T = 109°$
(v) $d = 115$ m, $E = 40.7°$ (vi) $R = 59.4°$ or $120.6°$

2 $BC = 4.68$ cm, $BD = 3.96$ cm

3 53.5°, 6.2 cm, 126.5°, 1.4 cm; 3.21 cm

4 (i) 4.8 cm (ii) 1:200 000, 9.6 km (iii) 045.4°

5 (i) 8.3 cm (ii) 14.9 cm (iii) 17.4 cm^2 (iv) 2.3 cm

6 7.9 ha

Exercise 14(c)

1 (i) 23.5°, 156.5° (ii) 120°, 240°
(iii) 37°, 323° (iv) 217°, 323°
(v) 33°, 147°, 213°, 327° (vi) 3°, 87°, 183°, 267°
(vii) 38°, 152°, 218°, 332° (viii) 61°, 169°, 241°, 349°
(ix) 53.5°, 186.5° (x) 112.5°, 292.5°

2 (iv) 73°, 211° (v) 0.76

3 (ii) 8.8 m (iii) 3.48 a.m. (iv) 2

Exercise 15(a)

1 (i)

$$S = \begin{pmatrix} 20 & 75 & 59 & 36 & 19 & 13 \\ 19 & 80 & 49 & 24 & 21 & 17 \\ 16 & 91 & 55 & 24 & 15 & 19 \\ 10 & 46 & 25 & 10 & 7 & 10 \\ 21 & 83 & 70 & 0 & 30 & 21 \end{pmatrix}$$

(ii)

$$C = \begin{pmatrix} 20 \\ 15 \\ 25 \\ 12 \\ 8 \\ 17 \end{pmatrix}$$

(iii) (a)

$$\begin{pmatrix} 3805 \\ 3550 \\ 3791 \\ 1861 \\ 4012 \end{pmatrix}$$

(b) 5×1

(c) The takings in pence on each day

(iv) (a) 21p (b) 10p (v) decrease of £2.24

Exercise 15(b)

1 (i) $\begin{pmatrix} 60 & 80 & 40 & 40 \\ 0 & 40 & 20 & 60 \\ 80 & 80 & 50 & 0 \end{pmatrix}\begin{pmatrix} 0.6 \\ 0.9 \\ 1.2 \\ 2.5 \end{pmatrix} = \begin{pmatrix} 256 \\ 210 \\ 180 \end{pmatrix}$

(ii) $(30 \quad 20 \quad 40)\begin{pmatrix} 60 & 80 & 40 & 40 \\ 0 & 40 & 20 & 60 \\ 80 & 80 & 50 & 0 \end{pmatrix} = (5000 \quad 6400 \quad 3600 \quad 2400)$

total = 17 400

2 (i) $\begin{pmatrix} 0 & 1 & 2 \\ 1 & 0 & 1 \\ 2 & 1 & 0 \end{pmatrix}$

(ii) 5 two-stage routes from C back to C

3 (i), (ii)

4 (i) $\begin{pmatrix} 1 & 0 & 1 \\ 1 & 1 & 0 \end{pmatrix}$

(ii) $\begin{pmatrix} 1 & 1 & 0 \\ 1 & 0 & 1 \\ 1 & 1 & 0 \end{pmatrix}$

(iii) (a) *NLP*, *NSP* (b) $\begin{pmatrix} 2 & 2 & 0 \\ 2 & 1 & 1 \end{pmatrix}$ (c) 4

Exercise 15(c)

1 (i) $\begin{pmatrix} 7 & 4 \\ -3 & 2 \end{pmatrix}$ (ii) $\begin{pmatrix} 6 & 6 \\ -4 & -2 \end{pmatrix}$ (iii) $\begin{pmatrix} 9 & -3 & -3 \\ 1 & 5 & 1 \end{pmatrix}$

2 $\begin{pmatrix} 0 & 2.5 \\ 1 & 2 \end{pmatrix}$

3 (i) $\begin{pmatrix} -2.5 & 2 \\ 1.5 & -1 \end{pmatrix}$ (ii) $\begin{pmatrix} \frac{1}{3} & -\frac{1}{3} \\ -\frac{1}{3} & -\frac{2}{3} \end{pmatrix}$ (iii) $\begin{pmatrix} 0.2 & 0.6 \\ 0 & 0.5 \end{pmatrix}$

4 $a = -2.93$, $b = 5.4$

5 (i) (a) 5 (b) 11

(ii) $\begin{pmatrix} 28 & 42 \\ 112 & 168 \end{pmatrix}$

(iii) (a) $p = 3x + 2$ (b) $q = b + ax$

6 (i) $-10, 7$ (ii) $-2, -3$ (iii) $-3\frac{1}{4}, 2\frac{3}{4}$

8 3, 11, 4

9 $x = 1, y = 1, t = 2$ or $x = -1, y = -1, t = 2$

Index